Sustainable Development Goals Series

The **Sustainable Development Goals Series** is Springer Nature's inaugural cross-imprint book series that addresses and supports the United Nations' seventeen Sustainable Development Goals. The series fosters comprehensive research focused on these global targets and endeavours to address some of society's greatest grand challenges. The SDGs are inherently multidisciplinary, and they bring people working across different fields together and working towards a common goal. In this spirit, the Sustainable Development Goals series is the first at Springer Nature to publish books under both the Springer and Palgrave Macmillan imprints, bringing the strengths of our imprints together.

The Sustainable Development Goals Series is organized into eighteen subseries: one subseries based around each of the seventeen respective Sustainable Development Goals, and an eighteenth subseries, "Connecting the Goals", which serves as a home for volumes addressing multiple goals or studying the SDGs as a whole. Each subseries is guided by an expert Subseries Advisor with years or decades of experience studying and addressing core components of their respective Goal.

The SDG Series has a remit as broad as the SDGs themselves, and contributions are welcome from scientists, academics, policymakers, and researchers working in fields related to any of the seventeen goals. If you are interested in contributing a monograph or curated volume to the series, please contact the Publishers: Zachary Romano [Springer; zachary.romano@springer.com] and Rachael Ballard [Palgrave Macmillan; rachael.ballard@palgrave.com].

Yuheng Li
Editor

Global Perspectives on Building Resilience for Sustainable Rural Development

Institutional Innovation and Policy

Springer

Editor
Yuheng Li
School of Humanities and Law
Northeastern University
Shenyang, Liaoning Province, China

ISSN 2523-3084 ISSN 2523-3092 (electronic)
Sustainable Development Goals Series
ISBN 978-981-96-6148-0 ISBN 978-981-96-6149-7 (eBook)
https://doi.org/10.1007/978-981-96-6149-7

This work is supported by the National Natural Science Foundation of China (No. 42471240; No. 42171208).

This Springer imprint is published by the registered company Springer Nature Singapore Pte Ltd.
The registered company address is: 152 Beach Road, #21-01/04 Gateway East, Singapore 189721, Singapore

Foreword: Enhancing Rural Resilience for SDGs

Geography is a global science and an international discipline. As a science, it contributes to the globalization of knowledge, often, depending on the subjects, in interaction with other sciences. Such is the case of climate change and sustainable development. As a discipline, geography tends to focus on case studies at different scales in various parts of the world. The study of rural (agricultural) resilience and land use issues led by the commission on Agricultural Geography and Land Engineering (AGLE) of the International Geographical Union (IGU) is a good case in point. Rural areas play a crucial role in achieving the United Nations Sustainable Development Goals (SDGs), which aim to address global challenges such as poverty, inequality, climate change, environmental degradation, peace, and justice. Nevertheless, the halfway evaluation of SDGs in 2023 showed limited progress but big challenges across the world. As a result, more attention needs to be paid to rural areas which are integral to the achievement of the SDGs.

Globally, rural areas are facing a range of complex challenges that can hinder their development and well-being. These challenges are often interconnected and can vary significantly depending on the geographical, cultural, and economic context of each region. The challenges facing rural areas are multifaceted and require comprehensive solutions that take into account the unique contexts of each community. Addressing these challenges is essential for promoting sustainable development, reducing inequalities, and improving the quality of life for rural residents. Collaborative efforts among governments, NGOs, and local communities are crucial to developing effective strategies that empower rural populations and foster resilience.

The book titled *Global Perspectives on Building Resilience for Sustainable Rural Development* provides valuable perspectives on the factors that influence rural resilience and explores diverse strategies that can ensure the long-term sustainability and empowerment of rural populations worldwide. Contributors from 16 countries showcase the successful application of innovative approaches to enhancing rural resilience. I do trust the book will help to deepen our understanding and enrich the knowledge to the critical aspects of rural resilience and inspire innovative solutions for promoting SDGs in the world.

Prof. Yuheng Li has been active in the International Geographical Union (IGU) where he jointly established the Commission on *Agricultural Geography and Land Engineering* (AGLE) in 2016. The commission's mission targets both the theoretical and practical progress in improving the relationship between agriculture, rural and land where land engineering is introduced to help increase the agricultural productivity which lays solid foundation for securing food supply, maintaining peasants' livelihood, and sustaining rural development. Being the secretary (2016–2024) and vice chair (2024–) of IGU-AGLE, Prof. Yuheng has devoted much to bringing the commission from young to mature. His years of experience and achievements have well demonstrated his good ability in rural (agricultural) resilience studies and practice and set good example for the peers. For a better and resilient world, I strongly believe that proactive and innovative efforts are needed to accelerate the progress of SDGs while IGU could play an important role in this cause.

President of International Geographical Union, Professor at Université Paris 8
Paris, France

Nathalie Lemarchand

Foreword: Understanding Rural Resilience

Rural communities around the world have long faced significant challenges, ranging from depopulation and aging, economic vulnerability to environmental degradation and climatic impact. Over the last couple of decades, understanding of these processes and the responses of rural communities has been greatly enhanced by the engagement of rural scholars with the concept of resilience. As the contributions to this book richly demonstrate, resilience refers not only to "bouncing-back" after a shock, but also to levels of preparedness and the capacity to withstand disruption. Applied in a social context to communities it leads to a holistic approach that encompasses the interlocking social, economic, cultural, and ecological aspects of rural communities, the appropriateness of governance and planning systems, and the impacts of power and resource distribution on capacities to act. While the concept itself has gained traction in rural studies, it remains underdefined and often misunderstood, particularly in terms of how it can be operationalized in different rural settings.

These different dimensions of resilience are all explored in the 18 chapters in *Global Perspectives on Building Resilience for Sustainable Rural Development* that have been ably pulled together by Professor Yuheng Li, forming an essential reference for anyone interested in studying perspectives on rural resilience and applying lessons in real-world contexts. The collection captures the way in which research on rural resilience has matured and expanded, reflecting the cutting edge of the field.

Understanding rural resilience is central to the evolving concern with the planetary rural. This approach started as a riposte to theories of planetary urbanization which in essence deleted the rural by asserting that whole of global society has been urbanized. As well as asserting the continuing importance of the rural as a category, work on the planetary rural has gone further by exploring how rural places, societies, and environments are intrinsically enmeshed in planetary processes. Resilience is core to this understanding. The resilience of rural communities to planetary dynamics such as earthquakes, wildfires, and extreme weather events was an early entry point for the application of ideas of resilience in rural studies. It quickly became apparent that the same capacities and resources that enabled rural communities to withstand and recover from natural hazards also inform their potential to cope with other forms of shock.

These include preparedness for future environmental changes, including the planetary threat of the climate crisis. Adaptability to climate change and the ways in which it is altering the human and biophysical composition of rural places and consequently their capacity to perform functions such as producing food and providing water supplies is a theme explored in several of the chapters in this book. The resilience of rural communities and places will be at the heart of humanity's ability to successfully negotiate planetary crisis such as climate change, biodiversity loss, and food security.

Yet, resilience is not just about the environment. Other chapters follow a rich vein of work in economic geography in examining the economic resilience of rural localities against pressures from globalization, deindustrialization, neoliberalism, recessions, and the impacts of the Covid-19 pandemic. Population dynamics and managing flows of migration are also significant, with resilience required in some contexts to survive depopulation and the erosion of human resources and in other contexts to adapt to inward migration and an increasing population. More intriguingly, contributors also look at cultural resilience and the sustainability of rural heritage. This is about the durability of rural identities even as the material assemblage of rural places changes with ecological, social, and economic transitions. There are many ways of being resilient.

One of the main messages of this book is that achieving resilience is not a one-size-fits-all approach. Instead, it requires a nuanced, context-sensitive understanding of local needs, resources, and circumstances. A planetary rural approach also means being open to multiple voices and perspectives from around the world, challenging the conventional dominance of western or Anglophone scholarship. An especially commendable feature of this book is that it includes contributors from 16 countries, mostly in the Global South, providing insights into how requirements for and approaches to resilience are shaped by varying environmental and economic processes, as well as by socio-political dynamics such as governance issues, demographic changes, and land tenure systems, and variability in access to technology and infrastructure.

A second key message is that for successful rural resilience an integrated approach is required, understanding interactions across social, economic, political, and environmental domains and ensuring that solutions are context-specific with local realities are accounted for in policy and practice. The book offers multiple ways to enhance rural resilience, including economic diversification and transformation, institutional innovation and integrated policies, knowledge synergy and community engagement, smart technologies, and infrastructure development.

This emphasis is not surprising given the oversight of Yuheng Li as editor. Yuheng's research in China and internationally has been at the forefront of advancing approaches to rural resilience, combining critical analysis with practical application. This volume will further consolidate his growing reputation as a global leader in the field.

Ultimately, rural resilience is not only about surviving adversity but also about thriving in an ever-changing world. Through thoughtful policies, innovative strategies, and active community participation, rural communities can overcome challenges and pave the way for a sustainable and prosperous future. The experiences from various countries highlight that rural resilience is not an abstract concept but a practical, actionable framework that can be tailored to meet the unique needs of different rural contexts. With the right mix of policies, local knowledge, and institutional support, rural communities can build resilience and ensure their continued prosperity in the face of uncertainty.

Professor FAcSS, Aberystwyth University
Aberystwyth, UK
Michael Woods

Contents

Building Resilience for Rural Sustainability: An Introduction

Yuheng Li

Abstract

The halfway evaluation of the UN's Sustainable Development Goals (SDGs), which was released in September 2023, was not optimistic as expected. Among all the 169 goals, only 15% of the SDGs are on track, 49% show minimal or moderate progress, while 36% of the goals show signs of stagnation or regression. Such warning report arose human society that the 2030 Agenda will become no more than an epitaph for a world that might have been unless we take urgent action now.

1 Why *Rural* Matters?

The halfway evaluation of the UN's Sustainable Development Goals (SDGs), which was released in September 2023, was not optimistic as expected. Among all the 169 goals, only 15% of the SDGs are on track, 49% show minimal or moderate progress, while 36% of the goals show signs of stagnation or regression. Such warning report arose human society that the 2030 Agenda will become no more than an epitaph for a world that might have been unless we take urgent action now.

Looking at the detailed content of all 17 SDGs, it is obvious to find that 9 SDGs are directly linked to rural sector in terms of households' income, poverty reduction, food security and health, gender empowerment, and environmental sustainability. Rural areas offer basic livelihood means and employments to the majority of poor people around the world (SDG 1 and SDG 8), while agricultural production is the key source of food and nutrition for the people (SDG 2 and SDG 3). Farms are also large consumer of land and water resources, with further impacts on terrestrial ecosystems, forests, grasslands, wetlands, and biodiversity (SDG 12 and SDG 15). In the social domain, rural areas are also the places where gender inequality widely exists especially to those marginalized female groups (SDG 5). The social and economic inequalities and disadvantages in people's early life can limit opportunities for the realization of their rights and full potential (SDG 10). In the meantime, villagers' *Climate Action* (SDG 13) in the context of meaningful mitigation actions and transparency on implementations will exert long-lasting impact on local ecosystems and in particular agriculture, which further influences the above-mentioned SDGs.

Rural areas are playing important role in the socio-economic fabric of nations worldwide. Despite often being overshadowed by urban centers, rural areas contribute significantly to various

Y. Li (✉)
School of Humanities and Law, Northeastern University, Shenyang, Liaoning Province, China
e-mail: liyuheng@wfxy.neu.edu.cn

Y. Li (ed.), *Global Perspectives on Building Resilience for Sustainable Rural Development*, Sustainable Development Goals Series, https://doi.org/10.1007/978-981-96-6149-7_1

aspects of society, economy, and culture. Understanding the importance of rural areas is essential for several reasons.

1.1 Economic Contributions

Rural regions are vital to the agricultural sector, which remains a cornerstone of many economies. They produce a significant portion of food, fiber, and raw materials that sustain both local and global markets. Agriculture not only provides livelihoods for millions but also underpins food security, making rural areas essential in the fight against hunger and malnutrition.

In addition to agriculture, rural economies often encompass other industries such as forestry, mining, and tourism. These sectors contribute to national GDP and provide employment opportunities, helping to diversify local economies and reduce dependency on single industries.

1.2 Cultural Heritage and Identity

Rural areas are often home to rich cultural traditions, languages, and lifestyles that reflect the history and identity of a nation. They preserve unique customs, crafts, and practices that contribute to cultural diversity. This cultural heritage is invaluable, offering insights into sustainable living, community cohesion, and resilience.

Moreover, rural places often foster strong social ties and networks, promoting a sense of belonging and community spirit. These social dynamics are essential for maintaining cultural identity and heritage, which can be lost in the rapid urbanization and globalization of modern society.

1.3 Environmental Stewardship

Rural regions are often the custodians of natural resources and ecosystems. They play a critical role in biodiversity conservation, land management, and environmental sustainability. The practices employed in rural areas, such as sustainable agriculture and forestry, can have a profound impact on environmental health.

Additionally, rural landscapes contribute to ecosystem services such as clean water, air quality, and carbon sequestration, which are essential for combating climate change. Protecting and promoting sustainable practices in rural areas is vital for ensuring the health of the planet.

1.4 Population Dynamics and Urbanization

As urban centers continue to grow, rural areas face significant demographic changes, including depopulation and aging populations. Understanding the dynamics of rural populations is crucial for effective policy-making and resource allocation. Addressing the challenges faced by rural communities, such as access to healthcare, education, and employment opportunities, is essential for maintaining balanced regional development.

Moreover, rural areas can serve as a counterbalance to urbanization by offering alternative lifestyles and living conditions. Promoting rural development can help alleviate some of the pressures faced by urban centers, such as overcrowding, housing shortages, and environmental degradation.

1.5 Food Security and Resilience

In an increasingly interconnected world, rural areas are essential for ensuring food security. They provide the agricultural base needed to feed growing populations, especially in the face of climate change and environmental challenges. Resilient rural economies are better equipped to adapt to shocks, such as natural disasters or market fluctuations, thereby contributing to overall national stability.

Supporting rural development enhances the capacity of communities to respond to crises, ensuring that they can continue to provide essential services and resources. This resilience is critical for both local and global food systems.

1.6 Innovation and Sustainability

Rural areas are often sites of innovation, particularly in sustainable practices and technologies. Many rural communities are at the forefront of developing and implementing sustainable agricultural techniques, renewable energy solutions, and conservation strategies. These innovations can serve as models for urban areas and contribute to broader sustainability goals.

By investing in rural innovation, societies can harness local knowledge and creativity to address pressing global challenges, such as climate change, resource depletion, and social inequality.

In short, rural areas play a crucial role in meeting the major parts of the SDGs. However, rural decline is today an undisputed fact and it has become a global issue as the world endeavors to promote urbanization and industrial development (Liu & Li, 2017). Both developed and developing countries have either experienced or are experiencing rural decline, which indicates the downward spiral of decreasing employment, depopulation, economic depression, and deteriorating quality of life in the countryside (Forth, 2000; Ono, 2005; Carr & Kefalas, 2009; Li et al., 2016). It is no doubt that rural decline will reverse the SDGs progress across the world while it is necessary to enhance rural resilience to revitalize the countryside which further promote the progress of SDGs.

2 Rural Resilience as a Key Attribute

Generally, rural area is a complex social-ecological system which consists of both material elements like resources endowment, location, population density, local economy and immaterial elements like people's behavior and action, trust, social norms, and network (Li et al., 2019). These elements have jointly formed the physical capital, human capital, social capital, cultural capital, institutional capital, etc., which support the operation of rural communities and help rural function well (Quaranta & Salvia, 2014). In the meantime, rural is not an isolated object, instead it interacts with the external context including urbanization, globalization, climate change, unexpected risks, etc. In this process, certain rural communities could endure and survive or tend to transform into a new development pattern, while some villages failed to go through the challenges and risks. That's why we often see some villages decline or vanish, while some others not. We attribute such rural evolutionary differences to resilience, which is the key capacity for rural areas to deal with external pressures and challenges.

Rural resilience has gained prominence and become a front edge in the field of sustainability as a response to various factors challenging rural communities in the recent decades. Unlike the concepts of "community-function" and "community vitality" which failed to clarify the interactions between rural areas and the external environment, resilience portrays the capacity of a system to absorb disturbance, undergo change, and retain the same essential functions, structure, identity, and feedbacks (Holling, 1973). Rural resilience is thus defined as a responsive capacity in the process as rural communities interact with pressures, shocks, or disturbances to maintain stability, minimize loss, or move further to create a completed new regime. Generally, rural resilience consists of three progressive capacities, namely resistance, adaptation, and transformation (Li, 2023). Then, resilient communities are believed to possess the capacity to prevent unwelcome challenges in the face of external circumstances and to adapt to the changing external environment in such a way that a satisfactory standard of living is maintained. Further, rural communities tend to transform into a completed new development pattern through thorough reform and innovation. The important role of rural resilience in maintaining sustainability is seen in the following aspects.

2.1 Adaptation to Climate Change

Rural areas, particularly those reliant on agriculture, are often on the front lines of climate change impacts, such as extreme weather events, droughts, and shifting growing seasons. Resilient rural communities develop adaptive strategies that enhance their capacity to cope with these changes. This can include diversifying crops, adopting sustainable farming practices, and implementing water conservation techniques. By fostering resilience, rural areas contribute to broader climate adaptation efforts, ensuring food security and ecosystem health.

2.2 Sustainable Resource Management

Rural resilience emphasizes the sustainable management of natural resources, including land, water, and biodiversity. Resilient communities often practice sustainable agriculture, forestry, and fisheries, which not only provide livelihoods but also protect ecosystems. By prioritizing sustainability in resource management, rural areas help maintain ecological balance, reduce environmental degradation, and promote biodiversity, all of which are crucial for long-term sustainability.

2.3 Economic Diversification and Stability

Economic resilience in rural areas is vital for sustainability. Communities that diversify their economies—by integrating agriculture with tourism, renewable energy, or artisanal crafts—are better equipped to withstand economic shocks. This diversification reduces dependency on single industries, enhances local job creation, and promotes economic stability. A robust rural economy contributes to overall national economic health and reduces the vulnerability of rural populations to market fluctuations.

2.4 Social Cohesion and Community Engagement

Resilient rural communities often exhibit strong social networks and a sense of belonging, which are key components of social sustainability. Community engagement and participation in decision-making processes lead to more effective and inclusive governance. When residents are involved in shaping policies and practices, they are more likely to support sustainable initiatives and collaborate on local solutions. This social cohesion fosters resilience, enabling communities to respond collectively to challenges.

2.5 Preservation of Cultural Heritage

Rural resilience is closely tied to the preservation of cultural heritage and traditional knowledge. Many rural communities have developed sustainable practices over generations, rooted in local customs and ecological understanding. By valuing and integrating this traditional knowledge into modern sustainability efforts, rural areas can promote practices that are both culturally relevant and environmentally sound. This preservation of cultural identity contributes to the social fabric and sustainability of communities.

2.6 Innovation and Technological Advancement

Rural resilience often drives innovation, particularly in sustainable technologies and practices. Many rural communities are experimenting with renewable energy sources, precision agriculture, and eco-friendly practices that not only enhance their resilience but also serve as models for other regions. By fostering an environment that encourages innovation, rural areas can contribute to the development of sustainable solutions that address global challenges, such as climate change and resource scarcity.

2.7 Food Security and Nutritional Sustainability

Rural resilience is critical for ensuring food security, which is a cornerstone of sustainability. Resilient rural communities produce diverse crops and livestock, enhancing nutritional security and reducing reliance on imported food. By promoting local food systems and sustainable agricultural practices, rural areas can contribute to healthier diets and reduce the environmental impact associated with long-distance food transportation.

3 Building Resilience for Rural Sustainability

Given the potential risks and unexpected challenges, it is quite necessary to maintain rural stability through enhanced resilience, a reaction to the notion of rural decline and associated with improved villagers' livelihood by way of changes in their behavior and adaptation to new circumstances, rather than being passively dictated at the mercy of unmanageable external forces. So far, interdisciplinary studies have identified various factors which are influencing rural resilience (Murphy, 2007; Wilson, 2010; Steiner & Markantoni, 2013; Li et al., 2016; Yang et al., 2020). As rural resilience is a multifaceted concept, gaining resilience can be achieved through diversified ways. Most importantly, any resilience studies should be carried out in a context-based way to understand the particulars of local settings and social, economic, environmental, and institutional compositions of rural communities as well as the specific challenge that is pressing rural communities.

This book synthesizes various studies which are focusing on the topic of rural resilience. The overarching theme highlights the interconnected challenges faced by rural communities and the strategies proposed to address these challenges and enhance rural resilience in different context. The editor of the book invited world-known rural scholars from 16 different countries of which 5 are from the Global North (USA, United Kingdom, Australia, Sweden, Poland) and 11 are from the Global South (Sudan, Senegal, Zimbabwe, Iran, India, Nepal, Thailand, China, Mongolia, Brazil, Slovakia). Multidisciplinary research background has enabled all authors to contribute 18 chapters with integrated approaches to identify current challenges facing rural communities and diversified ways to improve rural resilience.

3.1 Categorization of Challenges in Rural Development

3.1.1 Economic Vulnerabilities and Poverty

A primary challenge faced by rural economies is their heavy reliance on agriculture, particularly rain-fed farming. Regions like Zimbabwe (Vol. 2, Chapter "Enhancing the Resilience of Rural Economies in Zimbabwe") exemplify this vulnerability, where economic activities are limited to small-scale agricultural practices and crafts. Similar situation is also seen in rural areas of Mongolia (Vol. 2, Chapter "Rural Poverty and Responsive Social Protection to Resist Climate Shocks in Mongolia"), where many households rely on low-income sources like agriculture and husbandry, making it difficult to escape the poverty cycle. High poverty rates in turn pose significant barriers to rural development.

3.1.2 Climate Change and Environmental Degradation

Climate change poses a significant threat to rural resilience, impacting agricultural productivity and resource availability. For instance, in Senegal (Vol. 2, Chapter "Exploring the Challenges and Strategies for Enhancing Resilient and Sustainable Rural Development Tailored to Thiès Region, Senegal") and Brazil (Vol. 1, Chapter "Embeddedness of Environmental Politics in Tropical Rural Resilience: Analytical Innovation and Application in Brazil"), traditional subsistence farming is increasingly challenged by climatic variations. In China's rangeland (Vol. 1, Chapter "Vegetation Rehabilitation for Rangeland Resilience in China: Comparative Effectiveness Between Grazing Control and Ecological Engineering"), climate change and intense encroachment by human activities lead to vegetation degradation, which affect local herders' livelihood. In India (Vol. 1, Chapter "Planning for Smart Village Initiatives Towards Climate Resilience in Natural Resource Management: Case Study of Udaipur District, India"), Thailand (Vol. 2, Chapter "Community-Based Social Enterprises for Rural Revitalization and Community Resilience to Social Environmental Change in Rural Thailand"), and Iran (Vol. 1, Chapter "Resilience for Rural Sustainable Development in Iran: Challenges and Approaches"), water scarcity has not only significantly disrupted local economy, but also driven migration from rural areas to cities and reduces agricultural productivity. The lack of diversification also leads to susceptibility to climate change and natural resource depletion, which exacerbate socio-economic challenges, highlighting the interconnectedness of environmental factors and rural livelihoods. This is also seen in remote areas of Australia (Vol. 2, Chapter "Three Faces of Rural Resilience in Australia"), where countryside is facing poverty and biodiversity threats due to climate change.

3.1.3 Weak Infrastructure and Services

Infrastructure development and services provision are crucial for rural growth, yet many countries, such as Nepal (Vol. 1, Chapter "Smart and Compact Rural Settlements for Resilient and Sustainable Development in Nepal: Policy and Implementation Perspectives"), Mongolia (Vol. 2, Chapter "Rural Poverty and Responsive Social Protection to Resist Climate Shocks in Mongolia"), and Poland (Vol. 1, Chapter "Smart Villages—An Instrument for Building Rural Resilience in the European Context: The Example of Poland"), face challenges due to difficult terrain and limited funding. Rural areas suffer from inadequate infrastructure and services, including transportation, healthcare, education services, social protection, and disaster relief services. Historically, rural China (Vol. 1, Chapter "Smart and Compact Rural Settlements for Resilient and Sustainable Development in Nepal: Policy and Implementation Perspectives") experienced disfavoring environment due to the urban-bias policy, which leads to insufficient investment in rural infrastructure and public services. Such imbalanced infrastructure and services not only hamper rural economic activities but also lead to rural depopulation as individuals migrate to urban areas for better opportunities. As a result, entrenched socio-economic disparities between rural and urban areas are seen.

3.1.4 Social Dynamics and Demographic Changes

Besides those Global South countries which are suffering rapid rural depopulation, the Global North countries such as England (Vol. 2, Chapter "The Emergence of a Diverse and Age-Differentiated English Countryside: Changing Rural Populations and Resilience-Building in the Twenty-First Century"), Poland (Vol. 1, Chapter "Smart Villages—An Instrument for Building Rural Resilience in the European Context: The Example of Poland"), and Australia (Vol. 2, Chapter "Three Faces of Rural Resilience in Australia") are experiencing rural depopulation with communities facing loss of multiple services. In particular, those rural peripheries which are outside the growth influences of the city-regions in countries such as Sweden and Slovakia (Vol. 1, Chapter "Rural Development Challenges and Resilience Building in Northern and Central Europe: A Comparison of Sweden and Slovakia") require tailored responses to depopulation and necessitate a reevaluation of strategies to ensure community cohesion and resilience in the face of evolving socio-economic landscapes. Certain big incident like Brexit has induced significant social changes in UK with distinct rural populations depleted, and key rural economic sectors, such as agriculture, horticulture, hospitality and dairies, experienced a lack of workers due to the return of previous waves of immigrants back to their European places of origin. Similarly, the Pasak Jolasid Dam's construction in Thailand (Vol. 2, Chapter "Community-Based Social Enterprises for Rural Revitalization and Community Resilience to Social Environmental Change in Rural Thailand") not only caused land loss (agricultural land, forests, and residential areas) but also jeopardized villages' unique culture, traditions, and traditional livelihoods in the affected areas.

3.1.5 Political and Governance Issues

Environmental politics in tropical countries such as Brazil (Vol. 1, Chapter "Embeddedness of Environmental Politics in Tropical Rural Resilience: Analytical Innovation and Application in Brazil") is influenced by the international trade and global environmental issues such as climate change regimes, biodiversity negotiation, and deforestation-pertinent regulation. Such politics in turn exert great impact on the human–ecosystem interrelationship and rural continuity and transformation. Land governance remains a critical barrier to rural development. In Sudan (Vol. 2, Chapter "Land Governance for Enhancing Resilient Livelihood in Rural Sudan"), land ownership is characterized by inequality, with traditional land tenure systems influenced by government policies, conflict, and climate change. The lack of secure land rights for local communities, particularly pastoralist communities, hampers effective governance and resource management in the countryside.

3.2 Strategies for Enhancing Rural Resilience

3.2.1 Institutional Innovation and Integrated Policies

China's rural development experience has proved the efficiency and effectiveness of institutional innovation at different stages (Vol. 1, Chapter "Institutional Innovation and Its Effect on Rural China's Socioeconomic Resilience"). The institutional innovation and associated policies have jointly coordinated human–land relationship and promoted production relations in the countryside. The development of environmental politics in Brazil effectively strengthens the compatible interrelationship between ecology and social economy and improves tropical rural resilience (Vol. 1, Chapter "Embeddedness of Environmental Politics in Tropical Rural Resilience: Analytical Innovation and Application in Brazil"). Sweden and Slovakia have also adopted integrated rural policies that address the specific needs of both city-close and peripheral rural areas. This approach provides equitable access to services and promotes economic opportunities across diverse rural landscapes (Vol. 1, Chapter "Rural Development Challenges and Resilience Building in Northern and Central Europe: A Comparison of Sweden and Slovakia"). In Iran (Vol. 1, Chapter "Resilience for Rural Sustainable Development in Iran: Challenges and Approaches"), integrated sectoral-spatial policymaking is introduced and emphasized to invest in local communities over sectoral approaches. By focusing on spatial policies that cater to the unique needs of rural societies, policymakers can promote rural development, improve living standards, and enhance rural resilience. The practice in Sudan also indicates the importance of land policy reforms that recognizes customary land rights. The National Land Policy integrates marginalized communities into land governance frameworks, which can help ensure equitable access and user rights, empower local communities, and foster resilient livelihoods (Vol. 2, Chapter "Land Governance for Enhancing Resilient Livelihood in Rural Sudan").

3.2.2 Economic Diversification and Transformation

To combat rural economic vulnerability to climatic and resource shortage impact, regions such as Zimbabwe and Senegal are exploring pathways for diversification. By investing in social overhead projects and rural industries, intensified rural–urban linkages with stimulated rural economic activity are seen. These changes can help to reduce dependence on agriculture and create sustainable growth opportunities (Vol. 2, Chapters "Enhancing the Resilience of Rural Economies in Zimbabwe" and "Exploring the Challenges and Strategies for Enhancing Resilient and Sustainable Rural Development Tailored to Thiès Region, Senegal"). The case in Australia shows how multi-functional development with tourism, food processing, and a growing service sector added to rural vitality in those peri-urban areas, where loss of prime agricultural land, farm fragmentation, and inadequate infrastructure widely exist (Vol. 2, Chapter "Three Faces of Rural Resilience in Australia"). Agroecological practices in the semiarid regions of Brazil (Vol. 2, Chapter "Agroecology and

Resilience in Brazil from a Southern Perspective") are considered as a path, which leads to complex social, economic, and ecological transformations rather than adaptations. In this process, agricultural production with lifestyle, culture, local knowledge, biological diversity, and the management of natural resources are integrated, helping to cope with water access problems in the countryside. Cases in Sweden and Slovakia also show that balanced development of tourism facilities and the preservation of natural and cultural heritage are both environmentally sustainable and beneficial to the local communities in the peripheral places (Vol. 1, Chapter "Rural Development Challenges and Resilience Building in Northern and Central Europe: A Comparison of Sweden and Slovakia").

3.2.3 Knowledge Synergy and Community Engagement

In developing countries, integrating local knowledge into agricultural practices and resource management has proven effective in enhancing resilience. This approach not only respects traditional practices but also empowers communities to adapt to changing environmental conditions. By synergizing local and modern knowledge systems, rural communities can enhance their socio-ecological systems and develop diverse livelihood strategies. The case in China shows that the collision, synergy, and interdependence of village local knowledge and external modern knowledge help to transform rural socio-ecological systems from fragile to more resilient system (Vol. 2, Chapter "Logic of Synergizing Knowledge Systems for Rural Resilience: Case Study on Cultural Revitalization Projects in Three Villages in China"). In Thailand, social enterprises have emerged as vital actors to address the socio-environmental challenges and market failures facing the countryside. Social enterprises leverage local knowledge and values to foster innovation within communities, preserve cultural roots, enhance economic security, and build rural community resilience (Vol. 2, Chapter "Community-Based Social Enterprises for Rural Revitalization and Community Resilience to Social Environmental Change in Rural Thailand").

Engaging communities in decision-making processes enhances local capacity and fosters a sense of ownership, which is crucial for building resilience. Participatory processes led by local communities have demonstrated significant improvements in agricultural production and resource management. In the United States (Vol. 2, Chapter "Considering the Role of Community Engagement in Rural Resilience and Sustainability Narratives in the USA") and other regions such as Brazil (Vol. 1, Chapter "Embeddedness of Environmental Politics in Tropical Rural Resilience: Analytical Innovation and Application in Brazil"), community engagement principles are being utilized to foster partnerships between local communities, governments, and organizations. Such strategy seeks to empower local communities through recognition of local community knowledge, finding local solutions, and dialog, which builds social capacity and allows a range of perspectives to be heard. All these efforts tend to address complex socio-economic and environmental challenges and contribute to rural resilience.

3.2.4 Smart Technologies and Infrastructure Development

Integrating smart technologies and infrastructure development into rural socio-economic development can enhance rural connectivity with outside world and improve resource management. In Nepal, the government is promoting smart villages and compact rural settlement policies that integrate technology with local resources. This initiative with supportive policy frameworks leads to enhanced socio-economic connectivity, access to services, and sustainable resource management (Vol. 1, Chapter "Smart and Compact Rural Settlements for Resilient and Sustainable Development in Nepal: Policy and Implementation Perspectives"). Similar things are seen in India, smart village initiatives are emphasized for effective natural resource management and conservation on the condition of financial, technological, and marketization input as well as infrastructure development (Vol. 1,

Chapter "Planning for Smart Village Initiatives Towards Climate Resilience in Natural Resource Management: Case Study of Udaipur District, India").

In Poland, the concept of smart villages is being promoted to enhance rural resilience through digital innovation and local resource utilization. This initiative aims to address issues such as depopulation and limited access to services by fostering community engagement and sustainable development practices (Vol. 1, Chapter "Smart Villages—An Instrument for Building Rural Resilience in the European Context: The Example of Poland"). Australia even initiated climate smart agriculture (CSA) to deal with climate variability, increase agricultural productivity, enhance adaptation to climate change, and reduce greenhouse gas emissions. The initiative includes diversified technology-supported content such as carbon farming, precision agriculture, and utilizing accurate weather forecasts in farm management (Vol. 2, Chapter "Three Faces of Rural Resilience in Australia").

4 Rural Resilience in the Future

Rural resilience is poised to play a pivotal role in shaping sustainable futures as global risks and challenges intensify. The challenges facing rural regions are complex and multifaceted, requiring innovative strategies and collaborative efforts among stakeholders. By recognizing the interconnectedness of these challenges and the importance of local context, governance, and community engagement, we can develop comprehensive approaches to enhance rural resilience. As we move forward, the importance of building resilient rural communities becomes increasingly clear, particularly in the context of climate change, economic uncertainty, and social transformation. Here are several key aspects of how rural resilience will evolve and influence the future.

4.1 Adapting to Climate Change

As climate change continues to pose significant threats, rural areas will need to develop innovative strategies to adapt to its impacts. This includes implementing sustainable agricultural practices, enhancing water management systems, and investing in climate-resilient infrastructure. Future rural resilience will hinge on the ability of communities to anticipate changes, mitigate risks, and recover from climate-related shocks.

4.2 Technological Integration

The future of rural resilience will be greatly influenced by advancements in technology. Innovations such as precision agriculture, remote sensing, and data analytics will empower farmers and rural businesses to optimize resource use, increase productivity, and reduce environmental impact. Access to digital tools and platforms will also enhance connectivity, enabling rural communities to engage in e-commerce and access vital information and services.

4.3 Economic Diversification

To bolster resilience, rural areas will increasingly focus on economic diversification. Beyond traditional agriculture, communities will explore opportunities in renewable energy, tourism, and local manufacturing. By creating a more varied economic base, rural regions can enhance their stability and reduce vulnerability to market fluctuations, ensuring sustainable livelihoods for residents.

4.4 Strengthening Social Networks

Social cohesion will be a cornerstone of rural resilience in the future. As communities face shared challenges, fostering strong social networks and collaborative governance will be essential. Engaging residents in decision-making processes and promoting community-led initiatives will enhance collective action and empower individuals to contribute to resilience-building efforts.

4.5 Sustainable Resource Management

Future rural resilience will prioritize the sustainable management of natural resources, recognizing their critical role in community well-being. This includes preserving biodiversity, maintaining healthy ecosystems, and implementing sustainable land-use practices. By valuing and protecting natural resources, rural communities can ensure their long-term viability and contribute to global sustainability goals.

4.6 Cultural Preservation and Adaptation

As rural areas evolve, the preservation of cultural heritage and traditional knowledge will remain vital. Future resilience strategies will integrate local customs and practices with modern sustainability approaches, creating a unique blend that honors the past while addressing present and future challenges. This cultural richness will enhance community identity and cohesion.

4.7 Policy Support and Investment

For rural resilience to flourish, supportive policies and investment are crucial. Governments and organizations will need to prioritize rural development, providing resources for infrastructure, education, and capacity building. Collaborative efforts between public and private sectors will be essential to create an enabling environment for resilience initiatives.

5 Summary

Given the increasing risks and unexpected challenges, the global discourse on rural resilience must continue to evolve, incorporating lessons learned from different regions to address the complex and interconnected challenges facing rural communities today. The book highlights the complexity of building rural resilience in the face of diverse challenges. The insights drawn from the diverse case studies emphasize the importance of tailored strategies that consider local contexts and integrate economic, social, environmental, and institutional conditions. By fostering diversification and transformation, institutional innovation and policy reforms, knowledge synergy and community engagement, and integrating smart technologies, rural regions can enhance their resilience and pursue a sustainable development pattern.

Rural resilience is not just a response to current challenges, while it is also a proactive approach to shaping a sustainable future of the countryside. Case studies from both the developed and developing regions of the world indicate that enhancing rural resilience is a multifaceted challenge that requires a comprehensive understanding of the various factors influencing rural development in terms of socioeconomic and environmental aspects. By investing in and prioritizing rural resilience, we can create vibrant, sustainable communities that thrive in the face of adversity and contribute to a more equitable and resilient world.

Two types of abilities need to be highlighted for rural survival and growth in an increasingly uncertain world. The first is the ability to anticipate future challenges. Effective anticipation ensures that rural communities remain vibrant, resilient, and capable of thriving in the face of an unpredictable future. The second ability is rural coexistence with changing circumstances. Such ability is driven by a combination of stakeholders' resilient thinking and decision making, community cohesion, and innovative practices. In short, proactive planning and actions are key for rural communities to overcome challenges and seize new opportunities for prosperity when addressing environmental risks, economic shifts, technological advancements, or demographic changes in the future.

References

Carr, P. J., & Kefalas, M. J. (2009). *Hollowing out the middle: The rural brain drain and what it means for America*. Beacon Press.

Forth, G. (2000). The future of Australia's declining country towns: Following the yellow brick road. *Regional Science Policy & Practice, 9*(2), 4–10.

Holling, C. S. (1973). Resilience and stability of ecological systems. *Annual Review of Ecology and Systematics, 4*, 1–23.

Li, Y. H. (2023). A systematic review of rural resilience. *China Agricultural Economic Review, 15*(1), 66–77.

Li, Y. H., Westlund, H., Zheng, X. Y., & Liu, Y. S. (2016). Bottom-up initiatives and revival in the face of rural decline: Case studies from China and Sweden. *Journal of Rural Studies, 47*, 506–513.

Li, Y., Westlund, H., Liu, Y. (2019). Why some rural areas decline while some others not: An overview of rural evolution in the world. *Journal of Rural Studies, 68*, 135–143.

Liu, Y. S., & Li, Y. H. (2017). Revitalize the world's countryside. *Nature, 548*(7667), 275–277.

Murphy, B. L. (2007). Locating social capital in resilient community-level emergency management. *Natural Hazards, 41*(2), 297–315.

Ono, A. (2005). *Sanson Kankyo Shakaigaku Jyosetsu [Introduction to Environmental Sociology of Mountain Villages]*. Nouson gyoson bunka kyokai.

Quaranta, G., & Salvia, R. (2014). An index to measure rural diversity in the light of rural resilience and rural development debate. *European Countryside, 6*(2), 161–178.

Steiner, A., & Markantoni, M. (2013). Unpacking community resilience through capacity for change. *Community Development Journal, 49*(3), 407–425.

Wilson, G. (2010). Multifunctional quality and rural community resilience. *Transactions of the Institute of British Geographers, 35*(3), 364–381.

Yang, B., Feldman, M. W., & Li, S. (2020). The status of perceived community resilience in transitional rural society: An empirical study from central China. *Journal of Rural Studies, 80*, 427–438.

Institutional Innovation and Its Effect on Rural China's Socioeconomic Resilience

Yuheng Li

Abstract

Agriculture, villages, and farmers are crucially important to China, which has transformed from an agrarian country to an urbanized country during the past four decades. However, villages which are situated at peripheral places and stay away from the positive impact of big city regions have generally declined due to depopulation, aging, and loss of industrial competitiveness. Historically, there have been a series of institutional innovations on rural issues. The chapter aims to comprehensively review China's institutional innovation toward rural development and to examine how rural areas respond to these changes with increased resilience. Through in-depth analysis, this chapter expects to draw a clear connection between institutional innovation and rural resilience building in China and offers a practical route for rural revitalization strategy.

Keywords

Institutional innovation · Rural resilience · Urbanization · China

Y. Li (✉)
School of Humanities and Law, Northeastern University, Shenyang, Liaoning Province, China
e-mail: liyuheng@wfxy.neu.edu.cn

1 Introduction

As the world has entered into a stage of ever-increasing challenges, uncertainties, and unexpected risks, rural decline, which is normally described as a drop in quality of life and socio-economic decay, has emerged and become a global tendency (Liu & Li, 2017). Among all the explanations for rural decline, policies which have distorted the urban-rural relationship and put the countryside in a disfavoring environment have received particular attention. For example, rural places in Canada used to be considered as a "bank" from which resources were directed to support infrastructure construction and the supply of public services in urban areas (Markey et al., 2008). As a result, sharp rural decline emerged with decreasing resilience against shocks and pressures. It is thus widely accepted that resilience is playing an important role in maintaining the rural socio-ecological system while the concept of resilience indicates how the rural internal system interacts with external environment (Li et al., 2019).

At present, there is increasing public attention paid to resilience studies, which have been widely applied in multidisciplinary subjects like ecology, physics, engineering, economy, etc. (Bruneau et al., 2003; Cimellaro et al., 2010; Tanner et al., 2015; Grafton et al., 2019; Chaigneau et al., 2022). According to Heijman et al. (2007), rural resilience describes the rural

Y. Li (ed.), *Global Perspectives on Building Resilience for Sustainable Rural Development*, Sustainable Development Goals Series, https://doi.org/10.1007/978-981-96-6149-7_2

capacity to deal with external shocks while maintaining rural function and a satisfactory standard of living. So far, various factors have been examined to reveal their role in facilitating rural resilience, including social capital, leadership, collective efficiency, individual cognition, community belief, etc. (Leykin et al., 2013; Cohen et al., 2016; Li et al., 2016a, b; Yang et al., 2020; Xiong & Li, 2024). Rural revitalization is thus viewed as a strategy and process to reverse the trend of rural decline by stimulating socioeconomic development that benefits local communities and residents while preserving and sustaining rural livelihood. In this process, rural resilience is expected to increase through self-reinforcing development capacities and intensified connections with urban areas.

Being a highly populous country of vast territory, China's rural areas have been playing an important role in agricultural production and maintaining peasants' livelihoods. Historically, China's urban-rural relationship has moved forward continuously as the country evolved from a centrally planned country to a market-oriented economy (Li, 2011). Ever since the People's Republic of China was founded in 1949, the new country has faced rather backward socioeconomic conditions, when the majority of its population lived in rural areas. To pursue a quick socioeconomic recovery, the Chinese central government adopted a long-term urban-biased policy and directed resources to support the capital-intensive industries and the development of cities. Then, a series of measures were taken to intensify such a distorted urban-rural relationship. China initiated its reform and opening-up in 1978, and the booming economy and ever-improved urban-rural relationship, due to the adjusted policies and institutions, have largely contributed to the urban-rural interactions. The rural revitalization strategy, which was coined and implemented in 2017, marked a new era when priorities were given to agriculture, villages, and farmers in China. In short, it is the institutional innovation and adjustments, which are embedded at different stages of China's socioeconomic transformation, that have reshaped the development of rural areas.

The term "institution" generally indicates formal rules like laws and regulations, and informal constraints like customs and sanctions, which are to restrict human behavior and activities (North, 1991). Looking back to human history, institutions are normally designed and devised to establish order and reduce uncertainty in responding to the changing social environment. As a result, institutions that are of high efficiency tend to replace those inefficient ones (Li & Hu, 2015). At the micro level, there are evidences that institutional design through collaborative governance has contributed to the creation of robust villages (Chu et al., 2023). However, how has institutional innovation has influenced rural socioeconomic development at the macro level and contributed to rural resilience in China remains to be fully analyzed. Thus, this chapter aims to comprehensively review China's institutional innovation toward rural development and to examine how rural areas respond to these changes with increased resilience. Through in-depth analysis, this chapter expects to draw a clear connection between institutional innovation and rural resilience building in China and offers a practical route for rural revitalization strategy.

2 Evolution of Major Institutional Innovations in Rural China

2.1 Centrally Planned Era Before 1978

In the early stages after P. R. China was founded, land reform was prioritized and implemented to stimulate agricultural productivity and stabilize the society when the state was extremely poor after years of war. China used to have the feudal land lease system which lasted for over two thousand years, where landlords and rich farmers rented out their land to poor tenant farmers who had to pay high rents. Such a system had exacerbated the divide between rich and poor and affected social stability in China. In June 1950, the Chinese government issued the land reform law, which abolished the feudal land system and established land ownership by farmers. In this new system, each household was given the ownership and operational rights of certain land that used to be owned and managed by the landlords. By 1953, the nationwide land reform was implemented, and over 300 million farmers obtained ownership and operational rights to approximately 700 million mu (15 mu equals 1 ha) of land and other essential agricultural production materials.

China's land reform in the 1950s had enabled farmers to secure their basic livelihood and significantly triggered agricultural production. As depicted in Fig. 1, during the land reform period, the total national grain output grew from 113.18 million tons in 1949 to 143.69 million tons in 1952, with an average annual increase rate of 8.9%. Additionally, the total national agricultural output value reached 48.39 billion RMB in 1952, an increase of 48.5% in comparison to 1949. The per capita grain output increased from 209 kg to 288 kg, the per capita cotton output increased from 0.82 kg to 2.29 kg, and the per capita meat output increased from 2.05 kg to 5.95 kg.

Generally, China is a country of numerous smallholder farmers. The 1950s not only saw a weak and backward country but also witnessed how fragile the smallholder farmers were in resisting natural disasters and other risks. Thus, the idea of promoting agricultural cooperation was coined to bring smallholders into a rural collective production system. In December 1953, the "Resolution of the Central Committee of the Communist Party of China (CPC) on the Development of Agricultural Production Cooperatives" was issued and implemented. Then, a collective farming system was introduced so as to stimulate and promote agricultural production by way of collective work. Smallholder farmers thus put their land into the cooperative and obtained land shares. All cooperative mem-

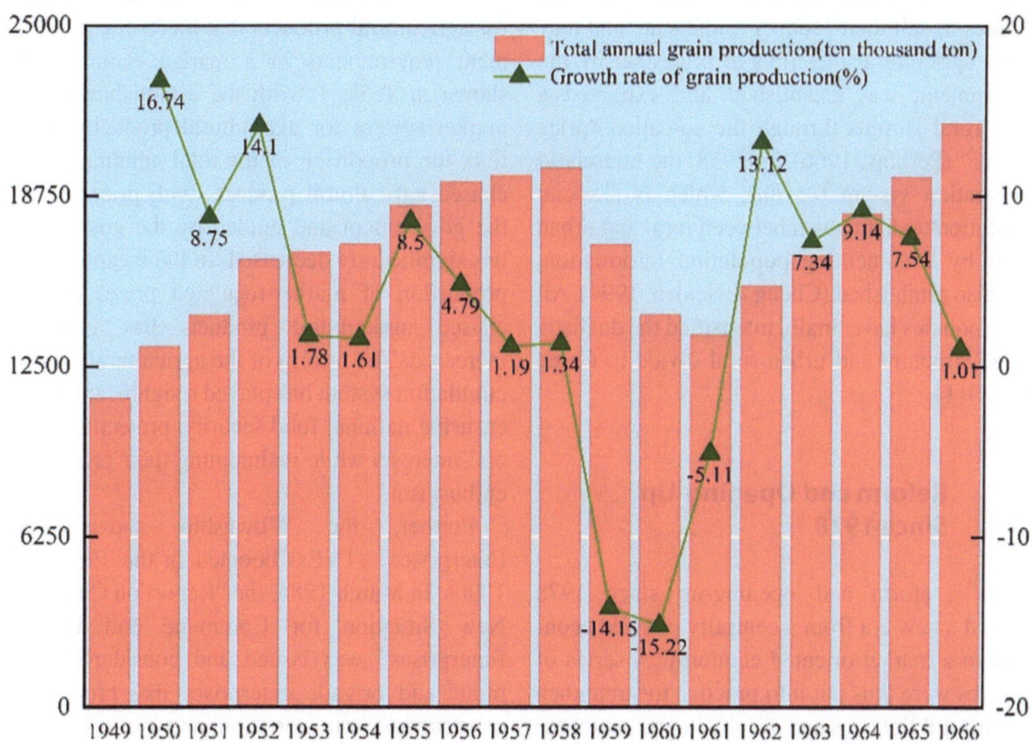

Fig. 1 The total grain production and growth rate in China: 1949–1966. Source: Comprehensive Statistics Department of the National Bureau of Statistics. China compendium of statistics 1949–2008. Beijing: China Statistics Press, 2010

bers work together through mutual assistance in production and unified management of the land.

For one thing, such reform had transformed rural land ownership from individually owned by smallholders to collectively owned by cooperatives. For another thing, the development of agricultural cooperatives had contributed to smallholders' resilience against disasters and risks. By the end of 1956, the number of agricultural cooperatives reached 540,000, with 96% of rural households joining the cooperatives. Statistics in 1956 showed that 326 tractor stations, 14,230 agricultural technology extension stations, 2257 animal husbandry and veterinary stations, 545 livestock breeding stations, and 207 new agricultural machinery stations were established across rural China.

It is noteworthy that the 1950s and 1960s were the periods of China's centrally planned system under which rural areas were the base to support cities and industries through extracting rural surplus (Yang & Cai, 2000). In 1953, the compulsory procurement policy, which requested farmers to sell their product to the state and purchase agricultural materials at prices set by the government, was established and extracted a huge rural surplus through the so-called "price scissor" (Perkins, 1966). In 1958, the household registration system (*hukou*), which is a social institution to distinguish between rural and urban areas by documenting population information, was also established (Cheng & Selden, 1994). All these policies have finally intensified the dualistic social structure and urban-rural divide in China (Li, 2011).

2.2 Reform and Opening-Up Since 1978

China's reform and opening-up since 1978 opened a new era from a centrally planned economy to a market-oriented economy. A series of policies were thus put into practice to strengthen urban-rural linkages and accelerate the socioeconomic development in rural China.

First and foremost, the household responsibility system replaced the collective farming system, whereby each rural household was given a certain plot of land, and farmers were responsible for both the profits and losses associated with managing their own land. By the end of 1983, over 98% of rural households in China had implemented the household responsibility system. This institution has highly triggered farmers' motivation and passion in agriculture by enabling them to deal with their agricultural surplus. As a result, the period from 1978 to 1989 witnessed the increase of China's grain output from 304.8 million tons to 446.2 million tons, a growth rate of 46.4 percent (National Bureau of Statistics of China, 1990).

Accompanying the rapid agricultural development, the compulsory procurement policy was replaced by a market-oriented system of agriproduct circulation. In 1985, the "Ten Policies for Further Activating the Rural Economy" were issued and marked an abolition of the 35-year history of compulsory procurement policy in China. This reform has liberalized the rural market and established a price formation mechanism for agricultural products that meets the development requirements in a market economy. As shown in Table 1, with the establishment of a market system for agricultural product circulation, the proportion of the total amount of purchased agricultural products with prices set by the government and guided by the government has significantly decreased. In the meantime, the proportion of market-regulated prices of purchased agricultural products has gradually increased. The reform of the agricultural product circulation system has played a significant role in ensuring national food security, protecting farmers' interests while maintaining their production enthusiasm.

Further, the "Township and Village Enterprises" (TVEs) boomed in the 1980s and 1990s. In March 1984, the "Report on Creating a New Situation for Commune and Brigade Enterprises" was issued and considered commune and brigade enterprises like processing, construction, transportation, and various service industries outside of agriculture as "township enterprises." In January 1997, the first "Law on Township Enterprises" was officially enacted to

Table 1 Proportion of purchased agricultural products of different prices in China, 1978–2002

Year	Government-set price (%)	Government-guided price (%)	Market-regulated price (%)
1978	92.2	2.2	5.6
1984	67.5	14.4	18.1
1988	37.0	23.0	40.0
1990	25.0	23.4	51.6
1992	12.5	5.7	81.8
1994	16.6	4.1	79.3
1996	16.9	4.1	79.0
1998	9.1	7.1	83.8
2000	4.7	2.8	92.5
2002	2.6	2.9	94.5

Source: Cheng Zhiping. Price Reform 30 Years: 1977–2006. Beijing: China Market Press, 2006

support and guide the rapid development of township enterprises. The number of TVEs reached 11.88 million by the end of 1988, and TVEs employed 95.5 million peasants and produced 29 percent of China's GDP (National Bureau of Statistics of China, 1999). In 1992, TVEs employed 106 million workers in 1992 (52 million in collective enterprises and 54 million in private enterprises), in comparison to 145 million in state-owned and urban enterprises (National Bureau of Statistics of China, 1993).

The market-oriented agriproduct circulation system and TVEs have significantly contributed to the rural economy and households' income. According to Fig. 2, the rural households' per capita net income increased from 133.6 RMB in

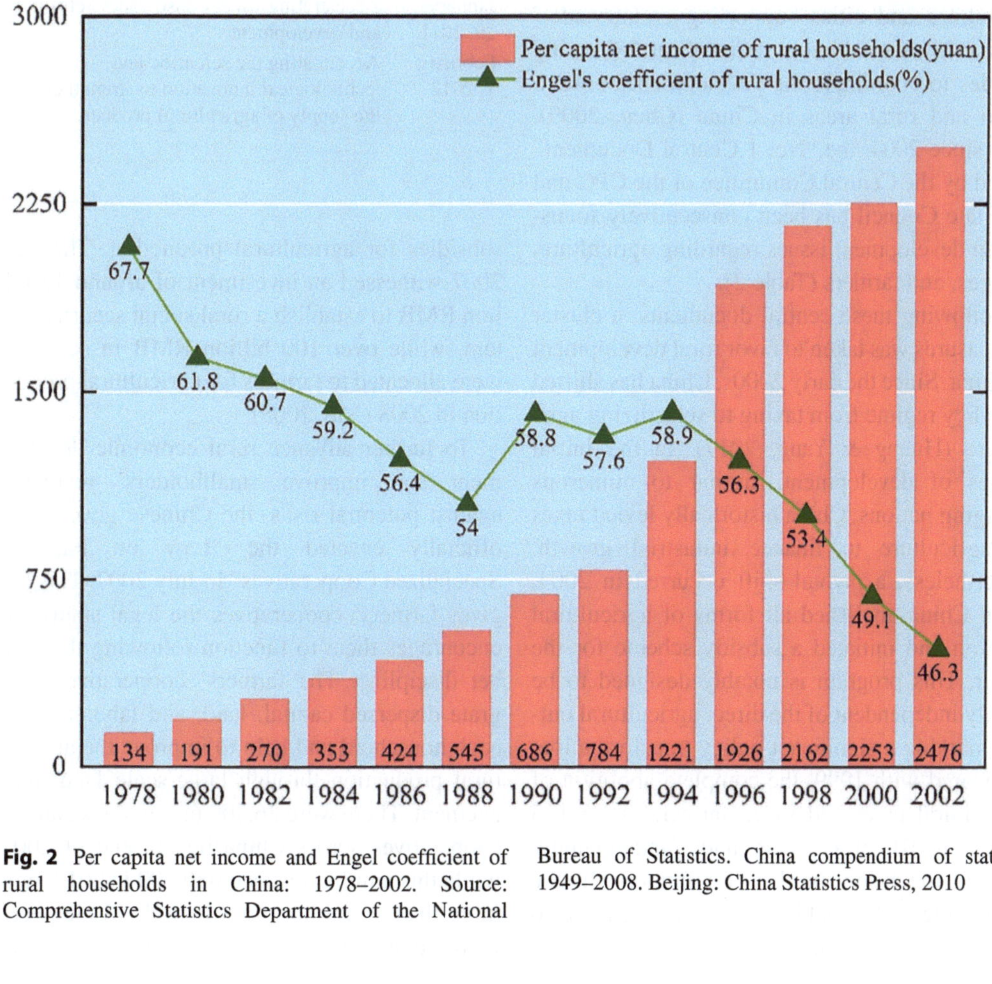

Fig. 2 Per capita net income and Engel coefficient of rural households in China: 1978–2002. Source: Comprehensive Statistics Department of the National Bureau of Statistics. China compendium of statistics 1949–2008. Beijing: China Statistics Press, 2010

1978 to 2475.6 RMB in 2002, with annual average growth at 7.1%. In the meantime, the Engel coefficient of rural households decreased from 67.7% to 46.3%. What's more, rural living standards also improved with the per capita housing area growth from 8.1 square meters in 1978 to 26.5 square meters in 2002.

2.3 Coordinated Urban and Rural Development Phase

A long-term urban-biased environment has enlarged urban-rural inequality which has undoubtedly hindered China's development plan to build a well-off society. The beginning of the twenty-first century has witnessed a shift of China's urban-rural relationship from urban bias to a stage characterized by "industry nurturing agriculture and cities supporting countryside." Such a shift turned into a series of rural-favoring policies to coordinate the development between urban and rural areas in China (Chen, 2009). Ever since 2004, the "No. 1 Central Document" issued by the Central Committee of the CPC and the State Council has been consecutively focusing on development issues regarding agriculture, villages, and farmers (Table 2).

Table 2 No. 1 Central documents of China, 2004–2012

Date	Highlights and theme
February 8, 2004	Boosting growth in farmers' income
January 30, 2005	Strengthening the comprehensive production capacity of agriculture
February 21, 2006	Constructing a new socialist countryside
January 29, 2007	Developing modern agriculture and steadily promoting the construction of a new socialist countryside
January 30, 2008	Strengthening the agricultural infrastructure to promote agricultural development and increase farmers' income
February 1, 2009	Achieving steady agricultural development and sustained income increases for farmers
January 31, 2010	Speeding up coordinated development between urban and rural areas and further cementing the foundation of agricultural and rural area development
January 29, 2011	Accelerating water conservancy reform and development
February 1, 2012	Accelerating the scientific and technological innovation to strengthen the supply of agricultural products
……	……

Following these central documents, a cluster of measures was taken to favor rural development in China. Since the early 2000s, China has shifted its policy regime from taxing to subsidizing agriculture (Huang & Yang, 2017). In the initial phases of development, similar to numerous emerging nations, China historically levied taxes on agriculture to finance industrial growth. Nevertheless, a pivotal shift occurred in 2004, when China abolished all forms of agricultural taxation and initiated a subsidy scheme for the sector. This program is notably designed to be largely independent of the direct agricultural output, marking a significant policy transformation. Compared with 1999, the complete abolition of agricultural taxes had saved farmers about 125 billion RMB every year and fundamentally relieved the heavy burdens on farmers. In the meanwhile, the Chinese government also increased the investment in rural places and in subsidies for agricultural production. The year 2007 witnessed an investment of around 13 billion RMB to establish a rural social security system, while over 100 billion RMB in subsidies were allocated to farmers for agricultural production in 2008 (Xie, 2008).

To further advance rural economic development and improve smallholders' resilience against potential risks, the Chinese government officially enacted the "Law on Farmers' Specialized Cooperatives" in July 2007. This law gives farmers' cooperatives the legal status and encourages them to function following the market discipline. The farmers' cooperatives integrate dispersed capital, land, and laborers from each household and help to improve the agricultural production through large-scale land management. There were 26,400 farmers' specialized cooperatives across China by the end of 2007, while this number reached 680,000 in 2012, with more than 46 million rural household members. Developing farmers' cooperatives, which enhances farmers' market competitiveness, represents another major institutional innovation following the household responsibility system.

In such a rural-favoring environment, urban-rural interactions in China have become stronger than ever before, while much progress has been achieved in rural places. Grain production increased continuously from 469.5 million tons in 2004 to 530.8 million tons by 2009. The rural per capita net income of China increased from 2622 RMB in 2003 to 7917 RMB in 2012. In the meantime, the ratio of urban to rural per capita net income decreased from 3.23 to 3.10 in the same period (National Bureau of Statistics of China, 2012).

2.4 Modernization of Agriculture and Rural Areas

The 18th National Congress of the CPC in November 2012 marked a new era of China's building socialism with Chinese characteristics. The Two Centenary Goals were articulated by the congress, which encompassed the objective of achieving a comprehensively moderately prosperous society by the 100th anniversary of the Communist Party of China (CPC), established in 1921. Additionally, the goals aimed to transform China into a modern socialist nation characterized by prosperity, strength, democracy, cultural advancement, and harmony by the 100th anniversary of the People's Republic of China, established in 1949. It is clearly indicated that rural issues are still the "short boards" of China's socioeconomic development, and this weak area must be strengthened to accomplish the first Centennial Goal of China. In this period, targeted poverty alleviation and rural revitalization strategies were proposed and implemented to accelerate the progress of constructing a strong agricultural country and promoting modernization of agriculture and rural areas.

Boasting a populace of 1.4 billion, China, grappling with underdeveloped infrastructure and uneven growth, has historically endured poverty on an unprecedented scale and severity, a condition seldom paralleled globally. In November 2013, Chinese president Xi Jinping proposed for the first time the concept of "giving differentiated guidance for targeted poverty alleviation in line with local conditions by seeking truth from facts." Then, the Chinese Central Government initiated the targeted poverty alleviation strategy with the aim of taking targeted measures to lift people out of poverty. In contrast to prior poverty alleviation efforts, the current strategy underscores the significance of precisely identifying impoverished areas, strategically deploying suitable projects, and ensuring the precise execution of these initiatives. This approach is designed to guarantee that aid effectively reaches the most impoverished communities and families (Li et al., 2016a, b).

Following the criteria "Two Assurances and Three Guarantees" [This refers to assurances of adequate food and clothing and guarantees of access to compulsory education, basic medical services, and safe housing for impoverished rural residents], 70.17 million impoverished people were identified in rural China. In the meantime, 128,000 impoverished villages and 832 nationally designated poor counties were identified as well to implement targeted measures and assistance. Through targeted poverty alleviation, divergent from previous poverty reduction endeavors, the contemporary approach emphasizes the criticality of pinpointing areas of poverty accurately, the strategic allocation of fitting projects, and the meticulous fulfillment of these endeavors. Such a methodology is crafted to ensure that support is efficiently delivered to the communities and households most severely affected by poverty. The per capita disposable income of the rural poor increased from 6079 RMB in 2013 to 12,588 RMB in 2020, up by 11.6% per annum on average. The growth rate was 2.3 percentage points higher than the national rural average.[1] Over time, their earnings from salaries and operations have escalated in tandem with their income from transfers, indicating an enhanced capacity to overcome poverty through individual initiatives. Concurrently, this initiative

[1] http://english.scio.gov.cn/whitepapers/2021-04/06/content_77380652_4.htm

marked the cessation of the underdevelopment prevalent in impoverished regions and boosted overall rural development with much improved local infrastructure, economy, and public provisions like education, healthcare, housing, and drinking water supplies. By the end of 2020, regions designated as impoverished had seen the construction of 1.1 million kilometers of revamped highways and the addition of 35,000 kilometers of railway lines. Villages, townships, and towns within these areas that possessed the requisite conditions were connected by paved roads and equipped with bus and postal services, thereby promoting economic growth. From 2016 onwards, the land area effectively irrigated expanded by over 5.35 million hectares, and the water supply capacity rose by 18.1 billion cubic meters. Enhanced water infrastructure has been instrumental in bolstering the development of these economically disadvantaged regions.

In 2017, a rural revitalization strategy was proposed as a key move for the development of a modernized economy at the 19th National Congress of the CPC. Chinese president Xi Jinping stressed that addressing the issues pertaining to agriculture, rural areas, and farmers should remain the foremost task on the government's agenda. In line with this, the government report explicitly highlighted the need to focus on agricultural development, aligning it with the overarching objectives of fostering prosperous industries, sustainable living environments, civilized rural communities, affluent lifestyles, and efficient governance.

China released a package of policies in 2018, charting the roadmap for rural revitalization. The schedule mandated the establishment of an institutional and policy framework for the rural revitalization strategy by 2020. At that juncture, it was projected that no individual in China would remain below the poverty line, with a significant enhancement in rural productivity and agricultural output. By 2035, the strategy was expected to achieve "decisive" advancements, with the foundational modernization of agriculture and rural regions completed. Universal access to basic public services for all Chinese citizens, irrespective of urban or rural residence, would be ensured, with enhanced urban-rural integration. By 2050, the vision was for China's rural areas to boast a thriving agriculture sector, picturesque landscapes, and affluent farmers.

Ever since 2012, a range of measures have been implemented to optimize the agricultural sector, improve the living standards of farmers, and enhance rural infrastructure and the environment. In 2020, China experienced its 17th consecutive year of bumper harvests, with grain output rising by 0.9 percent year-on-year to nearly 670 billion kg. This marked the sixth consecutive year that the country's total grain production exceeded 650 billion kg. Rural residents' income growth outpaced that of urban residents for the 11th consecutive year in 2020, further reducing the urban-rural income gap. Additionally, the country achieved its goal of doubling the average per capita income of rural residents by 2019 compared to 2010 levels. A new round of rural grid transformation and upgrades was completed in 2019, reaching a supply reliability rate of 99.8 percent and an integrated voltage qualification rate of 97.9 percent, ensuring stable and reliable power supply services across all rural areas. In terms of the rural living environment, over 65 percent of rural households now have access to sanitary toilets, with more than 30 million households having renovated their toilets since 2018. Furthermore, sewage and waste management in rural areas has improved, with waste collection, transportation, and disposal systems now covering over 90 percent of administrative villages.

3 How Institutional Innovation Contributes to Rural Socioeconomic Resilience

Resilience, as a key component under the social-ecological system framework, emerged in the 1980s (Carpenter et al., 2012) and has been developed for the rural context, which faces increasing challenges and risks. Rural places, because of the existence of people, are proactive and dynamic objects which possess the ability to mobilize resources and make necessary changes to minimize the impact of shocks so as to sustain a satisfactory standard of people's livelihood (Wilson, 2010; Quaranta & Salvia, 2014). Thus, local people's decision-making and action, which are interwoven with their access to available resources, determine how rural communities respond to shocks and disturbances (Li, 2023). Besides, the institutional innovation and deployments play an important role in sustaining rural development, providing essential support, resources, and infrastructure to empower rural communities (Li and Hu, 2015). The abovementioned institutional innovation over time in China has proved this positive role (Fig. 3).

China's experience of accelerating rural development through institutional innovations could be explained by the logical relationship between productivity and production relations. It can be seen that institutional innovation emerged each time the old production relations constrained productivity. In the early stage after the P. R. China was founded, the feudal land ownership, which used to exist for over two thousand years, had greatly restricted agricultural productivity. Farmers who had no land resources were exploited with high land rent charged by the landlords. The elimination of feudal land ownership and land redistribution to all farmers had highly triggered their enthusiasm for agricultural production. Then, the introduction of the household responsibility system had intensified the relationship between farmers and their land and further motivated farmers' agricultural production.

Reform and opening-up and the transition from urban bias to "industry nurturing agriculture and cities supporting countryside" have reshaped the urban-rural relationship in China. The consecutive "No. 1 Central Documents" since 2004 have created a rural-favoring environment to support issues related to agriculture, villages, and farmers. In this circumstance, the improved production relations have witnessed intensified resource flows and linkages between urban and rural areas. The rural economy thus became diversified, and households' income increased through flexible employment in different sectors. As a result, there was improved rural livelihood and development capacity in comparison to the era before reform and opening-up in 1978.

The rural revitalization strategy which emphasizes the advancement of agriculture and rural areas has established a long-term policy environment which favors rural China. A series of policies and institutional innovation are thus expected to be created and implemented to support this strategy. These are believed to further advance

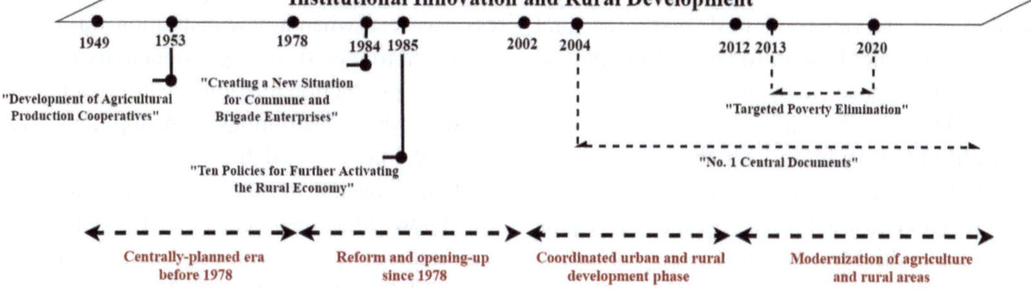

Fig. 3 Diagram of institutional innovation in China

the socioeconomic development of the countryside. By the time when the rural revitalization strategy is to be accomplished, it foresees building modernized and resilient rural areas of China with thriving businesses, pleasant living environments, social etiquette and civility, effective governance, and prosperity.

Besides the compulsory institutional changes that rely on law and government power, which were top-down implemented by the governmental agents, it is worth highlighting the institutional changes driven by grassroots actors, who, motivated by the pursuit of their own benefits, have spontaneously initiated bottom-up transformations. The household responsibility system, which replaced the former collective farming system, was originally driven by 18 farmers in Xiaogang village, Central China's Anhui Province. They agreed to distribute land to each household which was accountable for both the gains and the losses linked to its respective land. Such initiative not only triggered local farmers' production enthusiasm but also finally turned into a national institution which was formally established and implemented across rural China.

With the ability to independently manage their land, an increasing number of farmers shifted away from agricultural work, opting instead to pursue non-agricultural employment in urban areas or to establish new businesses. For one thing, the *hukou* system, which used to strictly control population mobility, became loose and enabled farmers to move to cities and engage in industrial or commercial activities. The provision of public services (including compulsory education, employment services, pensions, healthcare, and housing) also extended to those rural immigrants. For another thing, farmers were encouraged to start and run TVEs to diversify the rural economy after the "Law on Township Enterprises" was officially enacted by the Chinese government. What's more, the "Law on Farmers' Specialized Cooperatives," which was enacted in July 2007, was also an official response to farmers' intention to establish cooperatives to gain group competitiveness and maximize profits. The law gives farmers' cooperatives the legal status to mobilize capital, land, and laborers to manage large-scale agricultural production.

In short, China's rural development can be attributed to both the top-down compulsory institutional changes and bottom-up induced institutional changes. These two forms of institutions have jointly transformed the socioeconomic patterns of the villages through history. These institutional innovations had improved production relations and further advanced productivity, which laid a solid foundation to build socioeconomic resilience in rural China.

4 Concluding Remarks

Agricultural, rural, and farmer-related issues are fundamental to China, as they directly impact the nation's stability and the well-being of its people. Addressing these issues has a central place on the work agenda of the CPC and has remained a top priority for China. There has been significant progress in rural development in China as the state moved forward from a poor and vulnerable country to an energetic and resilient country. Nevertheless, the phenomenon of rural decline owing to rapid depopulation and related factors has been widely existing across China. The COVID-19 pandemic, climatic disasters, and the unexpected risks further added to the vulnerability of those declining villages. Thus, improving rural resilience against potential challenges and risks has become significantly necessary and an important contributor to sustainable rural communities in China in the long run.

Historical review has unveiled the evolution process of China' institutional innovation toward rural socioeconomic development. Much progress was seen when a new institution was established and triggered soaring productivity through improved production relations in the countryside. Although these institutions were top-down created and implemented by the Chinese government, the proactive initiatives from the grassroots who know what they really need shouldn't be neglected. And they are playing an important role in facilitating the institutional innovations in rural China. In this process, innovative institutions have empowered farmers with autonomy and contributed to the socioeconomic connections between urban and rural regions. As a consequence, improved livelihood and development capacities were seen and finally led to increased rural resilience against various challenges in China.

Rural revitalization strategy which is ambitious with definite goals demands a well-established institutional system which is expected to release its potential and reap the benefits of reform in the future. As rural households' livelihood rely more on non-agricultural sectors and more villagers tend to migrate to cities, the rural human-land relationship changes rapidly in China. Who is to cultivate the farmland? How to deal with the hollowing-out villages? How to govern the villages with more aged people and declining vitality? Thus, comprehensive institutional innovation is needed to induce behavioral changes and adaptation to new circumstances so as to properly cope with these problems. What's more, institutional innovation for promoting the rural revitalization strategy necessitates the involvement of multiple agents, especially the grassroots. How to efficiently motivate local stakeholders and have their voices heard by the decision-makers is crucially important to initiate practical and feasible institutions and policies for rural China.

References

Bruneau, M., Chang, S. E., Eguchi, R. T., Lee, G. C., O'Rourke, T. D., Reinhorn, A. M., Shinozuka, M., Tierney, K., Kathleen, W., William, A., & Von Winterfeldt, D. (2003). A framework to quantitatively assess and enhance the seismic resilience of communities. *Earthquake Spectra, 19*(4), 733–752.

Carpenter, S. R., Folke, C., Norström, A., Olsson, O., Schultz, L., Agarwal, B., Blavenera, P., Campbell, B., Castilla, J. C., Cramer, W., DeFries, R., Eyzaguirre, P., Hughes, T. P., Polasky, S., Sanusi, Z., Scholes, R., & Spierenburg, M. (2012). Program on ecosystem change and society: An international research strategy for integrated social–ecological systems. *Current Opinion in Environmental Sustainability, 4*(1), 134–138.

Chaigneau, T., Coulthard, S., Daw, T. M., Szaboova, L., Camfield, L., Chapin, F. S., III, Gasper, D., Gurney, G. G., Hicks, C. C., Ibrahim, M., James, T., Jones, L., Matthews, N., McQuistan, C., Reyers, B., & Brown, K. (2022). Reconciling well-being and resilience for sustainable development. *Nature Sustainability, 5*, 287–293.

Chen, X. W. (2009). Review of China's agricultural and rural development: Policy changes and current issues. *China Agricultural Economic Review, 2*(1), 121–135.

Cheng, T. J., & Selden, M. (1994). The origins and social consequences of China's hukou system. *China Quarterly, 139*, 644–668.

Chu, V. H. Y., Lam, W.-F., & William, J. M. (2023). Building robustness for rural revitalization: A social-ecological system perspective. *Journal of Rural Studies, 101*, 103042.

Cimellaro, G. P., Reinhorn, A. M., & Bruneau, M. (2010). Framework for analytical quantification of disaster resilience. *Engineering Structures, 32*(11), 3639–3649.

Cohen, O., Geva, D., Lahad, M., Bolotin, A., Leykin, D., Goldberg, A., & Aharonson-Daniel, L. (2016). Community resilience throughout the lifespan – the potential contribution of healthy elders. *PLoS One, 11*(2), e0148125.

Grafton, R. Q., Doyen, L., Bene, C., Borgomeo, E., Brooks, K., Chu, L., Cumming, G. S., Dixon, J., Dovers, S., Garrick, D., Helfgott, A., Jiang, Q., Katic, P., Kompas, T., Little, L. R., Matthews, N., Ringler, C., Squires, D., Steinshamn, S. I., Villasante, S., Wheeler, S., Williams, J., & Wyrwoll, P. R. (2019). Realizing resilience for decision-making. *Nature Sustainability, 2*(10), 907–913.

Heijman, W., Hagelaar, G., & Heide, M. (2007). *Rural resilience as a new development concept*. EAAE Seminar Serbian Association of Agricultural Economists.

Huang, J. K., & Yang, G. L. (2017). Understanding recent challenges and new food policy in China. *Global Food Security, 12*, 119–126.

Leykin, D., Lahad, M., Cohen, O., Goldberg, A., & Aharonson-Daniel, L. (2013). Conjoint community resiliency assessment measure-28/10 items (CCRAM28 and CCRAM10): A self-report tool for assessing community resilience. *American Journal of Community Psychology, 52*(3), 313–323.

Li, Y. (2011). Urban-rural interaction in China: Historic scenario and assessment. *China Agricultural Economic Review, 3*(3), 335–349.

Li, Y. (2023). A systematic review of rural resilience. *China Agricultural Economic Review, 15*(1), 66–77.

Li, Y., & Hu, Z. (2015). Approaching integrated urban-rural development in China: The Changing Institutional Roles. *Sustainability, 7*, 7031–7048.

Li, Y., Su, B., & Liu, Y. (2016a). Realizing targeted poverty alleviation in China: People's voices, implementation challenges and policy implications. *China Agricultural Economic Review, 8*(3), 443–454. https://doi.org/10.1108/CAER-11-2015-0157

Li, Y., Westlund, H., Zheng, X., & Liu, Y. (2016b). Bottom-up initiatives and revival in the face of rural decline: Case studies from China and Sweden. *Journal of Rural Studies, 47*, 506–513.

Li, Y., Westlund, H., & Liu, Y. (2019). Why some rural areas decline while some others not: An overview of rural evolution in the world. *Journal of Rural Studies, 68*, 135–143.

Liu, Y., & Li, Y. (2017). Revitalize the world's countryside. *Nature, 548*(7667), 275–277.

Markey, S., Halseth, G., & Manson, D. (2008). Challenging the inevitability of rural decline: Advancing the policy of place in northern British Columbia. *Journal of Rural Studies, 24*, 409–421.

National Bureau of Statistics of China. (1990). *China Statistical Yearbook*. China Statistics Press.

National Bureau of Statistics of China. (1993). *China Statistical Yearbook*. China Statistics Press.

National Bureau of Statistics of China. (1999). *50 Years of P.R. China: 1949–1999*. China Statistics Press.

National Bureau of Statistics of China. (2012). *China Statistical Yearbook*. China Statistics Press.

North, D. C. (1991). Institutions. *The Journal of Economic Perspectives, 5*, 97–112.

Perkins, D. H. (1966). *Market control and planning in communist China*. Harvard University Press.

Quaranta, G., & Salvia, R. (2014). An index to measure rural diversity in the light of rural resilience and rural development debate. *European Countryside, 6*(2), 161–178.

Tanner, T., Lewis, D., Wrathall, D., Bronen, R., Cradock-Henry, N., Huq, S., Lawless, C., Nawrotzki, R., Prasad, V., Rahman, M. A., Alaniz, R., King, K., McNamara, K., Nadiruzzaman, M., Henly Shepard, S., & Thomalla, F. (2015). Livelihood resilience in the face of climate change. *Nature Climate Change, 5*(1), 23–26.

Wilson, G. (2010). Multifunctional quality and rural community resilience. *Transactions of the Institute of British Geographers, 35*(3), 364–381.

Xie, X. (2008). *Report on China' central and local budgets*. Available at: http://english.gov.cn/official/2008-03/20/content_925090.htm (accessed March 10, 2010).

Xiong, A., & Li, Y. (2024). The role of social capital in building community disaster resilience – empirical evidences from rural China. *International Journal of Disaster Risk Reduction, 110*, 104623.

Yang, D. T., & Cai, F. (2000). *The political economy of China's rural-urban divide*. Working paper, Center for Economic Research on Economic Development and Policy Reform, Stanford Institute of Economic Policy Research, Stanford University, Stanford, CA, August 1.

Yang, B., Feldman, M. W., & Li, S. (2020). The status of perceived community resilience in transitional rural society: An empirical study from central China. *Journal of Rural Studies, 80*, 427–438.

Smart and Compact Rural Settlements for Resilient and Sustainable Development in Nepal: Policy and Implementation Perspectives

Sunil Babu Shrestha, Rojee Pradhananga, and Krishna Bhattarai

Abstract

Nepal, being a mountainous country with a significant rural population, faces numerous challenges in achieving sustainable development and building resilience. With a predominantly rural landscape and a history of agriculture-based livelihoods, Nepal has prioritized the development of rural areas and agriculture programs. However, the scattered and dispersed nature of rural settlements presents challenges in terms of infrastructure development due to high costs and limited funding. Moreover, the difficult terrain of mountains and hills adds to the complexity of providing essential services and infrastructure in an eco-sensitive manner. The recent restructuring of the state has resulted in 753 local units, with the majority categorized as rural municipalities. While more than 66% of the population now resides in urban municipalities, these areas still retain rural characteristics and face socio-economic challenges due to the lack of investment in infrastructure. Recognizing that today's rural areas will become tomorrow's municipalities, the Government of Nepal aims to promote resilient pathways to sustainable development through smart and compact rural settlement policies and implementation strategies. In this context, this book chapter examines the potential of smart villages and compact rural settlements as effective strategies for sustainable development in Nepal. Furthermore, it explores the need for supportive policy frameworks that incentivize the development of such settlements, including land-use planning policies, infrastructure guidelines, and governance mechanisms. It also emphasizes the importance of integrating smart technologies into rural development policies to enhance connectivity, access to services, and resource management. The research methodology employed in this study includes a comprehensive discussion on the review and analysis of the information obtained from secondary sources, such as literature reviews of relevant books, journals, websites, government budget

S. B. Shrestha (✉)
Nepal Academy of Science and Technology (NAST), Khumaltar, Lalitpur, Nepal

Nepal Development Research Institute (NDRI), Lalitpur, Nepal
e-mail: sunil.shrestha@ndri.org.np

R. Pradhananga
Pulchowk Campus, Institute of Engineering, Tribhuvan University, Lalitpur, Nepal

K. Bhattarai
Thapathali Campus, Institute of Engineering, Tribhuvan University, Kathmandu, Nepal

Y. Li (ed.), *Global Perspectives on Building Resilience for Sustainable Rural Development*, Sustainable Development Goals Series, https://doi.org/10.1007/978-981-96-6149-7_3

speeches, census data, periodic plans of Nepal, government policies and planning documents, as well as publications on rural and urban settlement development. The policy implications of this book chapter emphasizes the need for a holistic and forward-thinking policy framework that supports the adoption of smart and compact rural settlements as a pathway to sustainable development in Nepal.

Keywords

Compact rural settlements (CRS) · Smart villages · Resilient development · Policy implications · Implementation perspectives

1 Introduction

Nepal's population reached approximately 30 million people, distributed across 753 local units in 7 provinces following a recent state restructuring. Among these, 460 local units are classified as rural municipalities. The pictorial representation of rural municipalities is shown in Fig. 1. Koshi Province has the most rural municipalities (88), while Karnali Province and Sudurpashchim Province have the least (54 each). Even though most of the Nepal's population (66.2%) lives in urban municipalities (NSO, 2021a), rural characteristics persist in these areas. Geographically, Nepal is divided into Mountain, Hill, and Terai regions. The 2021 census highlights a big difference in the percentage of rural population in different regions. For example, in the mountains, 64.4% of people live in rural municipalities, much higher than the national average of 33.8%. This means that it is really important for Nepal to focus on developing rural areas sustainably, especially in the mountains, to help the people who live there.

Rural settlements in Nepal, particularly in mountainous and hilly regions, are often small and scattered, posing major challenges to sustainable rural development. This dispersion results in a low population density of 105 persons per square kilometer in rural municipalities, compared to 372 people per square kilometer in urban municipalities, dropping to as low as 34 people per square kilometer in mountainous regions (NSO, 2021b). Consequently, providing basic infrastructure and services becomes challenging, contributing to persisting issues such as low literacy rates and limited access to public services. The current statistics shows 28.15% of the rural population is still illiterate and 15% still do not have access to electricity. Similarly, still 75% of the rural households rely on firewood for cooking, and only 62.6% of the rural households have access to tap water (NSO, 2021b). Marked by inadequate infrastructure development and poor health services, together with increased cost and difficulty of disaster relief and recovery (Gao et al., 2007), scattered settlement causes serious repercussions such as low income, farmland abandonment, and rampant migration to urban areas. According to Census 2021, about 29% of the population is residing in places different from their place of residence prior to the census, with 61.6% moving from rural to urban municipalities for various reasons including work, business and education.

These challenges highlight the necessity to promote compact settlements in Nepal. International experiences highlight the benefits of compact settlements in enhancing accessibility to infrastructure and services, optimizing resource utilization, and bolstering resilience against natural disasters, particularly in the rural areas of mountains and hills. Smart and Compact Rural Settlements (SCRSs) hold particular promise for rural development by improving quality of life, economic well-being, and preserving cultural and historical values. Various governmental and non-governmental organizations have recognized the potential of SCRSs in Nepal, with the National Planning Commission (NPC) initiating policies and strategies for their implementation.

This book chapter examines the current structure, status, and challenges of the rural development in Nepal. It introduces the concept of SCRSs as potential solutions to sustainable rural development and presents some case studies on SCRS efforts both nationally and internationally. Furthermore, it delves into the necessary policy frameworks and strategies required for successful

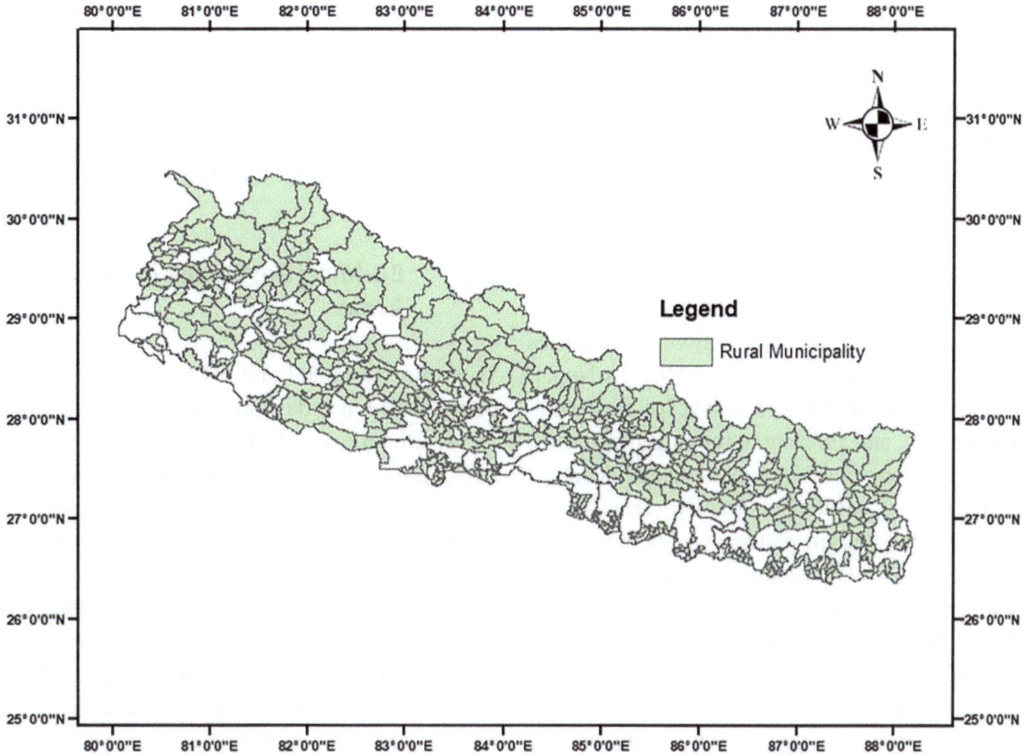

Fig. 1 Rural municipalities in Nepal

implementation, along with suggesting policy implications and strategies specifically tailored for Nepal.

2 Sustainable Development and Resilience in Rural Nepal

2.1 Challenges in Achieving Sustainable Rural Development

Sustainable Rural Development (SRD) is a concept aimed at improving the well-being of rural populations while preserving natural resources and the environment for future generations. It holds particular significance for rural areas due to the persistent issue of poverty, which significantly restricts access to both the quantity and quality of food (Montaldo, 2013). Workers in rural regions of developing countries often earn minimal wages, leading to reduced food production or distribution (Sheaffer & Moncada, 2012). Furthermore, poor infrastructure and healthcare services exacerbate the challenges faced by rural communities. According to the World Bank, the global rural population comprised approximately 43.94% of the total population in 2020, a percentage that has steadily declined from 57% in 1990 (World Bank, 2022). The trend is expected to continue as more individuals migrate to urban areas seeking better opportunities and services.

Rural communities encounter various challenges, including poverty, inadequate infrastructure, environmental degradation, and social exclusion. In Asia, the rural sector faces demographic shifts, diminishing local economies, a dearth of skilled labor, and a lack of entrepreneurship. The absence of fundamental life-sustaining amenities and economic opportunities leads economically active age groups to migrate to urban hubs within the country and abroad, seeking more promising prospects, significantly impacting rural development in Nepal.

Nepal, characterized by its multi-ethnic, multi-cultural, and multi-religious community, features vast rural territories with sparse populations and uneven regional economic development. Rural settlements serve as the regional conduits for natural, social, and economic composite systems. Spatial distribution and clustering of rural settlements across various scales and landforms exhibit distinct regional disparities (Yang et al., 2016; Luo et al., 2020). In Nepal, rural settlements exhibit a low level of clustering, limited scale, and substantial regional disparities, posing challenges to sustainable rural development. Notably, settlement density varies across regions, with denser patterns in economically advanced south plain regions compared to underdeveloped north hill and mountainous regions. That is, 460 people per square kilometer in south plain or Terai region, whilst only 34 people per square kilometer in Himalayan regions (NSO, 2021a). Poverty is particularly acute in the Himalayan region, with persistent economic disparities affecting education, health, and the environment.

Despite 66.2% of Nepalese residing in urban municipalities (NSO, 2021a), rural poverty remains significantly higher (28%) compared to urban poverty (12.3%) (NPC, 2021). Moreover, a majority (52.4%) of the poor people are concentrated in rural regions, emphasizing the continued necessity of prioritizing rural development initiatives. Human Development Index (HDI) indicators also reflect lesser progress in rural areas compared to urban areas, underscoring the need for targeted interventions.

Over the past few decades, Nepal's rural areas have witnessed a decline in population driven by urbanization. Different factors influence the rural settlement distribution that varies across different scales. At the national level, factors such as geomorphology, climate, and population distribution play pivotal roles, whereas at a more local scale, such as within wards, factors like slope, aspect, and availability of water resources become the primary influencers. Infrastructure development, government policies, ethnic customs, and cultural practices also exert an instrumental role in the rural settlement's spatial distribution, resulting in uneven economic development and unplanned expansion. For the rural economy and society to evolve rapidly, the scattered layout and lack of planning pose threats.

2.2 Building Resilience in Rural Areas

To ensure sustained and resilient rural development, policy agendas must prioritize several key areas: strengthening healthcare systems, supporting agricultural production, and emphasizing green, resilient, and inclusive development (World Bank, 2020). Prior to fortifying community resilience, it is imperative to identify both the at-risk and safe zones within rural villages. In Nepal, under the new three-tier government structure, there exist 460 rural municipalities, which are the lower administrative divisions of the country. Rural municipalities situated in remote regions are particularly vulnerable to natural disasters and often suffer from isolation and inadequate infrastructure, hindering accessibility. Thus, gaining a comprehensive understanding of the region's topography, environment, as well as the socio-economic and cultural traits of its inhabitants is crucial.

Economic growth driven by market forces can contribute to resilience and bolster resilience-building endeavors. However, it is crucial to acknowledge that traditional ways of life and livelihoods are not inherently resilient, despite claims made by certain populist figures. Moreover, certain forms of market-driven growth and the supporting policies may actually diminish resilience. Hence, it is imperative to critically analyze how such growth is achieved, its specific manifestations, and its impacts on diverse livelihoods (Oven & Rigg, 2016). Projects that leverage local resources and reflect the unique character of a locality should be prioritized for resilience-building efforts. This entails assessing existing settlements in rural regions to identify natural vulnerabilities, including dangerous zones. Additionally, identifying potential sites for resettling scattered communities from these hazardous areas is imperative. The selection of resettlement sites should consider various factors, including enhancing rural livelihood opportunities, improving infrastructure, facilitating access to energy and water resources, enhancing education and healthcare provisions, promoting on- and off-farm livelihood opportunities, cultivating high-value medicinal and aromatic plants, fostering tourism, nurturing rural enterprise development, and engaging the private sector.

2.3 Importance of Rural Settlements in Nepal

Rural settlements serve as the nexus of human interaction with the land and have generated considerable interest in recent years as the key hubs of rural activity (Li & Song, 2023). As the primary spatial unit for rural activities, these settlements are pivotal in shaping the rural economy and societal dynamics (Phillips, 1998). In Nepal, they are the primary form of habitat for 33.8% of the population, and 66.6% of this population relies on agriculture, forestry, and fisheries for their livelihoods (NSO, 2021a). Additionally, they hold significant religious and cultural importance as sites of ancient temples, monuments, and heritage, showcasing the diversity and richness of Nepalese culture. Moreover, rural settlements in Nepal are also important from tourism perspective, as they attract tourists seeking to immerse themselves in Nepal's natural beauty, rural lifestyle, and cultural diversity. However, as the urban economy and society continue to develop, the distribution attributes and spatial framework of rural settlements undergo continual change, leading to scattered layouts due to inadequate planning. This evolution sometimes poses threats to cultivated land areas (Whatmore, 1993; Zhang et al., 2014; Oven & Rigg, 2016). Moreover, there is a peak in youth migration to urban or foreign areas. According to NSO (2021b), 23.38% of rural households in Nepal have at least a person living abroad. Additionally, the recent COVID-19 pandemic has affected the rural areas more severely due to exorbitant prices and income shocks. Addressing these challenges requires creating better work and education environments in the rural areas, a goal that can be achieved through the development of smart and compact villages.

3 Smart and Compact Rural Settlements

3.1 Concept of Smart Village and Compact Rural Settlement

In Nepal, the NPC prepared guidelines and procedures for smart village initiatives in 2018. Smart villages are a rural development concept aimed at addressing local challenges and enhancing living standards. Rural areas often face issues such as poverty, inadequate education, and restricted technology access. The concept of the smart village emerged from the recognition of distinct differences between rural and urban areas (Aziiza & Susanto, 2020). The authors proposed a model of smart villages consisting of six dimensions: Resources, Technology, Living, Village Service, Tourism, and Governance. According to Singh and Patel (2015), technology, particularly driven by Information and Communication Technology (ICT) awareness, plays a pivotal role in establishing smart villages and fostering local economic development. Moreover, it facilitates infrastructure investment, business growth, skill development, and community empowerment (Guzal-Dec, 2018).

Smart and compact rural settlements embody innovative approaches to rural development, leveraging digital technologies, social innovation, and community engagement to elevate the standard of life, economic prosperity, and environmental sustainability in rural areas (Klenova et al., 2020). A SCRS doesn't always equate to a small village; instead, it refers to a rural area that maximizes resource utilization, minimizes ecological impact, and strengthens social cohesion (Cvar et al., 2020). Various models for SCRSs include eco-villages, emphasizing ecological governance; agri-villages, utilizing digital technologies for agricultural productivity; and smart cultural villages, dedicated to preserving and promoting cultural heritage. These models may also blend aspects of these approaches for comprehensive rural development strategies.

3.2 Benefits of Smart and Compact Rural Settlements

Considering the layout of rural settlements across the country, it is essential to establish a functional hierarchy among them. This hierarchical structure is crucial for ensuring adequate service provision in rural areas and optimizing the economic effectiveness of investments. It implies that the development of the SCRSs fundamentally relies on the premise that there must be a minimum population threshold for the diverse tiers of rural service/market centers to cater to the needs of the rural hinterland (NPC, 2013).

SCRSs offer significant benefits in enhancing the villagers' quality of life through technological innovation. This effort will help fulfill the lack of primary infrastructure needs, namely electricity, roads, water supply, and sanitation. By promoting economic sustainability in villages, it mitigates rural-urban migration pressures. Many rural residents, particularly in the western hill districts, face food shortages, compelling them to seek employment in India during the off-farming seasons. Additionally, Nepalese youth often opt to leave rural areas (Chaudhary, 2023). Therefore, fostering food security is imperative, achievable through rural economic development. Furthermore, adapting to climate challenges is also vital for rural livelihoods and productivity. The COVID-19 pandemic underscored the significance of local economies and food security, especially in areas experiencing insufficient food production and resulting malnutrition. Thus, the rationale for promoting SCRSs to support sustainable micro-economies is obvious. This approach also aids in addressing issues in the informal sector and rejuvenating agriculture and the rural economy through technology and innovation.

3.3 Role of Technology in Rural Development

Technology intervention has the potential to reduce the gap between urban and rural areas (Tiwari, 2023). It serves as a catalyst for improving education in rural areas, offering economic benefits. Developing appropriate and innovative technologies, particularly the utilization of Information and Communication Technology (ICT), is essential for the advancement of smart and compact rural development. This ensures a stable and adequate supply of basic needs and easy access to those supplies. It also entails creating jobs and generating income to alleviate poverty while safeguarding the environment and natural resources.

The development of the SCRS should emphasize building digitally skilled rural communities. This leads to income diversification, paving avenues for new employment and business opportunities for future generations, both on and off the farm. It also empowers individuals to develop these skills themselves. While significant advancements in agriculture historically centered on mechanical and genetic advancements, like enhancing the power and efficiency of machinery and boosting seed and fertilizer production, the next leap in the productivity requires the adoption of more sophisticated digital tools. These tools enable efficient and sustainable resource utilization. Additionally, new technologies offer improved decision-making capabilities, enabling better management of risk and variability to optimize yields and enhance economic outcomes. For instance, employing technologies in animal husbandry has the potential to enhance the well-being of livestock, effectively addressing the increasing concerns regarding animal welfare.

Furthermore, technology can stimulate rural tourism by showcasing local culture, heritage, and products, and by linking rural areas with urban and global markets (Monda et al., 2023). Likewise, Information and Communication Technologies (ICTs) have the capacity to drive development in underprivileged areas through the implementation of E-governance projects. E-Governance simplifies processes and enhances access to government information that benefits the rural population by improving service efficiency and delivery, standardization, accessibility, transparency, and accountability. It fosters direct connections between citizens and local government, encouraging their participation in governance processes. Eventually, it contributes to poverty alleviation, facilitates the direct involvement of women in government schemes and decision-making processes, and enhances democratic practices and citizen empowerment (Bhatnagar & Schware, 2000).

4 Case Studies

4.1 Examples from Asia

Smart and compact rural development has been studied and practiced in many regions of Asia, offering valuable insights into effective strategies and innovative approaches that can be replicated across the region. In a study titled 'Sikkim—Once a Captivating au *Naturel* Himalayan Kingdom in the Light of its Growing Urbanscape', the authors highlight the importance of compact development preserving ecological integrity while managing urbanization (Bandyopadhyay et al., 2020). As discussed previously, smart technology is an important constituent that facilitates smart agro-villages in rural areas. A study conducted by Hebsale et al. (2023) regarding the implementation of climate-resilient agriculture technologies among small-scale farmers in Gujarat discovered that emphasizing the significance of their involvement in the governance process related to the advancement of such technologies is crucial. This inclusiveness of stakeholders ultimately helps in promoting resilience in villages.

Liao and Huan (2015) discuss the development of smart and green rural settlements in rapidly urbanizing China, highlighting political-economic reform of the country in the 1980s aimed at protecting rural land from being transacted into the property market. However, these reforms also created barriers to rural development and investment, resulting in socio-economic discrepancies between urban and rural communities in terms of income gaps, economic stagnation, impoverishment, and excess labor in agriculture, stimulating rural-to-urban migration. The authors suggest encouraging farmers to consolidate their farmland, removing obstacles between urban and rural societies to facilitate the free movement of capital and people, and implementing transparent, pragmatic, and systematic strategies for planning and development control mechanisms regulating construction in the rural settlements. Likewise, a case study on rural settlement planning in Baidyerbazar Union, Bangladesh, suggests the immediate formulation of physical planning in response to decreasing agricultural land (Hanif et al., 2015). Furthermore, Alam et al. (2016) discuss a densification strategy in rural Bangladesh and caution against the direct adoption of foreign policies without a meticulous understanding of the local context. The study also analyzed the practicality of the draft Agriculture and Land Protection, and Land Use Bill. The finding from the study of two villages—Tetultala and Chhoygharia—suggest that the unique and evolving dynamics of rural settlements are leading to the erosion of farming practices and farmland within rural society, thereby disrupting traditional village morphology (Alam et al., 2016). This study serves as a good example of policy shortcomings that a developing country like Nepal can learn from.

4.2 Examples from Nepal

4.2.1 Potential Compact Rural Settlements in Kalikot, Humla, Myagdi, Rashuwa, Pyuthan, Achham, and Gulmi Districts

An initial step towards establishing CRSs in Nepal involved feasibility studies conducted under the initiation of DUDBC to identify potential CRS locations in Kalikot, Achham, Humla, Myagdi, Rashuwa, Pyuthan, and Gulmi districts. These studies assessed various criteria encompassing physical infrastructure, geography, social dynamics, economics, and environmental conditions and provided a ranking of the locations based on their suitability for CRS development. Parameters such as roads, water supply, geo-structure, potential for expansion, environmental factors, population, education, migration patterns, healthcare, security, community services, market services and facilities, domestic industries, livestock, forest resources, agricultural productivity, and tourism were key considerations in the evaluation process. In Kalikot district, for instance, the feasibility study carried out on 17 potential locations identified Vijayapur in Syuna VDC, Takulla and Chhapre settlements of Chhapre VDC, and Varta of Sanjalbada VDC as the top three choices for integrated and compact development (DUDBC, 2007). Similarly, in Gulmi district, the study identified Daha in Simichaur VDC, Wamibazar of WamiTaksar VDC, and Darbar Devisthan of Darbar Devisthan VDC as prime options for development (DUDBC, 2009), and also proposed compact settlement plans for these locations.

These studies are in line with development models outlined in DUDBC's National Concept Note (DUDBC, 2004), which recommends three distinct approaches: the welfare model, emphasizing targeted poverty alleviation programs for the marginalized groups in the dispersed settlements; the service-oriented model, aiming to enhance infrastructure and essential services to draw the residents from nearby areas; and the urbanization model, promoting urban growth within rural settlements based on hierarchical service distribution and centrality. This focuses to promote urbanization and foster urban processes within rural settlements, based on the settlement hierarchy and the centralization of services. The major goal is to improve services and livelihood opportunities for rural populations through compact settlements, while also reducing the per capita costs associated with the construction, operation, and maintenance of essential infrastructure and services.

4.2.2 Resettlement of Scattered Communities of Himalayan Region

The mountainous region of Nepal, encompassing 126 local units, is home to approximately 1.7 million people, constituting 6.09% of the country's total population (NSO, 2021a). Of this population, 64.42% reside in rural municipalities. Settlements in this mountainous area tend to be small and scattered, particularly as altitude increases in the Himalayan region. The 2021 Census reveals a population density of 34 people per square kilometer in this region, significantly lower than the national average of 198 people per square kilometer. The challenging terrain, coupled with a heightened risk of natural disasters and fragility, poses significant obstacles to infrastructure development and service delivery, adversely impacting the lives and livelihoods of communities in this region (Chidi, 2009). This situation results in adverse outcomes such as low income, farmland abandonment, and widespread migration to urban areas.

In response to the National Planning Commission's (NPC) initiative, the Center of Infrastructure Development Studies (CIDS) conducted a study to pinpoint potential resettlement areas for scattered communities in the Himalayan region (CIDS, 2020). This aligns with the government's objective outlined in the 15th Five-Year Plan to consolidate settlements in high-risk mountainous areas. The study meticulously assessed various factors such as slope, drainage buffer, relief, aspect, and elevation to identify settlements at risk within the region. Subsequently, 28 safe sites were identified, shown in Fig. 2, as potential growth centers for the relocation of these vulnerable communities. Recommendations for these growth centers were informed by considerations including risk sensitivity, post-earthquake scenarios, centrality, available infrastructure, social and economic prospects, agricultural and cultural significance, as well as land use and land cover patterns.

Central to the study's concept is the consolidation of scattered communities into planned settlements in safer locations, which offer improved access to health, education, job/business, and livelihood opportunities. This strategic approach not only enhances community resilience but also alleviates pressure on natural resources, fosters environmental conservation, and drives sustainable regional development. By consolidating settlements, resources can be efficiently directed towards fostering social and economic growth,

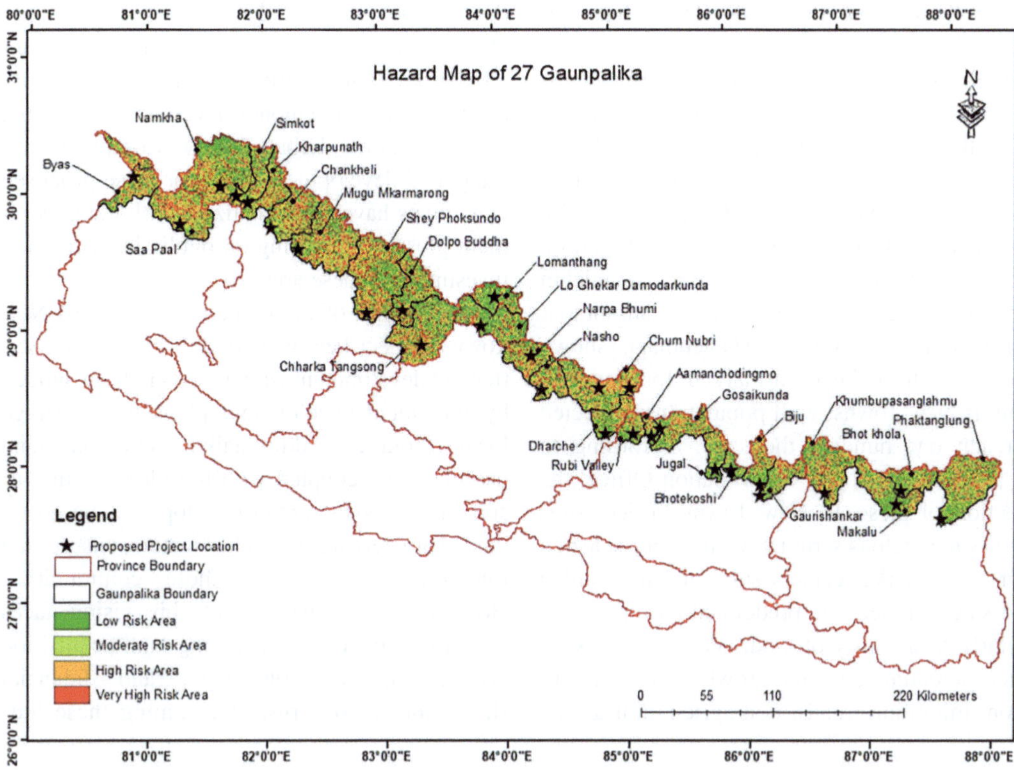

Fig. 2 Potential resettlement (proposed project) locations of the Himalayan region (CIDS, 2020)

thereby benefiting the broader community. Nevertheless, a major challenge in implementing this concept stems from individual's strong attachment to their original habitats and communities, which is more prominent in rural contexts. However, this obstacle can be overcome through judicious policy frameworks aimed at establishing CRSs nearby, enriched with job prospects, amenities, and other opportunities, all planned with a long-term vision. This process should begin gradually, with gradual transitioning of the current generation while targeting full integration of the new generation within a specified planning period. In summary, this study represents a holistic approach to addressing the challenges faced by scattered communities in Nepal's mountainous areas. Through appropriate detailed planning of growth centers, collaborative efforts, and long-term strategic interventions, this initiative seeks to create thriving and resilient communities capable of adapting to future challenges.

4.2.3 New Towns Project Along Mid-Hill Highway

New towns project along the mid-hill highway is a major national-level effort towards smart and compact settlement. The Government of Nepal is currently undertaking the construction of the Mid-Hill Highway, also known as the Pushpalal Highway, which will connect 24 hilly districts from east to west Nepal. Spanning 1776 km, this highway is expected to impact approximately seven million people (Dahal & Timalsina, 2017). Alongside this project, the Department of Urban Development and Building Construction (DUDBC) has been working on the "New Towns Project" along the highway. The completion of the highway, together with the establishment of the planned new towns, is anticipated to bring significant benefits to the remote and underserved communities in the mid-hill region. This initiative aims to reduce youth migration from mountainous regions to the Terai and larger cities such as Kathmandu.

The Department of Roads (DoR) has pinpointed 48 settlements poised for growth due to the highway's construction, with DUDBC identifying 23 settlements with high growth potential along the mid-hill highway (DUDBC, 2014). Out of these, 10 key areas—Patan (Baitadi), Safebagar (Achham), Rakam-Karnali (Dailekh), Chaurjahari (Rukum), Burtibang (Baglung), Dumere_Bhansar (Tanahun), Baireni-Galchhi (Dhading), Khurkot (Sindhuli), Basantapur (Terhathum), and Phidim (Panchthar), shown in Fig. 3—have been earmarked for development as New Towns, with populations projected to reach one hundred thousand. According to the New Town Project Coordination Office, the selection of these 10 New Towns Project sites follows meticulous scrutiny, with 54 pre-selected towns and market centers subjected to detailed assessment based on predefined criteria, particularly availability of land, access to drinking water, potential economic growth, current population, migration trends, non-agricultural activities, literacy rates, and electricity accessibility (NTPCO, 2016). The overarching aim is to foster a balanced national urban structure aligned with the principles of regional development outlined in the National Urban Policy. Comprehensive integrated development plans for these proposed new towns have been finalized, with the government presently focusing on public infrastructure investment in these areas.

At the core of the objectives of the 10 New Towns Project lies the adoption of compact settlement development with mixed land use, guided by integrated development plans that leverage local resources. Additionally, meticulous land use planning, coupled with the identification of suitable areas for urban development based on land characteristics and other factors, ensures the long-term resilience of these communities. However, the realization of this vision faces numerous challenges, including land acquisition issues, capacity constraints, and financial sustainability concerns. Overcoming these hur-

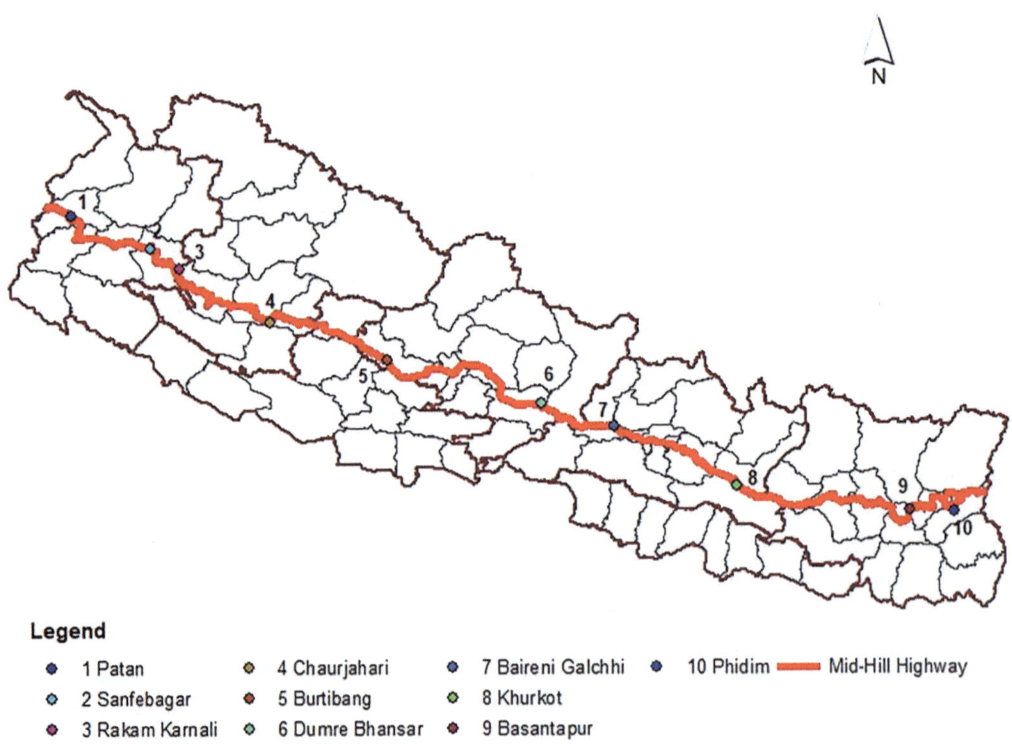

Fig. 3 Location of New Towns along Mid-Hill Highway

dles demands concerted efforts, innovative financing mechanisms, and strategic partnerships among government entities, the private sector, and civil society stakeholders. Nevertheless, as mentioned in Dixit and Shaw (2023), the establishment of a smart city in Nepal will be an evolutionary process and should be viewed as a continuous process of improvement. Despite the challenges, the 10 New Towns Project along the Mid-Hill Highway corridor symbolizes a beacon of hope for fostering inclusive, livable, and environmentally sustainable smart settlements that catalyze Nepal's resilient and sustainable development trajectory.

5 Policy Framework for Smart and Compact Rural Settlements

5.1 Land-Use Pattern of Rural Settlement and Their Planning Policies

Land use policy governs the allocation, management, and protection of land resources in rural areas, impacting the distribution, density, and variety of rural settlements. It also establishes the equilibrium between agricultural and non-agricultural sectors, addressing concerns like land degradation, fragmentation, and conversion, and promoting the preservation of natural and cultural heritage. Consequently, urban areas are predominantly characterized by non-agricultural activities, whereas agriculture primarily dominates rural regions, effectively distinguishing between the two. In the past, many urban planning approaches operated under the belief that cities and rural areas were entirely separate entities. This urban-rural divide portrayed cities as hubs of economic progress, contrasting with rural areas seen as lacking in development. This perspective treated cities and rural areas as disconnected entities, with no perceived connection or mutual relationship (Ntlhe, 2022). Nevertheless, contemporary rural areas are encountering challenges typically associated with urban regions, resulting in a heightened interconnection between these two zones (Strauss, 2019). The interlinkage arises as transition zones between rural, suburban, and urban areas, creating a continuum with no clear distinctions between rural and urban regions. The idea of an urban-rural division fails to support sustainable development because cities are not isolated entities; instead, they are interconnected and operate within various systems at multiple scales, spanning from regional to global levels. As per UN-Habitat (Arango et al., 2017), urbanization must transcend the obsolete dichotomy of urban and rural areas. Thus, planning needs to adopt an approach of good urban-rural linkage, considering the social, environmental, and economic interconnectedness between urban and rural areas. Addressing them separately is not sufficient. Sustainable development requires both metropolitan and rural areas to progress simultaneously. It is essential to minimize inequalities and bridge development gaps between and within these regions simultaneously.

The diverse environmental and cultural features of Nepal are evident in the structure, pattern, and type of rural settlements, which can be classified into three categories. Firstly, there are compact or dense settlements typical of the Terai and Himalayan belts. Secondly, there are agglomerated settlements found in the hilly areas between the Terai and the Himalayas. Lastly, there are isolated or scattered settlements located in specific areas of the inner Terai, and along steep and rugged hill slopes. These varied settlement configurations consist of an intricate hierarchy, including market towns, service centers, compact and large villages, hamlets, isolated clusters of houses, linear villages, and individual farms (NPC, 2021). Figure 4 depicts the settlement typologies. Considering the current arrangement of rural settlements in Nepal, functional hierarchy is essential among them. This is crucial for ensuring adequate service coverage in rural regions and maximizing the economic effectiveness of investments in infrastructure development and service delivery. CRSs represent both agglomerated and dense settlements. While a

Fig. 4 Scattered rural settlements

hierarchy exists among these settlements, at a fundamental level, a CRS could be established as a nucleated settlement characterized by concentrated development surrounding a public space.

5.2 Infrastructure Guidelines

CRSs provide numerous advantages for sustainable rural development. These benefits include improving accessibility to the high quality of infrastructure and services, reducing the environmental repercussions of rural sprawl, improving the resilience and safety within rural communities, and promoting the preservation and revitalization of rural heritage and culture. International experience demonstrates that well-planned and mixed-use neighborhoods within compact settlements offer diverse advantages. These neighborhoods often improve access to facilities and amenities, facilitate integration within extant infrastructure and public transport, optimize land usage, and support the transition to lower-carbon lifestyles.

The Government of Nepal has discussed various plans and guidelines related to infrastructure development for establishing CRSs. The Tenth Plan (2002–2007) recognized the need to develop rural infrastructure to uplift the rural populace, modernize agricultural practices, and implement pricing policies to increase agricultural production. Emphasis was also placed on rural energy development programs to boost the rural economy and alleviate poverty.

The three-year interim plan (2007–2009) aimed for a "Prosperous, Modern and Just Nepal." The plan focused on increasing investment in physical infrastructure, adopting an inclusive development approach, and implementing locality-targeted programs to establish CRSs. Similarly, the national development strategy paper in 2009 adopted a policy aimed at increasing investment in existing and new physical, social, and economic infrastructures. Mega, medium, and small projects were prioritized, with a focus on employment generation. Public-private partnerships (PPP) were also encouraged for construction projects. Strategic roads, north-south corridors, mid-hill highways, and rapid transit roads were given priority, along with exploring other transportation mediums such as cable cars, railways, and ropeways for mass transportation and tourism.

The Department of Urban Development & Building Construction has provided guidelines for the development of compact settlements nationwide. These guidelines cover various aspects such as geographical location and growth prospects, physical infrastructure, sanitation and waste disposal, environmental considerations, social services, safety and security, economic infrastructure and opportunities, efficient and inclusive local governance, and urbanization-friendly policies. DUDBC has initiated feasibility studies to identify potential CRSs in multiple districts, including Kalikot, Humla, Myagdi, Rasuwa, Pyuthan, Achham, and Gulmi. The studies focus on establishing a settlement hierarchy, particularly identifying settlements with higher-order centrality, excluding the district headquarters. Detailed environmental and socio-economic assessments are conducted for the identified settlements earmarked for CRS development. The study approach aligns with the urbanization model outlined in the DUDBC's National Concept Paper (DUDBC, 2004).

5.3 Governance Mechanisms

The National Urban Policy of 2007 advocates for the establishment of SCRSs. It promotes the development of compact settlements while discouraging dispersed settlements to protect natural resources and reduce investment costs associated with providing infrastructure and basic services. Additionally, it emphasizes the importance of implementing planning standards to prevent environmental degradation caused by scattered settlements.

Regarding the development of CRSs, the National Concept Paper recommends three potential models:

a. The Welfare Model, which prioritizes poverty alleviation programs for marginalized individuals, particularly those living in remote and scattered settlements.
b. The Service-Oriented Model, which emphasizes expanding and developing infrastructure and essential services to encourage people from surrounding areas to relocate to the compact settlements.
c. The Urbanization Model, which concentrates on the promotion of urbanization and urban growth processes within rural communities, based on their hierarchy and centrality of services.

The National Concept Paper also addresses cultural, social, and economic facets related to CRS planning and development, as well as the institutional and legal aspects relevant to the implementation of CRS programs.

5.4 Integrating Smart Technologies

The 2007 vision paper of the Ministry of Physical Planning and Works (MPPW) emphasized the development of integrated physical infrastructure to promote compact settlements (MPPW, 2007). It highlighted the need to develop physical infrastructure programs in the future provincial capitals as per the federal model to be embraced by the nation. Consequently, the integration of smart technologies into rural settlements is instrumental in creating sustainable rural development. In the context of Nepal's available resources, this can take the form of smart grid technology, climate-smart farming, digitization of local administration, and active citizen participation in local projects.

Nepal is making strenuous efforts to shift from a fossil fuel-dependent energy system to one based on renewable energy system. Smart grid technology (Bhattarai et al., 2023) plays a significant role in electrifying local communities with adequate power systems, integrating renewable energy sources, and harnessing bioenergy and solar energy. Climate-smart farming programs that provide subsidized inputs and services to farmers can contribute to sustainable rural development. There should be a strong focus on promoting the active participation of local governments through the use of digital tools and encouraging citizens to engage in the decision-making process. Additionally, administrative processes can be improved to enhance efficiency, transparency, and accessibility. This facilitates smoother service delivery, improved communication, and data-driven decision-making. In essence, crucial measures for establishing SCRSs include digitizing local administration, promoting citizen participation, and embracing smart technologies. These approaches empower communities, improve governance, and contribute to overall well-being.

6 Findings and Discussion

This book chapter presents an overview of the current status of rural settlements in Nepal, highlighting challenges in terms of geography, rural-urban migration, and proposed approaches and examples of initiations for the holistic development of SCRSs in Nepal. Challenges are imposed by rugged terrain, varying altitudes, and remote locations that hinder connectivity and infrastructure development, aided by scattered settlement which make it difficult to provide essential services such as health care, education and transportation. This has resulted in rural-urban migration, leading to reduced workforce, exacerbating the strain on urban resources while leaving rural settlements underserved. Despite the challenges of dispersed settlements and vulnerability to natural disasters, there is significant potential to create resilient rural areas and explore economic opportunities in tourism, agriculture, agro-industry, herbal industry, and more. National Urban Policy 2007 of the Government of Nepal has adopted strategies that discourage scattered settlements. Efforts from governmental and non-governmental organizations are underway to facilitate smart and CRSs.

Studies suggest that establishing 'growth centers' can play a vital role in the resettlement of dispersed households into compact settlements. These centers attract investment and technology, stimulating economic development in surrounding areas. For instance, the first two case studies discussed from Nepal demonstrate planning for CRSs by establishing economic centers, considering factors such as hazard mapping, post-earthquake scenarios, accessibility, and future social and economic growth potentials. The third case study example presents the effort of the Government of Nepal in creating new towns along major highways to mitigate migration to lowlands and major cities, balancing the national urban structure aligned with the principles of regional development outlined in the National Urban Policy. However, implementing such projects faces significant challenges, including the high capital requirement for the resettlement and land acquisition, as well as people's reluctance to relocate due to the pastoral values of their lands and homes.

Overcoming these challenges necessitates long-term planning and continuous effort. Resettlement projects can succeed with a well-prepared policy guide based on a long-term vision spanning 25–30 years, allowing for the adaptation of the new generation to new localities. The process of transitioning people from scattered settlements to compact ones can be best achieved through phased implementation over this extended timeframe. With decentralization, local governments have gained autonomy over resource management, supported by both central and federal governments, which enables them to more effectively achieve sustainable and resilient rural settlements.

7 Conclusion

This book chapter sheds light on the challenges and hurdles faced by rural areas in Nepal and explores the potential of SCRSs as a solution for promoting resilient and sustainable development. The chapter emphasizes the urgent need for economically feasible and environmentally sustainable investments in rural areas, considering the lack of basic infrastructure, essential services, and livelihood opportunities. The concept of SCRSs is advocated as a means to improve living standards, enhance social cohesion, boost economic productivity, ensure environmental sustainability, and mitigate risks from natural disasters, climate change, and urban sprawl. The policy framework, implementation strategies, and case studies illustrated in the chapter highlight the importance of inclusive and collaborative approaches among stakeholders and governance levels, as well as the utilization of modern technologies and digitization.

By leveraging advanced technologies, including Information and Communication Technology (ICT), and involving the community in the planning and implementation processes, resilient and self-reliant rural communities can be established in Nepal. This approach addresses the challenges faced by rural areas and paves the way for sustainable development. It emphasizes the significance of coordinated efforts, participatory planning, and the establishment of partnerships among various stakeholders. To sum up, the chapter advocates for the establishment of SCRSs in Nepal, which can serve as catalysts for resilient and sustainable development. By addressing the existing gaps in infrastructure, services, and livelihood opportunities, Nepal can create a more inclusive and prosperous future for its rural communities.

7.1 Policy Recommendations

Nepal can advance the development of SCRSs, leading to resilient and sustainable rural communities that improve the quality of life for the rural population, by implementing these policy recommendations:

1. Develop and implement supportive policy frameworks that prioritize the development of SCRSs. Provide incentives and guidelines for land-use planning, infrastructure development, and governance mechanisms to create sustainable and resilient rural communities.
2. Allocate sufficient funding for rural infrastructure development, focusing on improving road networks, electricity access, water supply systems, healthcare facilities, and educational institutions in rural areas. Pay special attention to areas with difficult terrain.
3. Develop sustainable and environmentally friendly infrastructure, including transportation, energy, and telecommunication systems, integrated into comprehensive development plans. Emphasize environmentally safe travel and transport infrastructure, such as roads, bridges, trails, and pedestrian bridges, while utilizing renewable energy sources where applicable.
4. Encourage public–private partnerships (PPPs) to mobilize resources and expertise for the development of SCRSs. Collaborate with the private sector, NGOs, and community organizations to leverage additional funding, technical knowledge, and innovative solutions.
5. Integrate smart technologies, such as digital connectivity, e-governance systems, and data-driven decision-making tools, to enhance connectivity, access to services, and resource management in rural areas. Improve service delivery efficiency, empower local communities, and bridge the urban-rural divide.
6. Strengthen capacity-building programs to enhance the skills and knowledge of local stakeholders involved in rural development. Provide training on sustainable agricultural practices, entrepreneurship, disaster preparedness, and technological literacy. Foster knowledge-sharing platforms to exchange best practices and innovative ideas.
7. Ensure inclusivity and social equity by meeting the requirements of marginalized groups, including youth, women, indigenous communities, and people with disabilities. Ensure their meaningful participation in decision-making processes related to SCRSs.
8. Promote the transition from scattered rural settlements to smart and compact ones through targeted policy interventions. One recommendation is to incentivize and motivate people to move towards these centralized settlements.
9. Encourage and support a range of economic activities within these settlements, such as tourism and agriculture.

7.2 Future Directions for Further Study

By conducting further studies and research in the following areas, Nepal can gather valuable insights and evidence to inform policy formulation, strategic planning, and implementation of SCRS:

1. Comprehensive studies and research: Further comprehensive studies and research are necessary to understand the various aspects affecting rural settlements and the development of SCRSs in Nepal. This includes investigating the interconnectedness between social and ecological dynamics, the interaction of local communities with their environment, livelihood practices, and natural resources.

2. Livelihood patterns and climate-sensitive sectors: There is a need to understand livelihood patterns in rural settlements and identify climate-sensitive sectors. Research should focus on sustainable agricultural practices, water management techniques, and income diversification opportunities for rural communities. This will help in developing strategies to enhance rural livelihoods and build resilience.
3. Traditional knowledge and practices: Recognizing the importance of traditional knowledge and practices in building resilience, further research should explore how traditional knowledge can be integrated into sustainable practices and contribute to the development of SCRS.
4. Government models for empowerment and sustainability: It is essential to identify appropriate government models that empower communities and promote sustainable practices. Further study should examine successful governance models, community participation mechanisms, and policy frameworks that support the development of SCRS.
5. Disaster Risk Reduction (DRR) strategies: Given Nepal's susceptibility to disasters, there is a pressing need for in-depth research to develop context-specific DRR strategies. This includes exploring early warning systems, evacuation plans, and community-centered disaster preparedness measures to enhance rural resilience.
6. Policy and institutional frameworks: Further study is needed to assess the existing policy and institutional frameworks related to rural development and SCRS in Nepal. This includes analyzing the effectiveness of current policies and identifying areas for improvement and alignment with sustainable development goals.
7. Long-term impacts and sustainability: Research should focus on assessing the long-term effects and sustainability of SCRS. This includes monitoring and evaluating the social, economic, and environmental outcomes of SCRS implementation and identifying strategies to ensure their long-term success.

References

Alam, A. F. M. A., Asad, R., & Enamul, K. M. (2016). Rural settlements dynamics and the prospects of densification strategy in rural Bangladesh. *SpringerPlus, 5*(1), 1–11. https://doi.org/10.1186/s40064-016-1883-4

Arango, J. et al. (2017). *Implementing the New Urban Agenda by strengthening urban-rural linkages – Leave no one and no space behind.* UN-Habitat, p. 90. https://www.unhabitat.org

Aziiza, A. A., & Susanto, T. D. (2020). The smart village model for rural area (Case Study: Banyuwangi Regency). *IOP Conference Series: Materials Science and Engineering, 722*(1), 12011. https://doi.org/10.1088/1757-899X/722/1/012011

Bandyopadhyay, S., Pathak, C. R., & Deninho, T. P. (2020). Urbanization and regional sustainability in South Asia. In *Contemporary South Asian Studies* (pp. 25–38). Springer. https://doi.org/10.1007/978-3-030-23796-7

Bhatnagar, S., & Schware, R. (2000). Information and communication technology in rural development: Case studies from India. *World Bank Institute, 3*(20), 1–192.

Bhattarai, T. N., Ghimire, S., Mainali, B., Gorjian, S., Treichel, H., & Paudel, S. R. (2023). Applications of smart grid technology in Nepal: status, challenges, and opportunities. *Environmental Science and Pollution Research, 30*(10), 25452–25476. https://doi.org/10.1007/s11356-022-19084-3

Chaudhary, D. (2023). The implication of "Smart Village" in Nepal: Sustainable rural development perspective. *Journey for Sustainable Development and Peace Journal, 1*(1), 1–20. https://doi.org/10.3126/jsdpj.v1i1.52595

Chidi, C. L. (2009). Human settlements in high altitude region Nepal. *Geographical Journal of Nepal, 7*, 1–6. https://doi.org/10.3126/GJN.V7I0.17436

CIDS. (2020). *Resettlement of the scattered communities/settlements in the dangerous zone of remote areas of the Himalayan Region of Nepal.* Center for Infrastructure Development Studies, Institute of Engineering, Tribhuvan University.

Cvar, N., Trilar, J., Kos, A., Volk, M., & Stojemenova, D. E. (2020). The use of IoT technology in smart cities and smart villages: Similarities, differences, and future prospects. *Sensors, 20*(14), 38–97. https://doi.org/10.3390/s20143897

Dahal, K. and Timalsina, K. P. (2017). New towns development: A wave of the future planning practices in Nepal. Tribhuvan University Journal, 31(1–2), pp. 139–152.

Dixit, A., & Shaw, R. (2023). Smart cities in Nepal: The concept, evolution and emerging patterns. *Urban Governance, 3*(3), 211–218. https://doi.org/10.1016/j.ugj.2023.08.003

DUDBC. (2004). *National concept paper on compact settlement 2061.*

DUDBC. (2007). *Feasibility study of compact rural settlement in Kalikot District.* In Nepali.

DUDBC. (2009). *Feasibility study of compact rural settlement in Gulmi District*. In Nepali.

DUDBC. (2014). *Business Plan 2014*. 2014. Government of Nepal, Ministry of Urban Development.

Gao, Y., Chen, G., & Shen, M. (2007). Scattered rural settlements and development of mountainous regions in Western Sichuan, China. *Wuhan University Journal of Natural Sciences, 12*(4), 737–742. https://doi.org/10.1007/s11859-006-0332-6

Guzal-Dec, D. (2018). Intelligent development of the countryside – The concept of: Assumptions, possibilities and implementation limitations. *Economic and Regional Studies/Studia Ekonomiczne Iregionalne, 11*(3), 32–49. https://doi.org/10.2478/ers-2018-0023

Hanif, M. A., Hossain, M. R., Hauque, A. M., Farhan, H. M., Akhter, M. S., & Sharmeen, N. (2015). A study on planning of rural settlement: A case study of Baidyerbazar Union, Bangladesh. *European Academic Research, III*(2), 1693–1722. https://www.researchgate.net/publication/359365163%0AA.

Hebsale, M., Pathak, V. K., & Tapan, B. (2023). Climate smart agriculture technologies adoption among small-scale farmers: A case study from Gujarat, India. *Frontiers in Sustainable Food Systems, 7*, 1–18. https://doi.org/10.3389/fsufs.2023.1202485

Klenova, T. V., Ivanov, A. S., & Koneva, D. A. (2020). Development of rural areas by means of "Smart Village" concept. In *Institute of Scientific Communicaitons Conference* (pp. 998–1006). Springer.

Li, J., & Song, W. (2023). Review of rural settlement research based on bibliometric analysis. *Frontiers in Environmental Science, 10*, 1–14. https://doi.org/10.3389/fenvs.2022.1089438

Liao, H., & Huan, H. (2015). *Developing smart and green rural settlements for the rapid urbanizing China: Experience learnt from England*. International Forum of Urbanism. https://doi.org/10.3390/ifou-D013.

Luo, G., Wang, B., Luo, D., & Wei, C. (2020). Spatial agglomeration characteristics of rural settlements in poor mountainous areas of Southwest China. *Sustainability (Switzerland), 12*(5), 1–13. https://doi.org/10.3390/su12051818

Monda, A., Feola, R., Parente, R., Vesci, M., & Botti, A. (2023). Rural development and digital technologies: A collaborative framework for policy-making. *Transforming Government: People, Process and Policy, 17*(3), 328–343. https://doi.org/10.1108/TG-12-2022-0162

Montaldo, C. R. B. (2013). *Sustainable development approaches for rural development and poverty alleviation & community capacity building for rural development and poverty alleviation* (pp. 2–17). Yonsei University. https://sustainabledevelopment.un.org/content/documents/877LRSustainableDevelopment v2.pdf

MPPW. (2007). *Vision Paper 2007 – New Physical Infrastructure Foundation of the New Nepal*, p. 1–20.

NPC. (2013). *Compact rural settlements development strategy for Nepal: A policy brief*.

NPC. (2021). *Nepal multidimensional poverty index, analysis towards action*. National Planning Commision. https://academicworks.cuny.edu/cc_etds_theses/599/.

NSO. (2021a). *National population and housing census 2021*. National Statistics Office. http://search.jamas.or.jp/link/ui/2019004882

NSO. (2021b). *National population and housing census 2021*. National Statistics Office. Urban Rural Municipality.

Ntlhe, D. (2022). The transformation of compact rural human settlements in South Africa-The Case of Elim in Limpopo. *International Journal of Sustainable Development & Planning, 17*(1), 313–322. https://doi.org/10.18280/ijsdp.170132

NTPCO. (2016). *Preparation of integrated development plan of ten new towns*.

Oven, K., & Rigg, J. (2016). Economic development, livelihoods and resilience in Nepal. *Earthquake Without Frontiers, 76*, 63–74. https://doi.org/10.1016/j.geoforum.2016.08.014

Phillips, M. (1998). The restructuring of social imaginations in rural geography. *Journal of Rural Studies, 14*(2), 121–153. https://doi.org/10.1016/S0743-0167(97)00056-9

Sheaffer, C. C., & Moncada, K. M. (2012). *Introduction to agronomy: Food, crops, and environment* (p. 588). Cengage Learning. https://books.google.com.np/books?id=aaIJAAAAQBAJ.

Singh, A., & Patel, M. (2015). Achieving inclusive development through smart village. *Journal of Energy and Management*. Pandit Deendayal Petroleum University.

Strauss, M. (2019). A historical exposition of spatial injustice and segregated urban settlement in South Africa. *Fundamina. University of South Africa, 25*, 135–168.

Tiwari, S. P. (2023). The role of technology in rural development in Asia: Opportunities and challenges. *Canadian Journal of Educational and Social Studies, 3*(3), 1–9. https://doi.org/10.53103/cjess.v3i3.139

Whatmore, S. (1993). Sustainable rural geographies? *Progress in Human Geography, 17*(4), 538–547. https://doi.org/10.1177/030913259301700408

World Bank. (2020). *Macro Poverty Outlook Zimbabwe – Key conditions and challenges*. p. 296–297. https://pubdocs.worldbank.org/en/750841492188177908/mpo-zwe.pdf

World Bank. (2022). *World population prospects: 2022 Revision*. https://data.worldbank.org/indicator/SP.POP.TOTL?end=2022&start=1960&view=chart

Yang, R., Xu, Q., & Long, H. (2016). Spatial distribution characteristics and optimized reconstruction analysis of China's rural settlements during the process of rapid urbanization. *Journal of Rural Studies, 47*, 413–424. https://doi.org/10.1016/j.jrurstud.2016.05.013

Zhang, Y., Li, X., & Song, W. (2014). Determinants of cropland abandonment at the parcel, household and village levels in mountain areas of China: A multilevel analysis. *Land Use Policy, 41*, 186–192. https://doi.org/10.1016/j.landusepol.2014.05.011

Planning for Smart Village Initiatives Towards Climate Resilience in Natural Resource Management: Case Study of Udaipur District, India

Bhanwar Vishvendra Raj Singh, Vijai Singh Meena, Ravi Mishra, Dushyant Thakur, and Devjaditya Singh Chouhan

Abstract

The effects of climate change and natural resource depletion are closely interconnected issues, as climate change significantly impacts these resources and, in turn, the conservation efforts in question play a pivotal role in the mitigation of the effects of climate change. Meanwhile, climate change, food security, and natural resource management are the initial stages in developing a feasible climate change mitigation and adaptation strategy. The objective is to devise a holistic and sustainable plan for smart village initiatives that enhance climate resilience in natural resource management. However, there is a gap between digital information on natural resource management and climate, aimed at mitigating and adapting to climate change, as well as social-environmental sustainability. This limits future research into the smart village's potential self-sufficiency in India. In this research work, a microanalysis of natural resources in rural regions using a variety of remote sensing indicators and then questionnaires to assess local people's perceptions and knowledge about climate change adaptation. The study areas have four Gram Panchayats in Rajasthan's Udaipur district, namely Bambora, Bori, Sulawas, and Sihad, to examine natural resources and climatic perceptions regarding the smart village initiatives. The findings indicate that the availability of water resources is a primary driver of the concentration of agricultural and related activities. As a result of water scarcity, people migrate to cities in search of jobs and essentials. The results indicate that natural resource depletion and agricultural productivity occur in all villages. The findings also show that this research area lacks environmental information, perspectives on natural resource management, and climate change adaptation. The findings of this research will assist in determining the necessity for climate change adaptation in the context of India's smart village initiatives, specifically in relation to natural resource management and conservation.

B. V. R. Singh (✉) · V. S. Meena · R. Mishra
Department of Geography, Faculty of Earth Sciences, Mohanlal Sukhadia University, Udaipur, India
e-mail: bhanwarsa28@mlsu.ac.in

D. Thakur · D. S. Chouhan
Udaipur, India

Y. Li (ed.), *Global Perspectives on Building Resilience for Sustainable Rural Development*, Sustainable Development Goals Series, https://doi.org/10.1007/978-981-96-6149-7_4

Keywords

Natural resources management · Climate change mitigation and adaptation · Smart village

1 Introduction

The fundamental challenge that modern society must answer is living in harmony with the environment. The social and economic issues faced by rural and urban residents as a consequence of climate change are of considerable magnitude. The issue is dealing with climate change with limited natural resources without damaging nature. Although technological advancements can improve access to energy, food, health care, education, and a variety of other services, climate change risk, and natural resource depletion have a detrimental influence on livelihood security, particularly in natural-resource-dependent communities. Natural resources are key sources of development in any industry (Broad, 1995). All people rely on natural resources, including water, land, and energy. The developing world faces a severe task in satisfying rising needs for land, food, water, and energy, which are exacerbated by climate change. Climate change's effects are already obvious, and they are going to have serious consequences for the supply of water, energy, and food unless proper steps are implemented. With huge populations, limited land resources, and increasing water scarcity, emerging countries face the common dilemma of producing more food on the same or less land, with less water, and at higher energy prices. "The effects of climate change on human health are manifold and include an increased incidence of mortality and morbidity from extreme weather events such as heatwaves, storms, and floods, disruption of the food system, an increase in the prevalence of zoonoses and food, water, and vector-borne diseases, and an elevated risk of mental health issues" (IPCC AR 6). As a result, climate change introduces new problems to the conservation and management of natural resources. To improve ecosystem services, natural resource management, and human well-being, climate change mitigation and adaptation must be recognized. The experiences and indigenous knowledge of local communities are also of significant importance in the context of climate change mitigation and adaptation. Thus, local people's perspectives and traditional knowledge must be considered at all stages of developing and implementing policies related to climate change, in order to mitigate its effects, adapt to it, and manage and conserve natural resources.

As indicated by data from the 2011 census, approximately 69% of Indian citizens reside in rural areas. This suggests that agriculture and related sectors are the key sources of livelihood security for Indian village dwellers. In contrast, climate change is expected to have a detrimental impact on agriculture, associated sectors, and other climate-dependent livelihoods. "It seems reasonable to posit that climate change will have an impact on the frequency and intensity of extreme events. Extreme occurrences are already having a large influence on agriculture" (Food and Agriculture Organization, 2015). "Climate change, along with land degradation, has the potential to significantly impact food security. Variations in temperature and precipitation patterns will cause disruptions in the food production process, which could result in higher food prices and less food available overall" (IPCC, 2019). One of the main concerns associated with climate change is the diminution of rural livelihoods and income (IPCC AR 5). Therefore, it is essential to address these challenges. The smart village concept has its roots in Mahatma Gandhi's vision of an ideal village and self-reliance. It has been accepted by India's governments as a holistic rural development project. The plan was an ambitious endeavor to transform rural regions into "economically, socially, and physically sustainable spaces," or "smart villages" that would "trigger overall regional development." Although the notion of a smart village is not new, it is a recurring theme in the agendas of numerous development agencies, both nationally and internationally, in both developed and developing countries. Many countries have adopted the smart village concept to address contemporary concerns confronting rural communities, such as climate change risk, natural resource depletion, economic disparity, and other infrastructure and technological requirements. However, financial resources, technology development and related skills, limited services, and marketization are important concerns in addressing the building of the smart village. Despite India's enormous potential for smart villages, there is a mismatch between the approaches to natural resource management and climate action, which is a major worry for agricultural and allied sectors in comparison to other sectors. This limits future research into the smart village's potential self-sufficiency in India. As a result, a comprehensive grasp of the interconnections between climate change, food security, and natural resource management is a necessary first step in developing a viable plan for climate change mitigation and adaptation. Natural resources and their values, resource utilization, and reliable environmental information are required for this purpose. However, there is a vacuum between accurate information on natural resource management and conservation and climatic initiatives to mitigate and adapt to climate change, as well as social-environmental sustainability.

The ability of people, communities, ecosystems, and societies to foresee, plan for, respond to, and recuperate from the negative effects of climate change is defined as resilience to climate change. It entails enhancing adaptive capability and lowering susceptibility to hazards associated with climate change. The following are some crucial elements of climate change resilience:

Adaptation Techniques: Building resilience entails putting adaptation techniques into practice to aid ecosystems and communities in adjusting to shifting climatic circumstances. This can involve the creation of infrastructure that is resistant to climate change, sustainable farming methods, and better water management.

Early Warning Systems: Reacting to climate-related events requires accurate and timely information. Communities can better prepare for extreme weather disasters like hurricanes, floods, droughts, and heatwaves by putting in place efficient early warning systems.

Community Engagement: Developing resilience frequently calls for the involvement and active participation of nearby communities. Communities can be better equipped to cope with and recover from the effects of climate change if they are allowed to actively participate in planning, strategy implementation, and decision-making processes.

Diversification of Livelihoods: Reliance on one revenue stream or way of life could make one more vulnerable. Diversifying economic endeavors is a crucial component of resilience since it mitigates the effects of climate-related disturbances on communities.

Management of Natural Resources: Managing natural resources in a sustainable and efficient manner is essential for enhancing climate resilience. This includes actions like soil preservation, reforestation, and sustainable water management techniques.

Infrastructure Development: It's critical to have a robust infrastructure that can endure severe weather occurrences. This involves enhancing the resilience of key infrastructure, like roads, buildings, and other structures, to climate change impacts.

Education and Capacity Building: It's critical to raise community knowledge and comprehension of climate change. Initiatives to increase capacity and education programs give people the tools they need to make wise decisions and adjust to their changing surroundings.

Policy and Governance: By implementing efficient policies and governance frameworks, governments may significantly contribute to the promotion of resilience. This covers the creation and implementation of laws that promote climate-resilient behavior as well as the incorporation of climate considerations into a range of industries.

Worldwide Cooperation: Addressing the global challenge of climate change necessitates worldwide cooperation. Knowledge sharing, technological transfer, and financial help for vulnerable areas are examples of collaborative activities.

Research and Innovation: Developing new technologies and approaches to mitigate the effects of climate change requires significant investment in research and innovation. This covers developments in climate modeling, sustainable agriculture, and renewable energy.

It takes concerted efforts at the local, national, and international levels to build resilience to climate change, which is a complex and continuous process. To build a more resilient and sustainable future, it combines adaptation, mitigation, and sustainable development strategies.

Timely and trustworthy data on natural resources in climate action is the foundation and necessary to transforming rural regions into smart villages. Natural resource variations can be efficiently monitored and controlled utilizing remote sensing technology (Goetz & Dubayah, 2011). Remotely sensed images are used in a variety of oceanic, terrestrial, and atmospheric applications, including environmental modeling, land use analysis, and geographical database updating (Kumar et al., 2015). Consequently, this study attempts to use remote sensing technology to assess natural resources using a variety of remote sensing indices, as well as to investigate perceptions and knowledge about the indicators, causes, effects, and implications of climate change in natural resource management through questionnaires and surveys in support of India's smart village initiative. We chose four Gram Panchayats in the Udaipur district of Rajasthan, India, for this purpose: Bambora, Bori, Sulawas, and Sihad. This research will aid in determining the need for climate change adaptation in natural resource stewardship and conservation in the context of India's smart village programs.

2 Study Area and Materials

2.1 Study Area

The study area, i.e., the village cluster of Bambora, Bori, Sulawas, and Sihad Gram Panchayat, falls under the Kurabad Tehsil of the Udaipur district of Rajasthan in India. The study area of the village cluster is around 62.67 km^2. The mean annual precipitation in the Udaipur district is 586 mm, with the majority of rainfall occurring between June and September. The mean diurnal temperature range in winter (December to February) is 5–30 °C, while in summer (May to July), it is 35–42 °C. Agriculture is the primary occupation of the people, and most of the agricultural activities are concentrated near water bodies (Fig. 1).

More than 10,000 families are living in these four Gram Panchayats as per the census of 2011. According to the Population Census of 2011, 11,423 people are living in these villages. There are a lot of children in villages between the ages of 0 and 6, which accounts for 20% of the hamlet's total population. These village's average sex ratio is greater than the state average of Rajasthan. However, the child sex ratio is less favorable than the state average.

The literacy rate of these panchayats is also higher than that of Rajasthan. Bori village's 2011 literacy rate was 70.22%, while Rajasthan's was 66.11%. Bambora village's 2011 literacy rate was 76.71%, while Rajasthan's was 66.11%. In Bambora, the rate of female literacy was 66.46%, while the rate of male literacy was 89.63% due to the small town and educational institutes.

The Schedule Tribe (ST) makes up the majority of the village's population in Bori, Bambora, and another gram panchayat. In Bori village, Schedule Caste (SC) made up 8.08% of the total population, while Schedule Tribe (ST) made up 36.65%. In Bambora village, the Schedule Tribe (ST) makes up 14.33% of the total population, while the Schedule Caste (SC) makes up 12.56%.

Thirty-five percent of the people living in Bori village were working. While 19.67% of workers engaged in marginal activities less than six months, 80.33% classified their work as their primary source of income (employment or earning more than six months). A total of 404 agricultural laborers and 1063 cultivators (owners or co-owners) made up the 3055 laborers involved in the main work.

Of the entire number living in Bambora village, 1445 were employed. While 33.91% of workers engaged in marginal activity that provided a living for fewer than six months, 66.09% of workers identified their work as main work (employment or earning more than six months). Of the 1445 workers involved in main work, 132 were agricultural laborers and 88 were cultivators (either owners or co-owners).

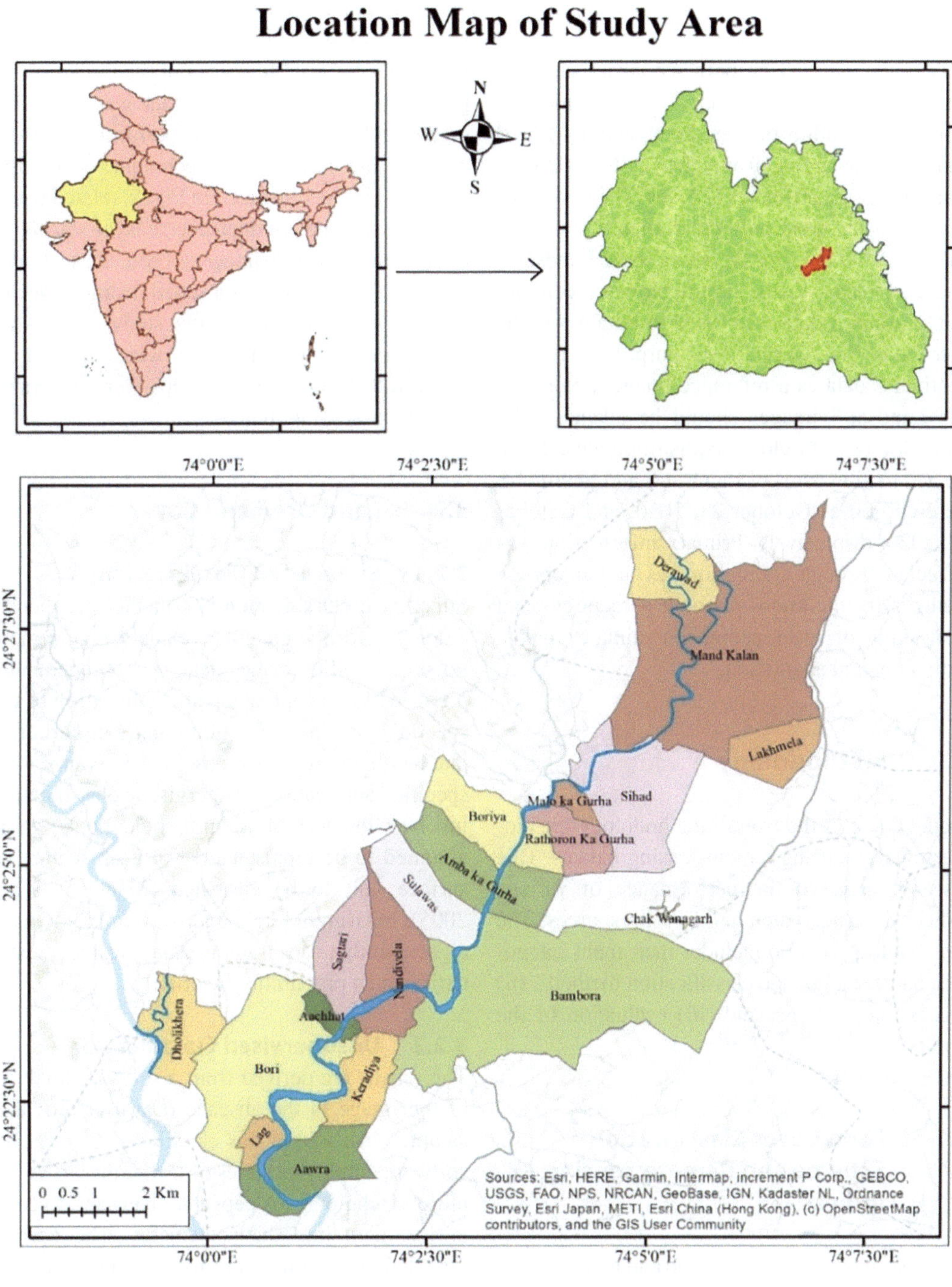

Fig. 1 Location of the study area. Source: Prepared by researchers

2.2 Database

This study is based on both secondary and primary data. The secondary data comprises cloud-free satellite images from Sentinel-2 and Landsat-8, sourced from the USGS (United States Geological Survey) server. The choice of data sets was based on their availability and the quality of the data for the study area. Landsat-8 was used to analyze the current biomass scenario, and Sentinel-2 data were used to obtain a reference map for comparison purposes. To use Sentinel-2 data as a reference image, Sentinel-2 and Landsat-8 images should be taken on the same day or with close acquisition dates. Thus, the acquisition dates of Landsat-8 and Sentinel-2 were chosen as October 23, 2020, and October 22, 2020, respectively. Primary information was collected through questionnaires and interview methods for the knowing about perceptions and knowledge of local people on climate change mitigation and adaptation.

3 Methodology

Figure 2 shows the overall methodology for biomass analysis using remote sensing indices. This research assessed the performance of various indices for land, water, and forest resources. The methodology section includes their main categories, namely (i) image classification methods, (ii) processing methods, and (iii) evaluation of the results.

3.1 Land Cover Classification Scheme and Remote Sensing Indices

To construct the land-cover map and many remote sensing indices such as NDVI, MNDWI, RDVI, SAVI, NDBI, BI, BSI, and SWMI using Landsat-8 and Sentinel-2, a classification approach was devised based on supplementary information, fieldwork, local knowledge, and visual interpretation. Maximum Likelihood Classification (MLC), ISODATA, and K-means classification were considered as classification methods. Landsat-8 satellite data were used to carry out the classification processes, and a Sentinel-2 image was utilized to create a reference map for the performance evaluation of the classification techniques (Al-Doski et al., 2013). As illustrated in Fig. 2, related bands of Landsat-8 and Sentinel-2 were layer-stacked to obtain multispectral images. Then, these multispectral images were subsetted using a vector boundary to differentiate the study area. After that, classification methods were applied to Landsat-8 data, and other maps were generated for remote sensing indices.

3.2 Land Use/Land Cover

3.2.1 Supervised Classification

Supervised classification is a method frequently employed for the quantitative analysis of remote sensing satellite image data. The fundamental premise of this approach is the delineation of the spectral domain into regions that can be correlated with ground cover classes of relevance for a specific application. Supervised classification involves the use of training field data that is assumed to be representative of each feature or surface unit to be classified (Mantero et al., 2005). Maximum likelihood is a commonly used algorithm that uses training data to calculate statistics and a probability function.

3.2.2 Unsupervised Classification

The results are derived from a software analysis of the image in the absence of a user-provided sample class within this classification. The GIS software employs a series of techniques to ascertain the relationships between pixels and to categorize them into distinct classes. The user is afforded the option of selecting the algorithm that the software will utilize, as well as the desired number of output classes. It is imperative that the user is aware of the area under consideration when the group of pixels with common features produced by the software is related to the actual features on the ground (Olaode et al., 2014).

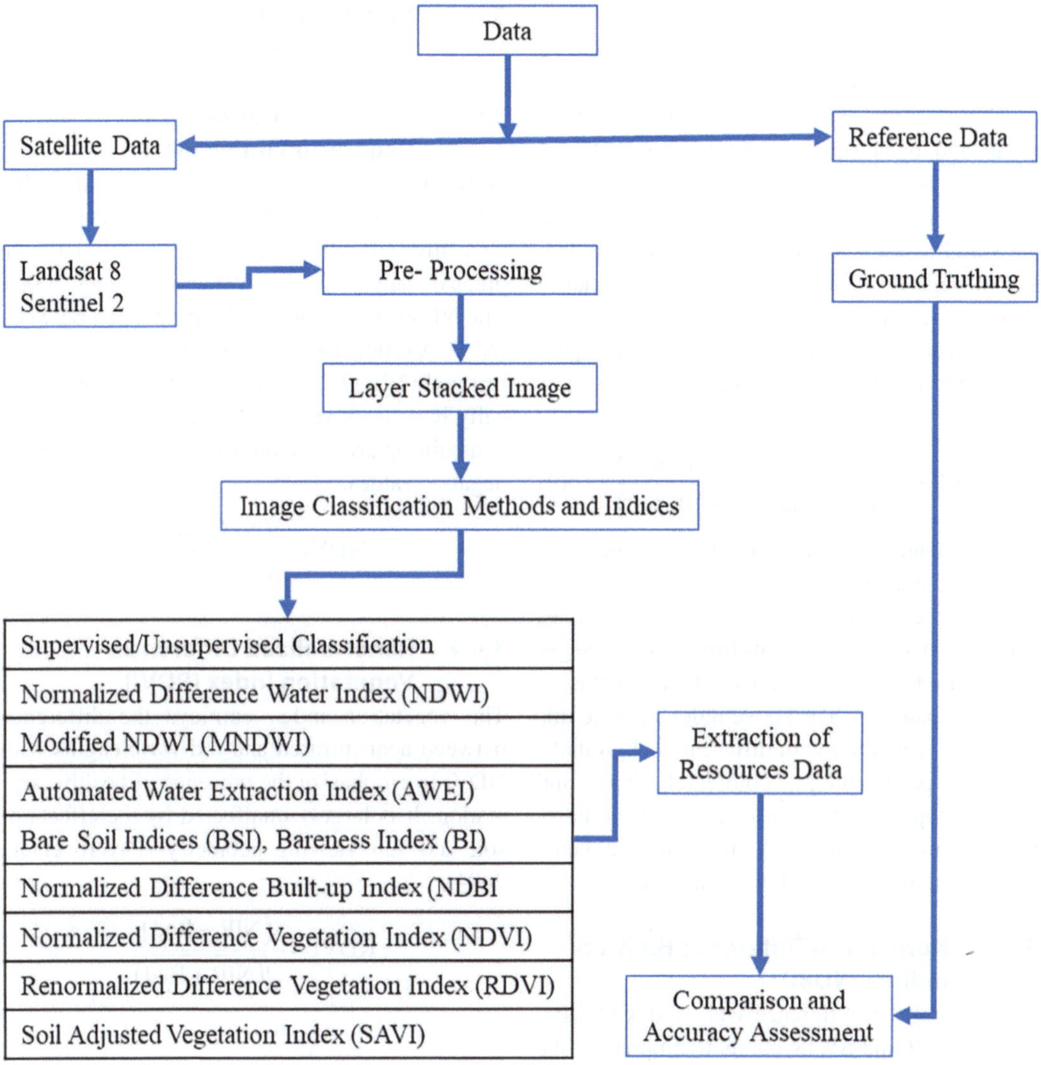

Fig. 2 Overall methodology framework to analyze the natural resources of the study area. Source: Prepared by researchers

3.3 Soil Resource Assessment

3.3.1 Bare Soil Indices (BSI)

The Bare Soil Index employs a mathematical approach to integrate spectral bands from blue, red, near-infrared, and shortwave infrared to measure soil variations. The spectral bands mentioned previously are utilized in a normalized fashion. The shortwave infrared and red spectral bands are employed for the measurement of soil mineral composition, whereas the blue and near-infrared spectral bands are utilized for the enhancement of vegetation (Nguyen et al., 2021). The BSI has a number of applications in remote sensing, including soil mapping and crop identification (in combination with NDVI), etc.

The BSI is calculated using the following formulas:

$$\mathrm{BSI} = \frac{(\mathrm{Band11} + \mathrm{Band4}) - (\mathrm{Band8} + \mathrm{Band2})}{(\mathrm{Band11} + \mathrm{Band4}) + (\mathrm{Band8} + \mathrm{Band2})} \quad (1)$$

3.3.2 Bareness Indices (BI)

In an ecosystem, the presence of bare soil plays a crucial role in sustaining its potential. It was developed primarily for the identification of urban and open surfaces. In our research, we focus on the bare surface devoid of vegetation. Consequently, this serves as a crucial indirect measure for assessing both the presence and density of vegetation in a given area. BI primarily encompasses the disparities between the inherent soil characteristics and their corresponding spectral responses. BI was calculated using the given Eq. 2:

$$\mathrm{BI} = \frac{(\mathrm{Band4} + \mathrm{Band2}) - (\mathrm{Band3})}{(\mathrm{Band4} + \mathrm{Band2}) + (\mathrm{Band3})} \tag{2}$$

Here, bands 2, 3, and 4 represent the blue, green, and red bands of Sentinel-2. Higher index values represent bare surfaces and urban areas, while lower values represent fully covered vegetation. The BI value increases with an increasing degree of bareness. The above index helps to differentiate vegetation with different backgrounds, such as sparse canopy, completely bare, and dense canopy. In addition, it facilitates comprehension of the physical and anthropogenic impacts on the ecological environment.

3.3.3 Normalized Difference Built-up Indices (NDBI)

The Normalized Difference Built Index (NDBI) uses the NIR and SWIR bands to emphasize the built-up areas. It is ratio-based to minimize the effects of atmospheric effects along with differences in terrain illumination. NDBI exploits the distinctive spectral properties of built-up areas and other land cover types. It has been extensively employed for extracting built-up areas from remote sensing imagery (Zha et al., 2003). The calculation of the NDBI was performed by employing the near-infrared (NIR) and short-wave infrared band 1 (SWIR1) from this formula.

$$\mathrm{NDBI} = \frac{(\mathrm{NIR} - \mathrm{SWIR})}{(\mathrm{NIR} + \mathrm{SWIR})} \tag{3}$$

where SWIR and NIR are images of band 11 and band 8 of Sentinel-2A imagery, respectively.

3.4 Vegetation Resource Assessment

3.4.1 Normalized Difference Vegetation Index (NDVI)

NDVI is a widely used vegetation index that measures the difference between visible and near-infrared reflectance, helping estimate the green density of an area (Weier & Herring, 2000). The NDVI is intended as the proportion between the TOA reflectance of the red band and the near-infrared (NIR) band. High vegetation areas typically have positive NDVI values, whereas water and built-up areas are represented by near-zero or negative values.

$$\mathrm{NDVI} = \frac{(\mathrm{NIR} - \mathrm{Red})}{(\mathrm{NIR} + \mathrm{Red})} \tag{4}$$

3.4.2 Renormalized Difference Vegetation Index (RDVI)

The vegetation index employs the difference between near-infrared and red wavelengths with NDVI to emphasize the presence of healthy vegetation. It is largely unaffected by the effects of soil and sun viewing geometry (Payero et al., 2004).

$$\mathrm{RDVI} = \frac{(\mathrm{NIR} - \mathrm{Red})}{\sqrt{(\mathrm{NIR} + \mathrm{Red})}} \tag{5}$$

3.4.3 Soil-Adjusted Vegetation Index (SAVI)

The Soil-Adjusted Vegetation Index (SAVI) is a vegetation index that employs a soil brightness correction factor to mitigate the impact of soil brightness on the index value. It is frequently employed in arid regions where vegetation cover is scarce (Huete, 1988).

$$\mathrm{SAVI} = \frac{(\mathrm{NIR} - \mathrm{Red})}{(\mathrm{NIR} + \mathrm{Red} + \mathrm{L}))} x(1 + L) \tag{6}$$

The terms NIR and red are used to refer to the bands related to specific wavelengths. The L value is subject to variation contingent on the extent of green vegetation cover. In areas devoid of green vegetation, L is typically equal to 1. In areas with moderate green vegetation cover, L is approximately 0.5, while in areas with a high degree of green vegetation, L is approximately 0. This index output value lies between −1 and 1.

3.5 Water Resource Assessment

3.5.1 Normalized Difference Water Index (NDWI)

The Normalized Difference Water Index (NDWI) (Gao, 1996) is a satellite-derived index from near-infrared (NIR) and shortwave infrared (SWIR) channels (Eq. 7). SWIR reflectance is indicative of alterations in both vegetation water content and spongy mesophyll structure within the vegetation canopy. In contrast, NIR reflectance is influenced by leaf interior structure and leaf dry matter content, but not by water content.

$$\mathrm{NDWI} = \frac{(\mathrm{NIR} - \mathrm{SWIR})}{(\mathrm{NIR} + \mathrm{SWIR})} \tag{7}$$

By integrating near-infrared (NIR) and shortwave infrared (SWIR) data, the effects of leaf internal structure and dry matter content on spectral reflectance are mitigated, thereby enhancing the precision of water content estimation in vegetation (Ceccato et al., 2001). The quantity of water present within the inner leaf structure exerts a significant influence on the spectral reflectance observed within the shortwave infrared (SWIR) interval of the spectrum. Consequently, SWIR reflectance is inversely correlated with leaf water content (Tucker, 1980).

3.5.2 Modified NDWI (MNDWI)

The Modified Normalized Difference Water Index (MNDWI) employs the green and SWIR bands to accentuate the attributes of open water, while simultaneously attenuating the effects of built-up features that are frequently associated with open water in other indices (Acharya et al., 2018).

$$\mathrm{MNDWI} = \frac{(\mathrm{Green} - \mathrm{SWIR})}{(\mathrm{Green} + \mathrm{SWIR})} \tag{8}$$

Green = pixel values from the green band

SWIR = pixel values from the shortwave infrared band

3.5.3 Sentinel Water Mask Index (SWMI)

The Sentinel Water Mask is a recently developed index that has been enhanced for use with Sentinel-2 data, which has undergone atmospheric correction. Following an analysis of TOA spectral reflectance curves for permitted mainland cover forms, the most appropriate spectral bands with spatial resolutions of 10 and 20 m were selected for use in the construction of the index formula (Robak et al., 2016).

$$\mathrm{SWMI} = \frac{(\mathrm{Blue} - \mathrm{Green})}{(\mathrm{NIR} + \mathrm{SWIR})} \tag{9}$$

For the other land cover classes, the sum of the blue and green light bands with the highest reflectance for water and relatively low reflectance was placed in the fraction of the equation. In the denominator, the sum of the NIR bands and SWIR bands was inserted, where the NIR bands are less reflective of water and the SWIR bands are more reflective of water. Such a combination of bands deepens the visual contrast and numerical separation between water and non-water regions, thus improving the ability to detect water.

3.6 Accuracy Assessment

The assessment of accuracy is an essential component of the validation process for digitally classified images. It is a method of evaluating the accuracy of a classification by comparing the results with geographical referenced data that is assumed to be correct (Batar et al., 2017). Accuracy assessment is performed by using 300 random points produced from remotely sensed land cover data with ground information obtained from other sources, e.g., Google Earth. This study performed an accuracy assessment for the 2021 image only. The accuracy assessment was conducted using a confusion matrix, which is generally used to evaluate classification performances.

3.7 Perceptions and Knowledge of Local People on Climate Change Mitigation and Adaption

A questionnaire survey at the village level was used to assess people's perceptions and understanding of climate change mitigation and adaptation. The random sample approach was used to choose 5% of the total population of each village for the survey. The questionnaire contained 11 questions about climate change and the environment. Consequently, around 600 survey replies were obtained. The sample number for each village is shown in Table 1.

4 Result and Discussion

4.1 Land Use and Land Cover

Figure 3 represents the land use and land cover area of the study area for 2021. The LULC contains seven classes: water bodies, fellow land, shrubland, dense forest, built-up area, barren land, and cropland.

Table 1 Village population and sample size of the study area

Sr. No.	Name	Total population	5% Sample size
1	Bambora	3964	198
2	Mand Kalan	1058	53
3	Malo Ka Gurha	286	14
4	Lakmela	226	11
5	Derawad	170	9
6	Chak Wanagarh	42	3
7	Boriya	346	17
8	Amba Ka Gurha	308	15
9	Rathoron Ka Gurha	164	8
10	Sulawas	206	10
11	Sagtari	458	23
12	Lag	158	8
13	Keradiya	292	15
14	Dholi Khera	507	25
15	Bori	903	45
16	Aawra	745	37
17	Aachhat	211	11
18	Nandivela	553	28
19	Sihad	840	42
Total		11,437	572

Water bodies, fellow land, shrubland, dense forest, built-up area, barren land, and cropland have areas of 1.74 km^2 (2.78%), 37.05 km^2 (59.12%), 5.46 km^2 (8.73%), 6.82 km^2 (10.88%), 1.86 km^2 (2.97%), 7.92 km^2 (12.64%), and 1.81 km^2 (2.88%), respectively, in Fig. 3 of land use land cover. In our LULCC research, we discovered that shrubland and dense forest are quite low in compared to fellow land in the study area (Table 2).

4.1.1 Accuracy Assessment

The quality of information that can be gained from remotely sensed data is determined by an accuracy assessment of a categorized satellite image. Overall accuracy is approximately 94.33%, which is quite good in terms of similarity to ground data (Table 3).

4.2 Soil Resource Assessment

In order to assess soil resources, a number of indices have been employed, including the bare soil index, the bareness index, and the normalized difference built-up index (Fig. 4).

Equation 1 was used to create the BSI map of the study area. It has been determined that when the BSI value is low, there is vegetation; when the BSI value is high, there is more bare soil. Equation 2 was used to create the Bareness Index map for the study area. A lower BI number indicates that the soil is covered with vegetation or water, whereas a higher BI value indicates that the soil is more open and less fruitful. The built-up or human encroachment region is depicted on the NDBI map. This is important for distinguishing between bare soil and settlement because, at the time of categorization, both characteristics seem identical in satellite photography. A final soil assessment map is created by merging all three indices using the weighted sum approach. Based on its fertility, soil is divided into five grades. Very High Fertility, High Fertility,

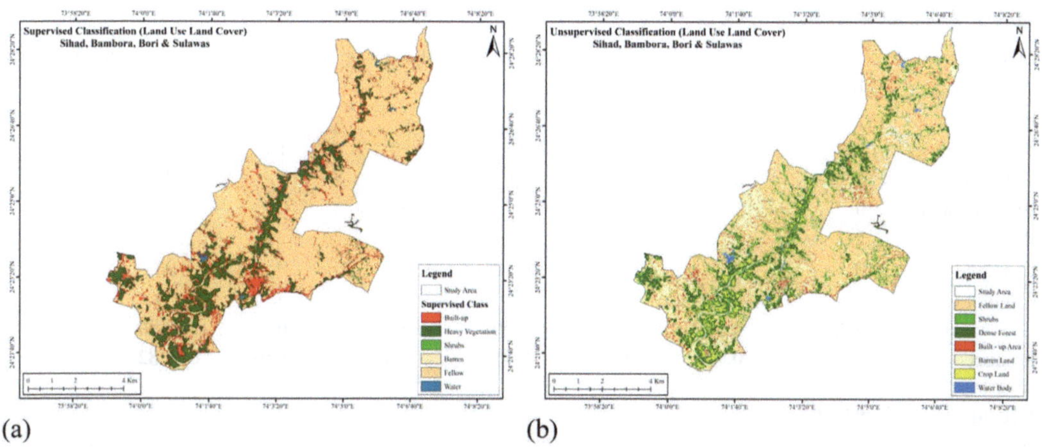

Fig. 3 Land use & land cover classification: (**a**) supervised classification method with six LULC classes and (**b**) unsupervised classification method with seven LULC classes. Source: Prepared by researchers

Table 2 Land use & land cover—2021

Sr. No.	LULC class	2021	Area percentage
1	Water bodies	1.74	2.78
2	Fellow land	37.05	59.12
3	Shrubland	5.47	8.73
4	Dense forest	6.82	10.88
5	Built-up area	1.86	2.97
6	Barren land	7.92	12.64
7	Cropland	1.81	2.88

Moderate Fertility, Low Fertility, and Very Low Fertility are the five categories.

4.3 Forest Resource Monitoring

Forest mapping is done by using overlay analysis of three major vegetation indices, namely NDVI, RDVI, and SAVI. The details are below:

The above-mentioned indices reveal that there is not a lot of vegetation cover; altogether, the same type of result is reflected in all indices. The minor increase is only visible in SAVI, which is the most widely used plant index and also includes soil types (Fig. 5c).

Overall, there is little forest resource in this research area; most regions have only small patches of forest cover. Sulawas, Bori, and Bambora have very little forest cover, but Sihad Gram Panchayat has the most. The high vegetation covers approximately 14.84 km^2 and the low vegetation covers approximately 27.13 km^2 (Fig. 5d).

4.4 Water Resource Assessment

Surface water resources are extracted through various indices:

Because there is not much surface water accessible in the region, water resource mapping is done using the indices indicated above, and the overall result is determined. There is only one seasonal river in the region behind it, the Nandi Vela River. In the surrounding areas, there are only a few small lakes. However, due to the ideal slope angles in this region, the watershed is very evident (Fig. 6).

4.5 Perceptions and Knowledge of Local People on Climate Change Mitigation and Adaptation

Figure 7 depicts local people's perceptions and knowledge about climate change mitigation and adaptation. The questionnaire survey is carried out in the research region to better understand people's opinions of climate change and to identify critical variables, obstacles, and impediments to environmental and climate-conscious behavior

Table 3 Accuracy assessment of LULC

LULC classes	Water bodies	Fellow land	Shrubland	Dense forest	Built-up area	Barren land	Cropland	Row total
Water bodies	18	2	1	0	2	1	0	24
Fellow land	1	73	2	0	2	2	1	81
Shrubland	2	1	58	2	1	1	2	67
Dense forest	0	2	1	43	0	0	2	48
Built-up area	0	1	2	0	34	0	0	37
Barren land	2	1	0	2	1	29	1	36
Cropland	0	1	2	0	0	0	28	31
Column total	23	79	65	47	38	32	34	300
User accuracy	**0.75**	**0.90**	**0.87**	**0.90**	**0.92**	**0.81**	**0.90**	
Producer accuracy	**0.78**	**0.92**	**0.89**	**0.91**	**0.89**	**0.91**	**0.82**	
Overall accuracy	**94.33**							

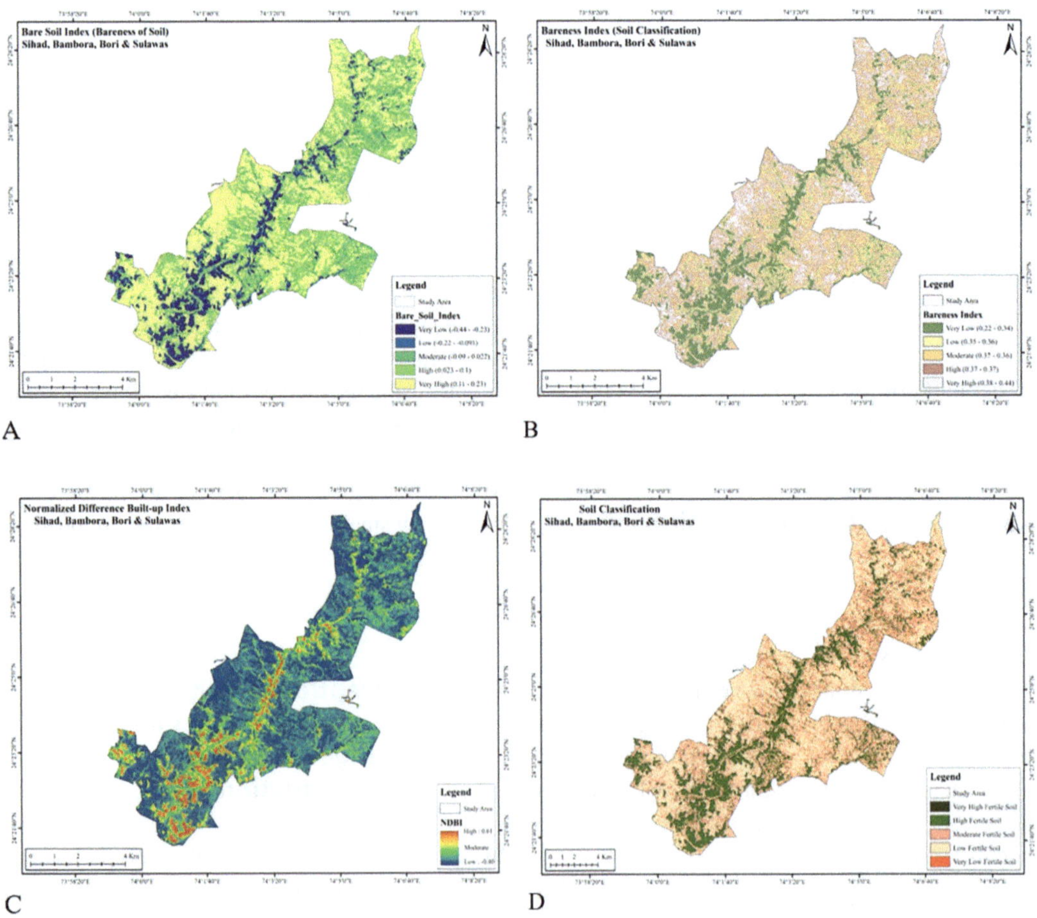

Fig. 4 Forest cover analysis: (**a**) BSI, (**b**) BI, (**c**) NDBI, and (**d**) weighted overlay of soil indices—2020. Source: Prepared by researchers

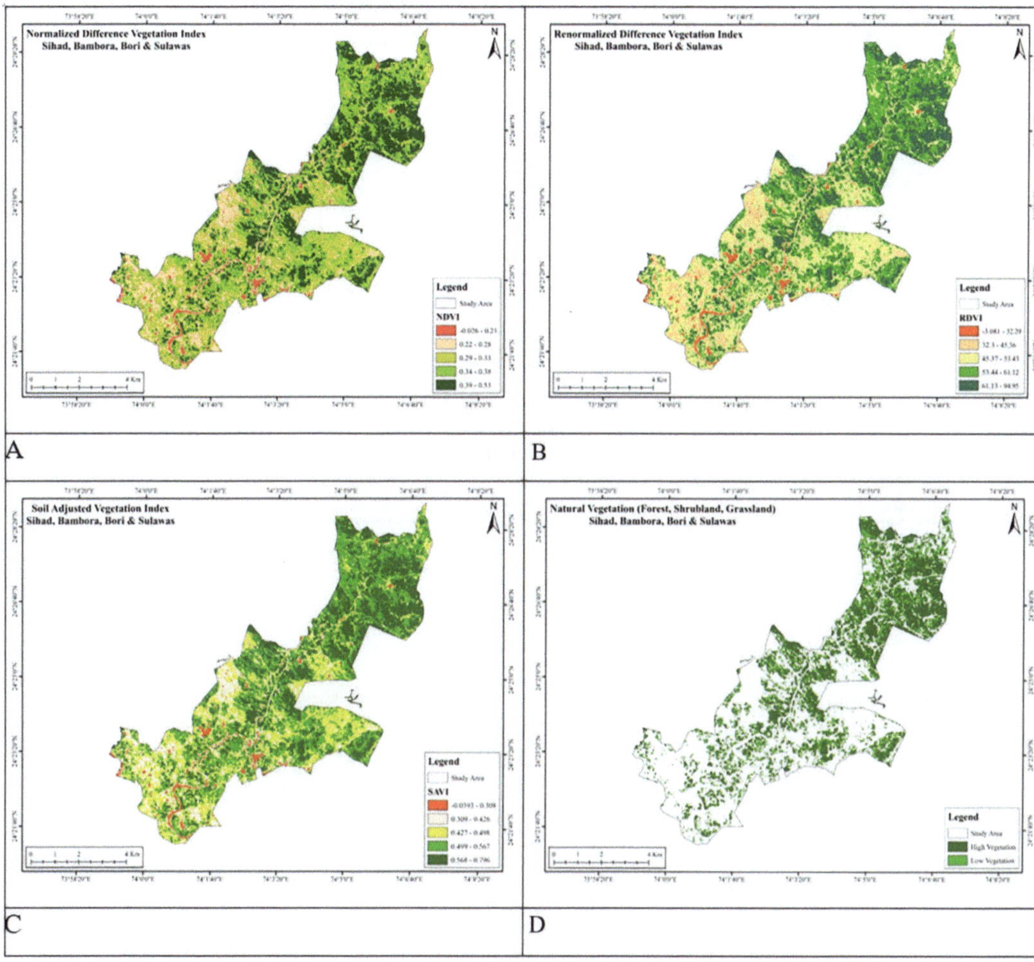

Fig. 5 Forest cover analysis: (**a**) NDVI, (**b**) RDVI, (**c**) SAVI, and (**d**) natural vegetation (forest, shrublands, grasslands, etc.). Source: Prepared by researchers

in smart village activities. The current survey identifies climate change as the most significant threat to local society. According to the findings, more than 40% of the population has never heard of climate change, yet they are aware of it. When asked about climate change, around 16% of the inhabitants in Sihad Gram Panchayat had no clue, 74% were in favor, and 10% did not respond. In the Sulawas Gram Panchayat, 27% denied it, while 73% agreed in the context of climate change. In Bori Gram Panchayat, 76% of residents are aware of climate change, while 24% are not. In Bambora Gram Panchayat, 94% of respondents support climate-related knowledge; 5% had heard nothing about climate change, and 1% did not respond.

The majority of people in these Gram Panchayats have witnessed various changes in climate change, such as an increase in temperature, a lack of rainfall, the onset of monsoons, an increase in the interval between cold and heat waves, a drop in water level, and so on. Water and food crises are two of the most obvious challenges caused by climate change; however, few responses have mentioned the soil catastrophe. This issue is most prevalent in Bambora Gram Panchayat. The majority of those present believed that natural and human forces were the primary causes of climate change. The percentage of persons who believe that social and natural causes are the primary causes is likewise relatively high. In Bambora and Bori Gram

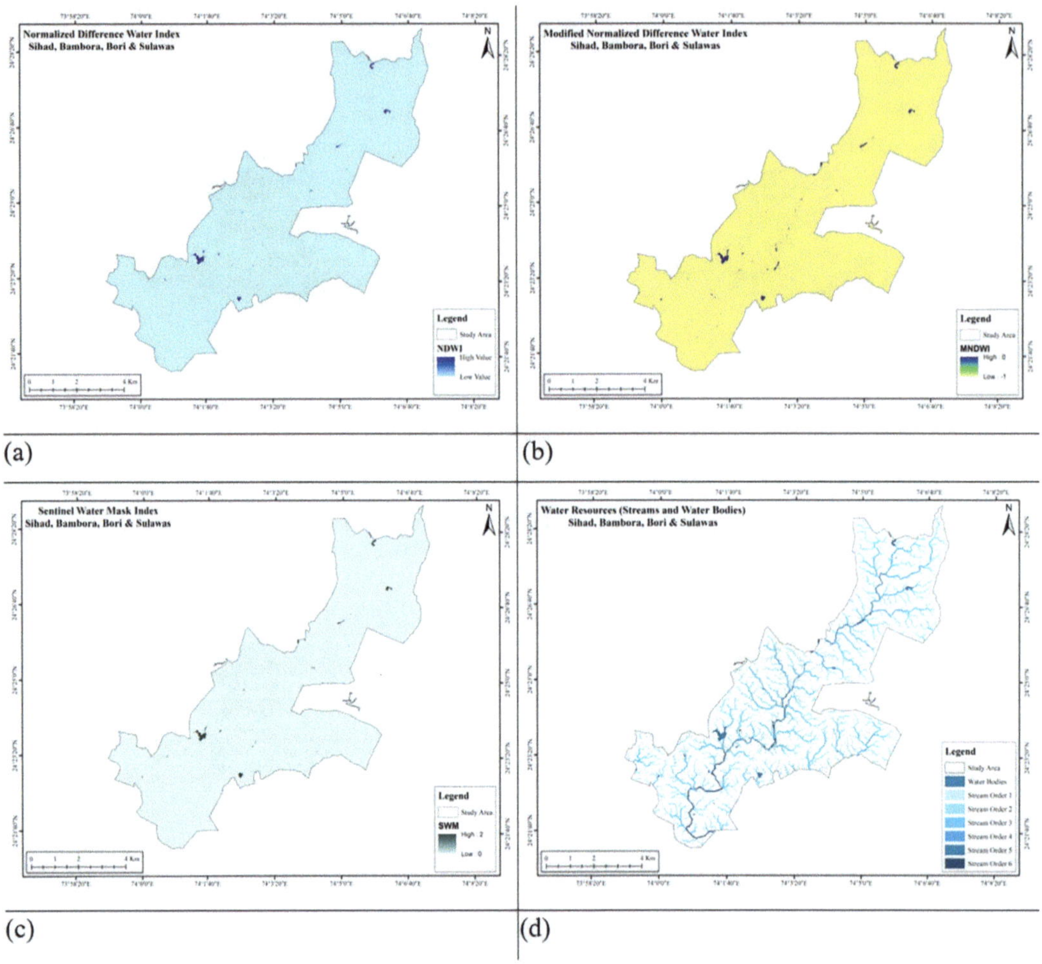

Fig. 6 Water resource: (**a**) NDWI, (**b**) MNDWI, (**c**) SWMI, and (**d**) drainage pattern with water bodies. Source: Prepared by researchers

Panchayats, these responses are extremely high. Climate change has had an impact on agriculture, water, the environment, and society (all at once). Almost all people have witnessed its huge influence in Bambora and Bori Gram Panchayat. Climate change has reduced agricultural activity, lowering production, modifying planting patterns, and even causing water scarcity in some circumstances. Because of the impact of climate change on water resources, the quantity of water has declined, as has the quality of water, and the quantity of groundwater has reduced dramatically. Some people have also expressed concern about the drought. Climate change, according to forest resource auditors, has resulted in a large increase in the amount of deforestation, degradation, and human-wildlife conflicts, with some even increasing the number of invasive species. People claim that the government has done nothing to address these issues, while others claim that the government has begun different management initiatives in this context. People believe that these issues can only be overcome if all resources are developed sustainably (Fig. 8).

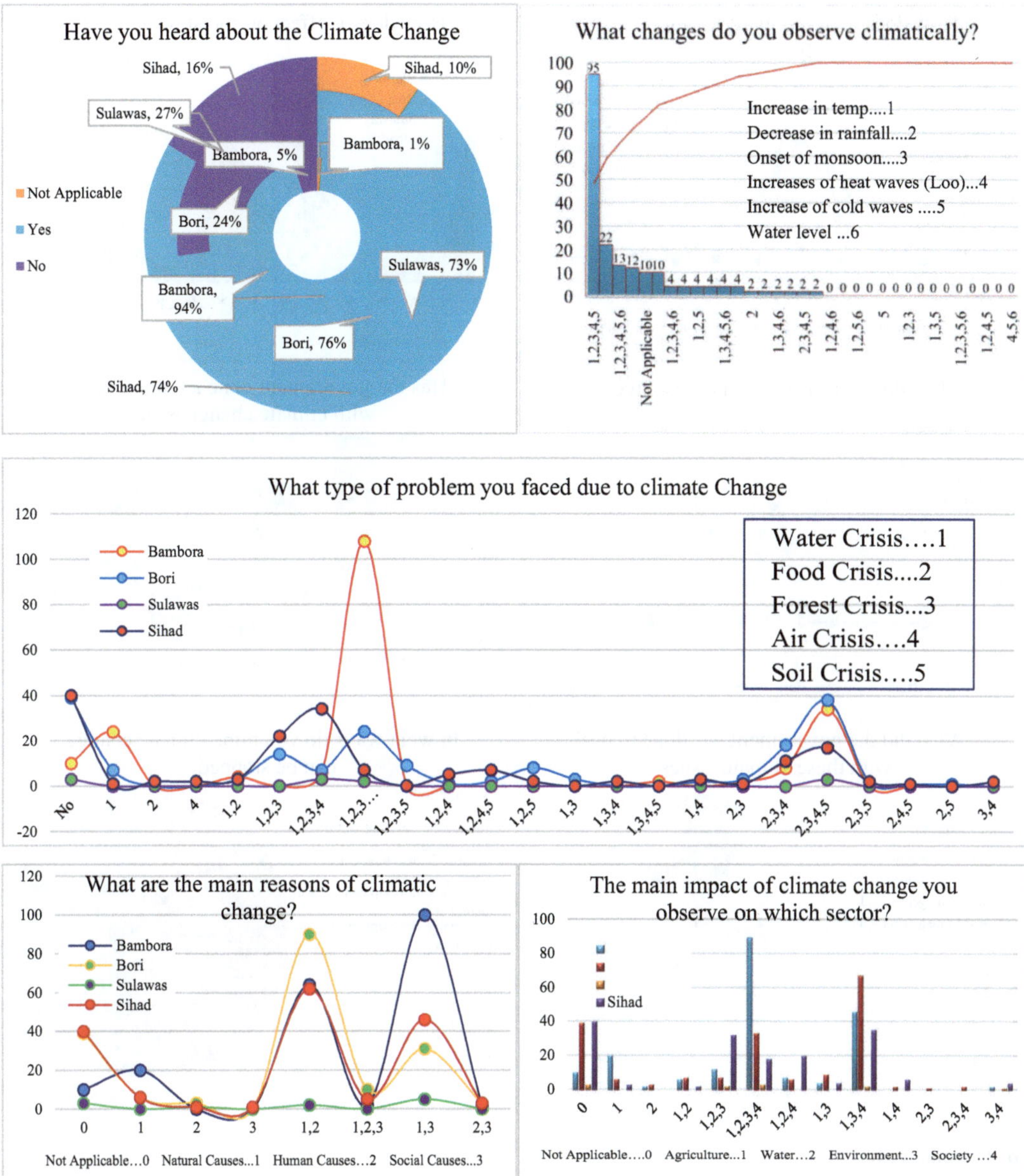

Fig. 7 2021 Survey response complied by the author using a questionnaire. Source: Compiled by researchers

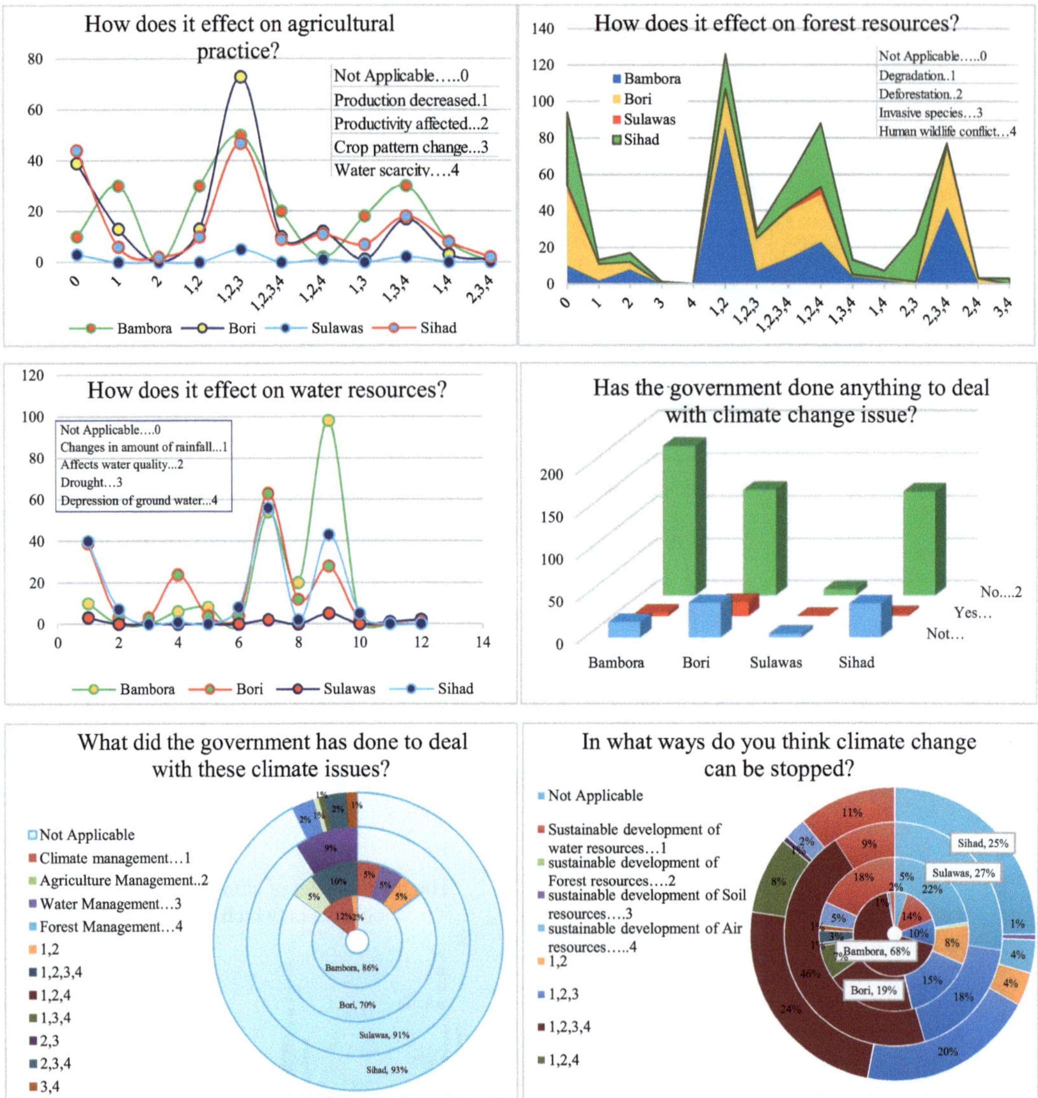

Fig. 7 (continued)

5 Discussion

The sustainability of a nation's economy, society, and culture largely depends to a large extent on its forest, land, and water resources. These natural resources are critical to a country's economy since they provide a source of raw materials and cash, as well as food, medicine, and energy for a variety of sectors. Countries need to devise strategies for the sustainable management of their natural resources to ensure that their benefits are accessible not only in the present but also for posterity. This is crucial because the failure to employ these resources in a manner that is both effective and efficient could lead to their depletion. The growth of the global population has led to a significant increase in the utilization of existing resources worldwide (Peizzy, 1992).

Human invasion has reduced the global coverage of important resources in the current situation. This has increased human-wildlife conflict and the appearance of desert-like environments.

Fig. 8 Baseline survey: (**a**) decreasing water quality, (**b**) decreasing water level, (**c**) water scarcity, (**d**) animal husbandry, (**e**) community level awareness, and (**f**) sustainable management of resources. Source: Photos taken by researchers

Ultimately, the exhaustion of natural resources has led to a substantial rise in living expenses, alterations in climate patterns, and a diminishment of the economic, social, and cultural advantages that were once derived from their use. Because of these limits, it is imperative to oversee the stewardship of these resources with care. A variety of strategies within the realm of natural resource management have been proposed to meet this objective. Groundwater is being overexploited as a result of rising demand and supply for drinking water and agricultural water, which is causing grave worry. Forests are being reduced daily as a result of human encroachment, increased settlements, agricultural land development, deforestation, and degradation. Disinterest in agriculture is growing as a result of environmental changes and the involvement of wild animals. Because of all of these factors, the food crisis has the potential to become severe shortly. Many other environmental challenges have begun as a result of increased population pressure on natural resources and excessive resource consumption, which can have catastrophic effects. People should be informed of all aspects of climate change to prevent and mitigate it. People should continue to share environmental knowledge by discussing these issues regularly. For more than three decades, land-based satellite remote sensing has been used in practically all disciplines of Earth research. Remote sensing and GIS technologies have consistently produced meaningful data that has led to long-term management.

6 Conclusion

The present study examined the phenomenon of climate change adaptation in the context of the smart village initiative, specifically in the domain of natural resource management, using four Gram Panchayats in Rajasthan's Udaipur district, namely Bambora, Bori, Sulawas, and Sihad. This study uses a microanalysis of natural resources in rural areas and questionnaires and surveys to evaluate local people's perceptions and understanding of climate change adaptation. The deterioration of natural resources and agricultural output was detected in all villages due to anthropogenic impact. Environmental information, perspectives on natural resource stewardship, and initiatives for climate change adaptation are limited in this study area, according to the findings. As a result, this study concluded that knowledge and awareness of climate change risk are required at all levels among rural inhabitants, farmers, and enterprises. This study has offered enormous potential for smart villages. Nevertheless, the outcomes of this research reveal the financial resources, services, marketization, technological development, and related skills required to address the building of the smart village.

Acknowledgments It is my great privilege to gratitude to RUSA M.H.R.D. 2.0 funding, Government of India to conduct this scientific research work on the project. My heartfelt thanks to all the authors, the fieldwork team, who worked and have referred during report writing, and also our colleagues and staff members of the Department of Geography, Faculty of Earth Science, Mohanal Suklhadia University, Udaipur, India.

References

Acharya, T. D., Subedi, A., & Lee, D. H. (2018). Evaluation of water indices for surface water extraction in a Landsat 8 scene of Nepal. *Sensors, 18*(8), 2580.

Al-Doski, J., Mansorl, S. B., & Shafri, H. Z. M. (2013). Image classification in remote sensing. *Department of Civil Engineering, Faculty of Engineering, University Putra, Malaysia, 3*(10).

Batar, A. K., Watanabe, T., & Kumar, A. (2017). Assessment of land-use/land-cover change and forest fragmentation in the Garhwal Himalayan Region of India. *Environments, 4*(2), 34.

Broad, R. (1995). The political economy of natural resources: Case studies of the Indonesian and Philippine Forest sectors. *The Journal of Developing Areas, 29*(3), 317–340.

Ceccato, P., Flasse, S., Tarantola, S., Jacquemoud, S., & Grégoire, J. M. (2001). Detecting vegetation leaf water content using reflectance in the optical domain. *Remote Sensing of Environment, 77*(1), 22–33.

Food and Agriculture Organization of the United Nations. (2015). Climate change and food security: Risks and responses. FAO.

Gao, B. C. (1996). NDWI—A normalized difference water index for remote sensing of vegetation liquid water from space. *Remote Sensing of Environment, 58*(3), 257–266.

Goetz, S., & Dubayah, R. (2011). Advances in remote sensing technology and implications for measuring and monitoring forest carbon stocks and change. *Carbon Management, 2*(3), 231–244.

Intergovernmental Panel on Climate Change. (2019). Climate change and land: An IPCC special report on climate change, desertification, land degradation, sustainable land management, food security, and greenhouse gas fluxes in terrestrial ecosystems. IPCC.

Huete, A. R. (1988). A soil-adjusted vegetation index (SAVI). *Remote Sensing of Environment, 25(3),* 295–309. (ESRI, 2018).

Kumar, N., Yamaç, S. S., & Velmurugan, A. (2015). Applications of remote sensing and GIS in natural resource management. *Journal of the Andaman Science Association, 20*(1), 1–6.

Mantero, P., Moser, G., & Serpico, S. B. (2005). Partially supervised classification of remote sensing images through SVM-based probability density estimation. *IEEE Transactions on Geoscience and Remote Sensing, 43*(3), 559–570.

Nguyen, C. T., Chidthaisong, A., Kieu Diem, P., & Huo, L. Z. (2021). A modified bare soil index to identify bare land features during agricultural fallow-period in southeast Asia using Landsat 8. *Land, 10*(3), 231.

Olaode, A., Naghdy, G., & Todd, C. (2014). Unsupervised classification of images: A review. *International Journal of Image Processing, 8*(5), 325–342.

Payero, J. O., Neale, C. M. U., & Wright, J. L. (2004). Comparison of eleven vegetation indices for estimating plant height of alfalfa and grass. *Applied Engineering in Agriculture, 20*(3), 385.

Pezzey, J. (1992). Sustainability: an interdisciplinary guide. *Environmental Values, 1*(4), 321–362.

Robak, A., Gadawska, A., Milczarek, M., & Lewiński, S. (2016). Wykrywanie wody na zdjęciach optycznych Sentinel-2((Sentinel Water Mask (SWM) – new index for water detection on Sentinel-2 images) na podstawie wskaźników wodnych. Teledetekcja Środowiska, 55.).

Tucker, C. J. (1980). Remote sensing of leaf water content in the near infrared. *Remote Sensing of Environment, 10*(1), 23–32.

Weier, J., & Herring, D. (2000). Measuring vegetation (NDVI & EVI). NASA Earth Observatory. https://earthobservatory.nasa.gov/features/MeasuringVegetation

Zha, Y., Gao, J., & Ni, S. (2003). Use of normalized difference built-up index in automatically mapping urban areas from TM imagery. *International Journal of Remote Sensing, 24*(3), 583–594.

Resilience for Rural Sustainable Development in Iran: Challenges and Approaches

Mostefa Abiyat and Morteza Abiyat

Abstract

The rural areas of Iran face various challenges, such as water scarcity, climate change impacts, economic fluctuations, limited access to services, migrations, and abandonment. Rural resilience is essential to maintain socio-economic and environmental systems while withstanding and recovering from challenges. Rural policy-making and planning focusing on rural revitalization, maintaining stability, and living standards can promote rural resilience and contribute to sustainable rural development. The centralized, sectoral, non-regional, and top-down policy-making system functions as the main phenomenon in the current pattern of rural development in Iran, which is a consequence of the weakness of intellectual and theoretical foundations in rural policy-making. Furthermore, it is recommended to achieve rural development and resilience, it is suggested that integrated sectoral-spatial policy-making should be the focus, and spatial policies should receive more attention with a relative reduction of sectoralism. Also, to achieve sustainable rural development, it is recommended to implement a spatial planning and policy system based on a network and hybrid approach to rural development. This system should cater to the unique needs of rural societies and be in harmony with their conditions. This can be achieved by emulating successful examples implemented by member countries of organizations like Economic Cooperation Organization (ECO), the Shanghai Cooperation Organization (SCO), and the Organization of Islamic Cooperation (OIC). In addition, it is important to continue formulating and implementing Iran's novel models of rural development that have been successful in recent years, such as the Rural Constellations Sustainable Development Plan (RCSD), the Economic Development and Rural Employment Plan (EDRE), the Underprivileged Foundation Development and Progress Plan (UFDP), the Comprehensive Model for the Progress of Deprived and Rural Areas with the Kowsari development approach of the Barakat Foundation (Sahab, Asman, Aftab, and Mahtab), and the Roostatish projects (economic, social, and socio-economic). These new patterns emphasize investment in places rather than sectors and improving socio-economic and environmental-physical status. The study's findings provide valuable insights for policy-makers, planners, and researchers on how to enhance the resilience and sustainability of rural communities in Iran.

M. Abiyat · M. Abiyat (✉)
Department of Geography and Rural Planning, Faculty of Geographical Science and Planning, University of Isfahan, Isfahan, Iran
e-mail: morteza.abiyat@geo.ui.ac.ir

Y. Li (ed.), *Global Perspectives on Building Resilience for Sustainable Rural Development*, Sustainable Development Goals Series, https://doi.org/10.1007/978-981-96-6149-7_5

Keywords

Rural resilience · Sustainable rural development · Policy-making · Planning · Iran

1 Introduction

Rural areas are often recognized as the birthplace of human civilization, as they have historically provided the necessary resources to meet the basic needs of societies (Leimgruber & Chang, 2019). They have always been full of authentic subcultures (Ahmadi & Mottaghi, 2022) and, as part of the geographical spaces, have played a fundamental role in regional and national development (Nilsen et al., 2023). These spaces have evolved through different ages, including the traditional agricultural civilization, collective economy, rural industrialization, rural differentiation, and rural revitalization (Ren & Wu, 2024). Today, rural areas provide crucial resources and ecosystem services such as food, raw materials, and energy to the cities (Haberman & Bennett, 2019). These settlements also contribute to the cultural and recreational opportunities provision, climate regulation, and biodiversity conservation (Mcneely, 2020; Welivita et al., 2021). Therefore, they are no longer just production surfaces but multifunctional spaces that provide possibilities and values of high interest to society (Quintá et al., 2018). However, considering climate change and globalization shocks, rural areas have lost their ability to reduce or prevent vulnerability to hazards in developing countries (Cupples et al., 2018). Local and regional institutions can be crucial in promoting the resilience and resourcefulness of rural areas against shocks (Pascariu et al., 2023). Nonetheless, the lack of structural adjustment and unfavorable performance in management can cause several problems in rural areas, such as poor demographic prospects, reliance on older industries, weak connectivity to the larger national economy (Kristensen et al., 2018), limited access to services, underdeveloped infrastructure, low-income levels, shortage of employment opportunities (Palmer, 2014), full of flaws transition from traditional to modern sectors, land-use changes, intensification of rural-urban migration (Maurel & Tuccio, 2016; Helbling & Meierrieks, 2021), rural hollowing, spatial shrinkage of villages, and vulnerability in front of economic headwinds. These factors negatively affect sustainable rural development (Cheng et al., 2023). Therefore, rural communities should increase their capacity, resistance, and adaptation to risks and disruptions while minimizing losses and maintaining stability (Li, 2023). Rural resilience approaches can help to achieve sustainable development by empowering rural areas to withstand economic challenges, social changes, and environmental crises, increasing the efficiency of the regional policy-making and management system, and preventing the evacuation of villages (Steiner & Markantoni, 2014; Yang et al., 2022).

With a population of 20 million, accounting for 26% of the country's population, and possessing more than 50% of the land area, Iran's rural areas have a considerable economic functionality, with an employment rate of 87.4% (higher than the national average of 85.2%) and an economic participation rate of 39.5% (higher than the national average of 38.1%). They contribute 26% of the value added to various economic sectors and support food security by producing 93% of agricultural products. Despite their active presence in the subsistence economy and their crucial role in the country's governance and planning system, rural areas in Iran still lack relative welfare, environmental quality, economic diversity, and employment opportunities (Badri et al., 2020). Also, the lack of proper regional planning and policy-making has resulted in numerous issues with population stability in rural areas. Surveys show that the share of the rural population in the country's population has decreased from 79% to 25.9% from 1900 to 2016. In addition, from 1920 to 2016, the number of cities experienced a growth of 14.3 times and increased from 78 to 1242 urban areas. This growth after the Islamic Revolution (1976–2016) was about 177%. The conversion of rural settlements into cities played a fundamental role in this growth (Varmazyari, 2021). Demographic estimates show that the country's rural population will face five challenges by 2030, such as the continuation of the rural-urban migration process, the absolute and relative reduction of the country's rural population, the aging of the rural population's age structure, and the increase in the supply of rural labor. In fact, in 2030, the country's rural population will decrease to 22.85% from 31.4% in 2006. In other words, more than 77% of the country's population will live in cities in 2030. Also, based on estimates, the country will witness the continuation of the aging process of the rural population in 2006–2030, so that the average age of the rural population will reach 37.1 years in 2030 from 23.5 years in 2015. Iran Statistics Center has predicted that the rural population share will be less than 20% in 2036. United Nations estimate in 2020 also shows that 24.1% of the country's population lived in villages, and this share will reach 16% in 2050. Thus, Iran will be the tenth country in the world based on the rate of rural population decline in 2050–2018. The decrease in rural population is not solely due to an increase in population threshold or the conversion of villages into urban areas but also due to migration caused by environmental and socio-economic factors (Kalantari et al., 2018).

Before the Islamic revolution, rural development policies were affected by the unilateral pursuit and implementation of modernization and urbanization policies, which could not curb the income gap between cities and villages, the heavy migration of rural people to cities, and the rapid growth of food and crop imports. At the outset of the Islamic Revolution's triumph, many intellectuals and policy-makers criticized these policies and their consequences, leading to high hopes for their revision (Shakuri, 2014). Although the revision of the organizational structure and the formation of revolutionary institutions in charge of rural affairs, such as Jihad of Construction in the early years of the Islamic Revolution, were evaluated in such an environment and in line with the mission of eliminating deprivation and expanding social justice (Fedai & Khalili, 2011), due to the lack of theoretical and intellectual foundations as a support for the rural development policy-making system, it has not enjoyed the comprehensiveness, effectiveness, abilities, and continuity necessary to realize the anticipated goals and the livability of rural areas (Badri et al., 2019). In the country's policy-making and planning process, the lack of a holistic and systemic perspective has resulted in cities and villages acting like separate islands, neglecting the principle of cooperation and complementarity between different places (Pourtaheri et al., 2014). Indeed, the country's current state reveals the weakness of the public sector due to the lack of capability, knowledge, and sufficient experience of agents and policy-makers. As a result, centralized policies are being adopted, which sometimes may not be compatible with the needs of rural shareholders. The review of development programs after the revolution, which was responsible for policy-making and implementation, indicates that these programs rarely achieved their goals. Also, the allocation of credits and investments in the agricultural sector has experienced significant fluctuations compared to other economic sectors (Varmazyari et al., 2018). Moreover, evidence indicates that Iran has primarily focused on construction activities in its rural development plans (Hosseini et al., 2015). Upon examining the components of rural development credits in the annual budgets, it is clear that 70% of these credits were allocated to infrastructure and service projects, and the remaining small amount was expended on rural economic and employment (Varmazyari, 2021). In another challenge, the survey of household expenditure and income evolution in the 2005–2015 decade found that income and consumption inequalities among rural households are higher compared to citizens (Noghani & Rahnama, 2020).

In addition to socio-economic issues, the rural areas of Iran are affected by various environmental hazards, such as drought, flood, storm, dust, landslide, and earthquake (Hejazi et al., 2022). The country faces pressing ecological challenges due to water mismanagement, leading to drought conditions in 97% of the country (Czulda, 2022). Iran's water problems are attributed to the over-abstraction of surface and groundwater, leading to drying lakes, declining groundwater levels, and water quality degradation (Madani et al.,

2016). The country's environmental crisis is further exacerbated by factors such as economic sanctions and fluctuations and nonoptimal water and natural resource use (Tahbaz, 2016). Efforts to tackle land degradation and climate change have been hindered by the lack of coordination between stakeholders and the absence of a comprehensive database (Amiraslani & Caiserman, 2018). The environmental hazards, have led to the desertion of villages and migration of rural populations, impacting agricultural production and leading to economic challenges (Shahi, 2019). Residents of Iran's arid and semiarid regions are experiencing respiratory, cardiovascular, cancer, and infectious diseases as a result of environmental issues (Rahimi & Dong, 2022). Increased vulnerability to environmental hazards is projected to rise by 27% by 2030, with a significant population exposed to high levels of vulnerability (Mafi-Gholami et al., 2022).

Thus, rural issues in Iran are multifaceted and include low development levels, depopulation, economic instability, environmental crises, and low quality of life. These issues stem from top-down, sectorial, and centralized rural policy-making, which have hindered the effectiveness of rural development policies (Badri et al., 2021). Additionally, rural communities in Iran have been more influenced by macro developments rather than state-specific rural reforms, leading to greater access to modern amenities and hindering the achievement of revolutionary objectives such as equal distribution of development benefits and poverty reduction (Shakoori, 2019). The lack of a coherent strategy for sustainable rural development, unavailability of proper infrastructure, weak management of governmental structures, and lack of diversification in economic activities are contributing factors to Iran's rural underdevelopment (Namdar & Sadighi, 2013; Riahi & Masoumi, 2020). Attention to the rural environment characteristics, organizing the spatial relationships of settlements and economic sectors (Darvishi et al., 2019), dealing with the consequences of previous policies, designing appropriate policies and patterns, and improving the efficiency of the planning and policy-making system are necessary to overcome the challenges of resilience and rural development in Iran (Afsharipour et al., 2021).

The study, conducted through qualitative content analysis, examines the challenges and resilience approaches required for sustainable rural development in Iran. This goal is achieved by addressing the following key questions: Firstly, what is the current social and economic situation of rural development in Iran? Secondly, how do rural communities respond to the challenges they face? And thirdly, what are the tools available to create rural resilience against future risks and challenges?

2 Characteristics of Rural Planning and Development System in Iran

2.1 Features of Rural Planning Structure

In Iran, it has been prevalent to pursue and promote development projects at the local scale by government institutions in line with the general upstream policies like the overall policies of the resistance economy. However, these 5-year development plans have not had the expected efficiency because they are centralized and do not define the indicators that determine the state of development, outsourcing, and the partiality of construction activities. In addition to the challenges of the planning structure, the governance structure regarding local and regional development has weaknesses and problems. Currently, managing rural development affairs is done by at least 16 ministries and the Vice President of Rural Development and Underprivileged Areas. This topic has made it impossible for the executive bodies responsible for rural development to coordinate and cooperate. In the meantime, the literature on modern management of rural development, upstream documents, and related domestic laws have emphasized integration in policy-making and planning. For example, Material 27 in the Law of the Sixth Plan of Economic, Social, and Cultural Development of the Country, according to which it is one of the government's duties to create unity in the management and policy-making of the three aspects of the country's agricultural, nomadic and rural development with a jihadist and monitorable approach until the end of the first year of the implementation of the Law of the Sixth Plan. The nonfulfillment of the mentioned task after about 7 years of the sixth plan has fueled the intensification of inconsistencies and organizational disorganization in the management of rural development affairs. At the local level, many organizations have significant influence and intervene directly in various programs. The existence of multiple organizations has led to a dispersion of decision-making channels and resources, making it challenging to coordinate and synergize programs effectively. As a result, sometimes governing institutions pursue similar programs with different names and domains, which has led to system inefficiency or organizational disputes. Also, in some cases, they may have opposing programs, which cause organizational differences (Dadashpour & Alidadi, 2020). Table 1 shows the challenges of the country's planning space on a local and regional scale.

Table 1 Challenges of the country's planning space on a local and regional scale

Approach-based challenges	Outcome-based challenges
Centralized planning	Parallel work of various public and government organizations
Authoritarian and top-down development	Spatial, functional, organizational, political, and planning division
Point approach to settlement systems	Weakness of the legal framework of active public participation
Partial view and lack of attention to internal capacities	Ignoring territorial capacities, functions, and activities in planning
Centralization in decision-making and distribution of resources	Lack of unanimity on issues related to different organizations
Focusing on exogenous factors and external sources	Conflict of organizations and time-consuming dispute resolutions
Inflexible planning in the context of change control	Ignoring the complementary role of socio-economic relations
Tendency to quick implementation actions outside of the planning and policy-making format	Weakness of management mechanisms and interaction between public and private sectors
Emphasizing the details of the programs and prolonging the approval process until their implementation	The effectiveness of local communities against global shocks caused by weak planning foundations

Source Varmazyari et al. (2023a, 2023b)

2.1.1 Recent Pattern-Oriented Plans for Rural Development in Iran

2.1.1.1 Rural Constellations Sustainable Development Plan

One of the measures to address the challenges of the local and regional planning system was the "Plan of Organizing Space and Rural Settlements." The Supreme Council of Urban Planning and Architecture approved the said plan in January 2000, based on Material 137 of the Law of the Third Plan of Economic, Social and Cultural Development and during the implementation of the approvals of the "Pattern for Determining the Development Level of Rural Areas and Centers." As per the regulations, it is necessary to provide a well-coordinated and balanced plan for the growth of agricultural, industrial, and service sectors. This plan should ensure an appropriate distribution of population and activities and an optimal provision of services in rural areas in sync with the urban system and network. Despite this, in 2015, the plan to organize space and rural settlements by the Islamic Revolution Housing Foundation was abandoned in practice and illegally, and citing Material 194 of the Law of the Fifth Plan of Economic, Social and Cultural Development of the Country and the November 2015 resolution of the Council for The Institute of Urban Planning and Architecture presents the "Rural Constellations Sustainable Development Plan (RCSD)." This institution has used the word constellation instead of the word collection in the mentioned title, which is a wrong reference. The RCSD Plan should have been subordinated to the management systems as it was born from the plan of organizing space and rural settlements. The RCSD Plan was implemented in 2015 as a pilot without approval from the Supreme Council of Urban Planning and Architecture of Iran. During this period, the plan witnessed changes in the description of services, implementation methods, and specific orientations. Later, due to content changes in the plan text, its name was changed from "Underprivileged Foundation Development and Progress" to "Development and Progress of Rural Constellations." According to the announcement of the Islamic Revolution Housing Foundation, RCSD Plan, for the sustainable development of rural settlements through the spatial regulation of rural systems in the form of local and regional networks (paying attention to urban-rural relationships). The geographical scope of the system is the territory of one or more political divisions. The primary goal of these plans is to create sustainable rural constellations based on the principles of spatial planning. Figure 1 shows the sub-objectives of the RCSD Plan.

The study of Varmazyari et al. (2023a, 2023b) revealed many reasons for providing a sustainable development plan for rural constellations. The plan is the first of its kind in the country, emphasizing comprehensive development and sustainability in all aspects, from planning to governance. These experts pointed out three necessities that are the basis of their argument on the presentation of the constellations plan:

- Continuation of past challenges in the development process of the country's villages and emergence of new challenges in the planning system: This section examines the fundamental challenges of economic and social development in the rural areas of the country, which have persisted despite the emphasis of experts and authorities, adopting policies, preparing plans, and implementing various development measures. The issue stems from a one-dimensional development approach and sectoral decision-making governance. Some of the significant challenges mentioned in this context include the destruction of natural resources and environment, indiscriminate exploitation of underground water resources and pollution of water and soil resources, the low rate of economic growth in rural areas, weakness of the infrastructure related to the development of social, economic, and physical activities in rural areas and the lack of appropriateness and complementarity of the created infrastructure with economic and environmental advantages, weakness of spatial balances, and the continuation of the unmanaged process of rural-urban migrations. On the other hand, along with the lack of rural

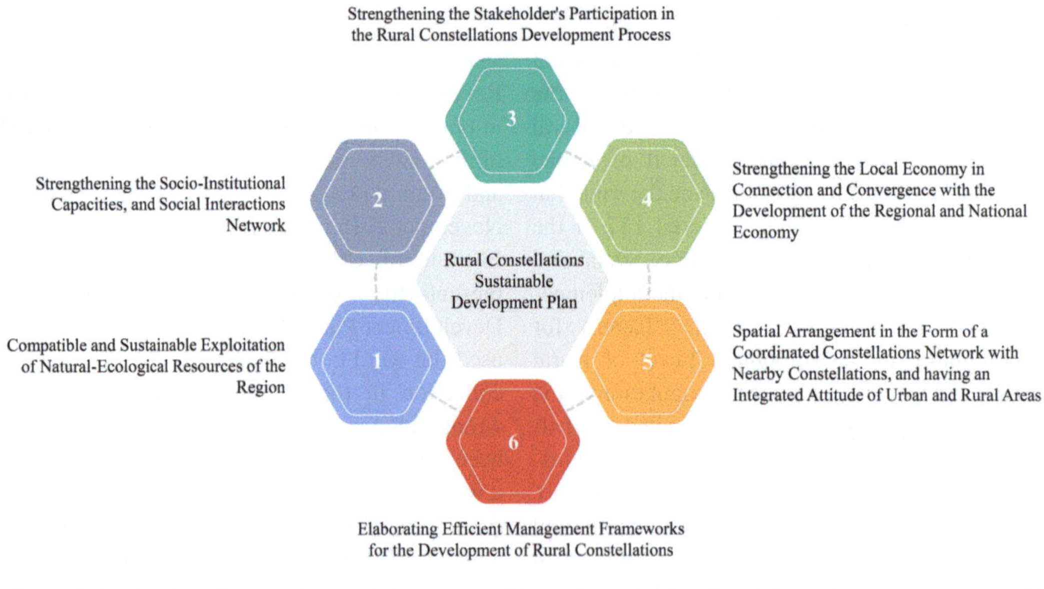

Fig. 1 Sub-objectives of the rural constellations sustainable development plan. Source https://www.bonyadmaskan.ir/pages/manzomerousta.html

development challenges in the country, new challenges have been added to this path over time, and the country's planning system has not been able to face them and take appropriate action. Those challenges that are sometimes rooted in unsustainable interventions in the environment and centralized planning system in the country include the emergence of natural and human hazards, the growth of consumerism and individualism, reducing the level of social security, the creation and development of second homes, marginalization of vulnerable groups, especially women, and the conflict between the interests of the government, people and local management. The elimination of rural development challenges and the emergence of new challenges have had consequences in the economy of space and socio-environment dimensions of rural development, including the occurrence of water stress, commodification of rural areas, population evacuation, and expansion of marginalization around central cities and metropolises, jeopardizing self-sufficiency and food security, increase in poverty spatial injustice and weakening of social capital and the instability of the rural ecosystem in the country.

- Neglecting the role and strategic position of villages in national development: Integrated and sustainable national development requires sustainable and balanced development of urban and rural areas. The evidence and opinions of regional and rural planning experts in the country show that the development of rural communities does not match their environmental assets and advantages. Also, the centralized planning system and single-product economy (carbon economy) have caused these spaces to remain open to development flows over time and through technological advances and related innovations. In this case, those involved in the RCSD Plan point out that with the structural changes created at the international and domestic levels, villages have particular importance in national development while accepting new functions for several reasons. Despite being the source of food supply and raw materials for industries, rural areas have not been able to utilize their resources and human capital for growth and development due to the lack of a comprehensive and integrated plan. Considering the new role of rural areas in the national economy, the trustees and experts of the RCSD Plan have emphasized the existence of a comprehensive development plan, which has given these roles a more central position.
- The requirement resulting from upstream macro policies affecting planning at local levels: Among the upstream policies and documents that the trustees of the RCSD Plan put forward as the necessity for the formation of such a plan is as follows: First, the RCSD Plan is in harmony with the general policies of a resistance economy. The RCSD Plan is one of the significant solutions to strengthen the local economy and realize the goals of resistance economy policies. The third paragraph of resistance economy policies refers to economic productivity by growth production, empowering the local workforce, strengthening economic competitiveness, and the basis of competition between the internal regions of the country. Second, the education of villagers and nomads has been considered a leading and facilitating element in local planning (Material 27 in the Law of the Sixth Plan of Economic, Social, and Cultural Development of the Country). Third, the rural industrial cluster development through the construction and development of competitive and export-oriented enterprises can assist the private and cooperative sectors (Material 27 in the Law of the Sixth Plan). However, the investigations show that the RCSD Plan lacks a clear and reasonable mechanism for the law's implementation. In addition, the mentioned laws implementation is fundamentally incompatible with the founding philosophy and missions of the Islamic Revolution Housing Foundation, and the entry of this institution into the implementation of the mentioned laws will lead to a waste of resources.

After a decade of the formulation and implementation of the RCSD Plan, rural development experts and thinkers in Iran raised four fundamental criticisms against the plan:

- The first part contains criticisms regarding the plan's content, principles, basics, and service description. It often refers to the fact that, in the implementation stage, the planning of the systems was not faithful to its principles and foundations (including the two principles of comprehensiveness and integration, as well as relying on the principles of sustainable development) or was less faithful to it. Criticisms related to the principles and foundations of sustainable development plans for rural systems are as follows: First, choosing the political scale of the district as the study area is not a suitable level for realizing the goals of the sustainable development plan of rural systems, because this plan does not have a mechanism to use the governance power of institutions such as the governorate and the county agricultural jihad department. Second, in the description of system planning services and the section titled Functional Recognition and Structural Analysis of Management-Institutional System, which pertains to institutions and institution building, there seems to be a misconception about the concept of institutions. The idea has focused solely on physical perception, which involves creating new institutions without considering the development of new procedures such as institution building, strengthening networks, organizing informal institutions, and convergence and divergence of trans-sectoral and sectoral policies. Third, spatial planning, which is the governing approach and the theoretical base of the RCSD Plan, cannot be considered alone as a complete approach for this comprehensive plan in the current situation. Another issue is that the spatial approach, proposed in the 1960s, does not incorporate the principles and concepts of regional planning, such as the regional innovation system. Therefore, it is essential to update the concepts of this approach with more modern perspectives on regional planning to address this issue.
- The second part includes criticisms that point to the low quality of the RCSD Plan in the process of preparing and compiling the program, such as the lack of proper adherence to the principles and foundations intended in the compilation of the RCSD Plan, inadequate matching of defined plans and projects with socio-economic capabilities and relative advantages of constellations, the predominance of a one-dimensional attitude on the nature of the plan, lack of sufficient expertise and working background of some consulting engineers designing programs with the nature and function of rural areas, the disproportion of the allocated funds for the preparation of plans with their extent and complexity, definition of micro-economic projects and plans in a structure other than plans with the nature and framework of rural constellations.
- The third part is criticism related to the implementation of the RCSD Plan. Experts have identified the following issues with the status of several RCSD pilot plans executed so far: the low flexibility of some plans with environmental changes over time due to the lack of implementation of a specific executive mechanism, limitation of financial resources for the full implementation of plans, and the lack of success in organizing the intervening institutions to implement the plan in some pilot areas.
- The fourth part discusses criticisms of the Islamic Revolution Housing Foundation's organizational position and governance. In this regard, the Islamic Revolution Housing Foundation lacks the legal support and technical capacity to implement the systems plan. The lack of this support has made this plan unable to bring the cooperation and support of government agencies in the implementation. Secondly, with the entry of the Islamic Revolution Housing Foundation into the RCSD Plan through a process, the position of this institution has been changed from a governing body responsible for physical affairs, carrying out the rural leadership plan, and providing housing for the deprived, to the trustee

of integrated plans for rural development, contrary to the principles of the correct administrative system.

The study by Varmazyari et al. (2023a, 2023b) reviewed and criticized the RCSD Plan, offering suggestions to improve the governance structure and strengthen the plan. They categorized suggestions into the following four sections for better organization:

- The proposals of the plan principles and basics category are as follows: Revising the concepts and principles of the RCSD Plan based on the latest approaches to rural development in the world, RCSD Plan's return to the philosophy of its formation, and avoiding a one-dimensional view and determining the city as the spatial and operational scale of RCSD Plan.
- The proposals of the plan preparation and adjustment process category are as follows: Focusing on the advantages of constellations, placement of infrastructure services and facilities, targeting and shortening the study section of the RCSD Plan, conducting cognitive studies based on the requirements and possible driving forces of each region, avoid going into business details, the definition of fundamental driving forces based on environmental advantages of constellations, and the integration of the RCSD Plan with the plan for economic development and employment in rural areas.
- The proposals for the plan execution and implementation category are as follows: Defining and determining incentives and financial facilities for the entry of private sector investors to carry out activities in the form of selected driving forces.
- The proposals of the responsibility for preparing and implementing the plan category are as follows: Assigning the responsibility of RCSD Plan formulation to the Planning and Development Council of the provinces, selecting the Vice President of Rural Development and Underprivileged Areas as the person responsible for developing and implementing the integrated plan of socio-economic development at the village rural area to the coordinated advancement of the RCSD Plan at the country level, and preparation and management of this project centered on the Ministry of Agricultural Jihad through a bottom-up collaborative process.

2.1.1.2 Economic Development and Rural Employment Plan

Based on Material 27 in the Law of the Sixth Plan of Economic, Social, and Cultural Development of the Country, in line with regional planning, strengthening the rural economy, and developing the export-oriented economy, the Iran Program and Budget Organization should prepare an Economic Development and Rural Employment Plan (EDRE) for 25,000 villages (5000 villages each year) by the end of the Sixth Plan. After receiving approval from the planning and development council of the province, this plan also must be implemented. The EDRE Plan aims to achieve objectives aligned with the Law of the Sixth Plan, such as:

- The scientific knowledge and strategic analysis of the advantages and challenges of economic development and employment in the target villages of the rural district/district.
- Creating a foundation for solving the bottlenecks and taking advantage of the economic development and employment opportunities in the target villages of the rural district/district.
- Integrating, targeting, and determining the priorities of economic investment to create employment in the target villages of the rural district/district.
- Strengthening existing capacities and increasing or developing a new production capability with an emphasis on completing the value chain and production chain with the participation of the nongovernmental sector in the target villages of the rural district/district.
- Institutionalizing public-private partnerships are being established to promote economic development and employment in the target villages of the rural district/district.

The EDRE Plan prioritizes local stakeholder participation and emphasizes employment opportunities for villagers. According to the provisions of this plan, annually for each province based on the defined quotas, out of 5000 rural areas approved in the Law of the Sixth Plan, the number of villages is determined to prepare an economic development and employment plan for the villages of one or more villages. Therefore, this plan preparing for districts or rural districts, each comprising several villages as the target community. This plan will become operational after preparation and will be a 5-year strategic plan for the target villages. The main focus of the EDRE Plan is to identify and address the priority issues of the rural economy. As per the announcement made by the Iran Program and Budget Organization, this plan aims to create development strategies that can help overcome economic underdevelopment. It does so by identifying the capabilities and challenges of rural employment and economic development and providing an overview of the state of the rural economy. Finally, it determines activities that stimulate the development of the local economy and provides employment creation plans and projects in the target villages and the district/district area. In other words, the final report of this program includes a strategic document at the rural district/district level and operational plans in the form of development measures such as investment activities, plans, and projects and their related requirements at the villages of the rural district/district (Islamic Parliament Research Center, 2018).

2.1.1.3 Underprivileged Foundation Development and Progress Plan

In 2014, by order of the leader of the Islamic Revolution and in line with the policies of the resistance economy, the Alavi Foundation presented the Underprivileged Foundation Development and Progress Plan (UFDP) in the direction of depriving rural areas. The implementation of this plan commenced in Qaleh-Ganj county of Kerman province as its initial phase, subsequently serving as a model for other counties. In the UFDP Plan preparation, features such as deprivation reduction, social justice, economic growth, and protection of natural resources have been included. This program has a comprehensive strategy that emphasizes knowledge-oriented, sustainable livelihood and poverty alleviation through people's participation and individual and institutional empowerment and promotes work culture and jihadi spirit. The Alavi Foundation emphasizes strengthening seven capitals for rural development of the UFDP Plan, such as cultural and religious, human, natural, financial, social, physical, and organizational. The main goals of the UFDP Plan are as follows: Deprivation and poverty alleviation, creation of equal opportunities by improving the business environment and business development, and capacity building in the capabilities of each region based on their relative advantages. It also includes macro strategies as follows:

- Branding in three scales of agency, participation, consultation and promotion, nonemployment, outsourcing, and community-oriented institutionalization.
- Focus on providing software services instead of hardware.
- The centrality of development and progress with the cooperation of local institutions.
- Attracting funds and cooperation of donors, jihadi groups, and volunteers.
- Applying government services in the region, attracting and consolidating micro-liquidity of each region in the form of funds.
- Carrying out projects based on the ability to delegate to other departments.
- Cultivating elites, specialized forces, and conscripts to keep them in the region and fulfill the responsibility of serving the people.
- Choosing the bottom three deciles of the society as the main audience of the plan.
- Development through the relative advantage of regions and completing the supply chain and value of products with export priority.
- Carrying out measures based on the comprehensive empowerment document in four dimensions livelihood, education, health, and culture.

According to Alavi Foundation's goals and strategies in disenfranchisement, the stages of selecting the target area until reaching the progress and sustainable development of that area are as follows: Choosing the deprived area, conducting studies of weaknesses, strengths, opportunities, and threats for the selected region, and presenting or editing the plan based on the Comprehensive Empowerment Document, implementing the plan, and developing the exit strategy. So far, the Alavi Foundation has started the process of implementing the UFDP Plan for 40 deprived areas of the country (The Islamic Republic News Agency, 2022).

The UFDP Plan model seeks to eliminate the deprivation of less developed areas by taking advantage of facilitation (empowerment) and studies. The UFDP Plan differs from the RCSD Plan in terms of how it selects study areas. While the RCSD Plan does not take the index of deprivation coefficient into account, the UFDP Plan considers it as the most fundamental key index for selecting study areas. Therefore, the UFDP Plan adopts a "Sustainable Livelihood" approach to improve livelihood capital. The UFDP Plan focuses on an integrated and practical management system. The study areas selected for this plan are the political divisions of "County". The role of the "governor" has been emphasized for the implementation of plans and projects. While the RCSD Plan emphasizes the spatial-functional coherence of settlements, the study areas are "Districts".

2.1.2 Comparison of Recent Rural Development Plans

After comparing the RCSD Plan, the EDRE Plan, and the UFDP Plan, it becomes apparent that all three plans aim to promote the growth of the local economy through regional development and environmental advantages. Although the three plans describe services and strategies differently, they share the same overall goal. The point approach was used in the EDRE Plan and the UFDP Plan, while the RCSD Plan used the polygon approach. But in practice and implementation, according to experts, all three plans have followed almost the same paths and have been successful in cases where government agencies have had the necessary support for them. As an example of the implementation of the RCSD Plan in Tarem city, one of the functions of Zanjan province, was supported by the management and planning organization of the province. In the framework of this support, the governorate, through the provincial planning committee, obliged the planning committees of the cities to allocate 25% of the budget of the local government departments to the RCSD Plan approvals. For example, the RCSD Plan was implemented in Tarom County of Zanjan province with the support of the Program and Budget Organization of the province. In the framework of this organization's support, the governorate, through the provincial planning committee, obliged the city planning committees to allocate 25% of the budget of local government departments to RCSD Plan approvals. However, since the government have not had enough support and cooperation, they have not been able to achieve much success (Table 2).

2.1.3 The General Issues of Planning and Policy-Making for Rural Resilience

In Iran, rural resilience faces challenges due to deficiencies in rural planning and policy-making. Centralized planning inhibits adaptability and responsiveness to diverse local needs, while the parallel efforts of various public and private organizations result in fragmentation and inefficiency. The authoritarian and top-down development approach marginalizes local voices and inhibits community participation, hindering the integration of local knowledge and priorities into decision-making processes. Furthermore, the challenges of discoordination are intensified by inequities in spatial, functional, organizational, political, and planning domains, impeding the formulation of coherent development strategies. The reliance on a point approach rather than a polygonal one in settlement systems fails to capture the complexities of rural landscapes and undermines comprehensive development planning. Moreover, the weakness of the legal framework inhibits active public participation, limiting

Table 2 Comparison of the RCSD plan with the UFDP plan

The RCSD plan	The UFDP plan
The district as a study area of plan	The county as a study area of plan
Study first, then implement	Emphasis on performance while studying
Targeting comprehensive development in all sectors	Aiming to solve fundamental and most effective issues
Lack of facilitation group during the study	Simultaneous activity of facilitation group and study group
Shallow knowledge and emphasis on questionnaire tools	Relatively deep knowledge and targeting local society participation
Ignorance of the local community about the preparation of the plan	Public awareness of plan preparation
Uncertainty of the effects despite the high history of plan preparation	Identification of positive effects in the field of microfinance despite the short duration of the first experience of the project in Qaleh-Ganj
Adopting the approach of physical-spatial in recent plan studies by strengthening the economic dimension	Adopting the approach of economic-livelihood in plan studies
Prolongation of obtaining information from executive institution	Maximum participation of executive bodies with the requirement of the governor to provide plan information
The lack of executive power of the district governor as the political institution of the district to make demands and advance plan	The appropriateness of the executive power of the governor compared to the district governor
Low executive guarantee of the plans due to the prolonged study process until approval and country's political-economic changes	The predictability of the indices and the plan's high implementation guarantee according to the implementation during the study
Having a memorandum of understanding between the Islamic Revolution Housing Foundation, the Program and Budget Organization, and the Governorates for the plan implementation	Having a memorandum of understanding between the Alavi Foundation and the Governorates for the plan implementation

Source Varmazyari et al. (2023a, 2023b)

the engagement of rural communities in shaping their resilience strategies. The lack of attention to internal capacities and the ignorance of territorial capacities, functions, and activities in planning overlooks valuable local resources and potential contributions to resilience efforts. Centralization in decision-making and resource allocation perpetuates disparities and alienates local stakeholders, while conflicts between organizations and time-consuming dispute resolutions further delay progress. The tendency to prioritize exogenous factors and external sources over local needs undermines the sustainability of resilience efforts, while inflexible planning fails to adapt to changing circumstances. The weakness of management mechanisms and limited interaction between public and private sectors hampers coordinated efforts, while the emphasis on procedural details prolongs approval processes, delaying implementation. Furthermore, the inability and instability of local communities to reduce global shocks comes more than anything else from the weak foundations of rural planning and management. Thus, to effectively deal with challenges and be resilient against them, it is necessary to use improved management mechanisms, embrace adaptive, flexible, and responsive planning approaches to changing circumstances, increase multi-sectoral collaboration, simplify decision-making processes until implementation, adopt localized solutions, empowering local communities, natural resource management, and make investments in infrastructure.

2.1.4 A Comprehensive Model for the Progress of Deprived and Rural Areas in Iran

A comprehensive model for the progress of deprived and rural areas with a Kowsari approach, in connection with the development of regions by the Barakat

Foundation, affiliated with the Executive Headquarters of Imam Khomeini's Order, was prepared and implemented. This model consists of a set of variables, communications, and activities that prioritize the fivefold capital resources and focus on the specific and proportional prioritization of each community's growth level (unique characteristics of each target area) and integrates these capital resources in a manner consistent with the principles of modeling. In determining the subset variables and the type and level of relationships between variables, theoretical foundations related to the level of analysis have been used, including stakeholders, living systems, empowerment, participation, and the business network, commensurate with the level of intervention.

Comprehensiveness, integration, and stimulation of advantages are the main features of this model, and mechanisms related to it are designed and implemented. It is emphasized that the comprehensive model of progress with the Kowsari approach seeks to facilitate the participation and role-playing of members of deprived communities in alleviating deprivation in material and spiritual dimensions, by Islamic values. This model is implemented and operationalized within the framework of network governance. Network governance is a complex system of interactions between players (actors) in the form of structures, values, functions, and processes that are defined by three fundamental values: responsiveness, transparency, and participation. The predicted mechanisms in this model are based on an approach that seeks to regulate the relationship between variables in such a way that individuals and institutions operate within it, and through collective action, the path of societal growth and prosperity is pursued from within and institutionalized. In other words, the Barakat Foundation, by designing a comprehensive model of progress with a Kowsari approach, seeks to empower less privileged communities through institutionalization in the framework of network governance, where all members of society play an effective role, and provide the groundwork for comprehensive internal growth (Barakat Foundation, 2021) (Fig. 2).

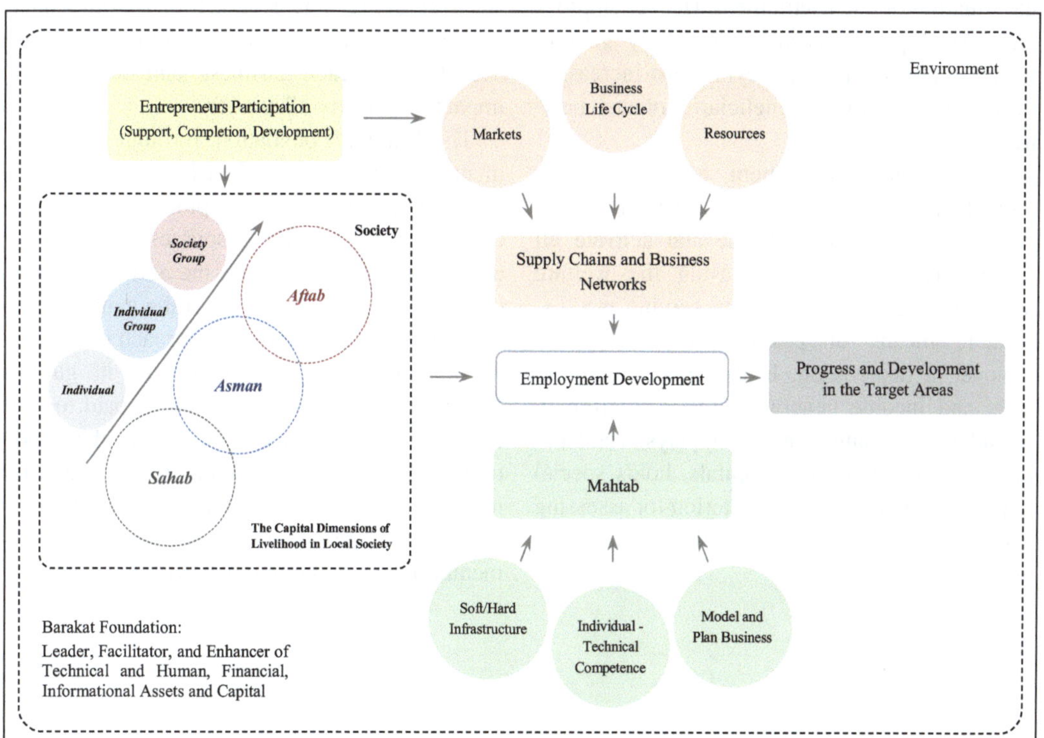

Fig. 2 Comprehensive model for the progress of deprived and rural areas with Kowsari approach of Barakat Foundation. *Source* Barakat Foundation, 2021

Comprehensive Model for the Progress of Deprived and Rural Areas with the Kowsari development approach emphasizes facilitating and guiding quintuple investments in target areas towards advancing progress and development with sustainability. In this regard, the intervention method, considering the situation of these investments, is carried out through three main regulations entitled: Regulation of Employment Support Investment (Sahab), Regulation of Public Investment and Employment of Human Resources (Asman), Regulations on Poverty Alleviation and Empowerment (Aftab), and Conductor Model of Economic Effort (Mahtab). These regulations are interconnected and each, with its evolutionary and exponentially increasing nature, covers a part of the development and growth process of rural communities. Therefore, achieving the goal is contingent upon the full implementation of the model at various levels of society. It's worth mentioning that each regulation addresses all five capitals. However, the manner of attention and intervention mechanisms for each in the community will vary according to the situation of each settlement. Hence, employment development as an intermediary goal in model implementation is beyond just increasing the income level of the beneficiaries of each business plan.

Employment development, accompanied by individual and social empowerment in every community, aims to mobilize and activate all community resources, increasing the welfare level and progress of the society, serving as a sustainable means of poverty eradication and removal of deprivation. In this model, employment and income generation resulting from it, considering the nature and role it plays in enhancing the level of the five capitals, holds special importance and serves as a criterion for assessing and ensuring the correctness of the goal achievement process.

According to the esteemed leader's statement: "Today, employment and income generation are very important and fundamental issues in our country, and if good and efficient planning and serious action are pursued, problems will gradually diminish." For him, the meaning of employment is that a capable young person is ready to work, bringing opportunities for effort and activity in both villages and cities, to fulfill their personal needs and contribute to the advancement and prosperity of the country and its economic growth (Barakat Foundation, 2021).

Based on the principles of the Comprehensive Model for the Progress of Deprived and Rural Areas with the Kowsari Development Approach and based on the business approach, efforts are made to use business plans aimed at growth and sustainability for employment development. In this regard, designing the model and business plans, assessing technical competence and individual eligibility, and providing soft and hard infrastructure on one hand, and providing a set of resources, influencing the nature of the business lifecycle, as well as identifying, creating access, and developing related markets on the other hand, are a series of actions that determine the correct definition of each business and its effective impact on poverty alleviation.

The Barakat Foundation, through the implementation of a comprehensive model with a Kowsari approach to development, aims to efficiently utilize existing resources and tap into unused opportunities, with the active participation of all stakeholders. One of the main features of the Barakat Foundation's approach to facilitating poverty eradication aligns with its guiding role. Therefore, special attention is paid to each of the main components, tailored to the conditions and capacities of the environment and the needs of applicants, both in terms of financial and human/social resources. The design and implementation of the comprehensive model are carried out at two levels of the Problem-Driven Iterative Adaptation (PDIA) approach. The first level oversees modeling and how to determine the main problem, creating opportunities for employing innovative methods, along with iteration, adaptation, and evolution accompanied by growth. The second level oversees the implementation of the model, where its executive elements, based on predefined mechanisms in the form of regulations and guidelines, practically apply four core problem-driven principles: acceptance, identification, and creation of innovative problem-solving methods, iteration, and adaptation in a continuously growing cycle. This cycle, in turn, becomes ingrained in the society, institutionalized, and transforms into sustainable behavior on the path of progress and excellence. In the following, the main components of the model and the predefined mechanisms are briefly explained (Barakat Foundation, 2021).

2.1.4.1 Regulation of Employment Support Investment: Sahab (Clouds)

The executive regulations of the supportive employment investment of Barakat Foundation (Sahab) focus on examining and identifying employment opportunities corresponding to environmental capacities and the characteristics of selected individuals. The main objective of the employment created through this path is to increase income and establish a cycle of income generation by sustainably establishing businesses in the target region.

In the implementation of the Sahab Plan, there is no expectation of a certain level of capital investment or the role of individuals, especially in initiating the process, and its implementation is recommended for communities with minimal capital resources to enter the employment generation cycle. Key actions in implementing the Sahab Plan, in addition to defining and selecting individual and group businesses appropriate to personal and environmental capacities, include:

- Identifying environmental capacities at the county level.
- Formation of the Employment Development Committee at the county level.
- Establishment of the Employment Development Task Force in the target area.
- Direct communication with the people and encouraging them to identify and implement business plans.
- Identifying the capabilities of applicants for employment generation and implementing business plans.
- Attracting participation, guiding, attracting, and utilizing the resources of the applicant.
- Creating appropriate skill training opportunities for the selected and determined businesses and obtaining licenses.
- Allocation of interest-free loans.
- Formation of specialized groups in the selected business area.
- Preparing and facilitating business networks.
- Setting up and establishing a business.
- Establishment of a specialized—guild-based or community-level (targeted) interest-free loan fund to sustain and continue businesses through monitoring and utilizing standards, social capital, and expert consultants.

The Sahab Plan, by raising the income level and fostering a culture of entrepreneurship and development, not only provides a platform for sustainable and effective income generation but also helps break the cycle of poverty and creates a suitable groundwork for the implementation of more advanced and multidimensional foundation plans, such as Asman, and Aftab (Barakat Foundation, 2021).

2.1.4.2 Regulation of Public Investment and Employment of Human Resources: Asman (Sky)

The regulations for public investment and human resource employment (Asman) focus on establishing an income generation cycle accompanied by promoting social responsibility and preserving and strengthening environmental resources. The Asman Plan, considering its characteristics and objectives, targets a different community compared to the target community of the Sahab Plan. This plan is implemented in communities where individuals, in comparison, are on a path of growth in human and financial capital and, alongside paying attention to personal needs, are sensitive to others and the environment. They recognize the importance of participation in personal and environmental development.

In this scheme, the subject of job creation and business establishment is pursued through the formation of development groups. These groups, formed autonomously and voluntarily by qualified individuals, serve as a means of socio-economic empowerment for their members. These groups, as centers for development and progress, not only create employment opportunities but also contribute to mutual problem-solving and the formation of common markets and supply chains up to the district level. The Asman Plan, regardless of national divisions, leverages regional development principles and considers the basis of the existence of environmental capacities and the level of access to resources and growth potential. The actions in implementing the Asman Plan include:

- Identifying environmental capacities from the district level to the target local community (region/urban area or village) and assessing the level of the fivefold sustainable livelihood capitals in that community.
- Establishing employment development committees in each target community and forming local advancement task forces (temporary or permanent).
- Engaging with the local community and forming local development and progress groups.
- Capacity-building training and enhancing the individual and social capabilities of group members to create self-confidence and attract public participation in solving common and individual problems.
- Capacity building training and enhancing the economic capabilities of group members focusing on home-based and small businesses and educating on business plan definitions through proportional joint actions according to capacities and needs.
- Identifying the capabilities of group members for employment in business ventures.
- Defining business models and plans considering capacities and growth and sustainability backgrounds.
- Mobilizing, guiding, attracting, and efficiently utilizing the personal resources of group members.
- Establishing support funds for business backing and allocating resources with internal management to support the business needs of each member (at two levels) and also addressing their personal needs.
- Allocating interest-free loans.
- Setting up and establishing a business.
- Providing inputs and facilitating access to sales markets with the aim of member participation and engaging stakeholders.
- Forming specialized trade groups.
- Sustaining and continuing businesses through monitoring and utilizing standards, social capital, and specialist experts.
- Establishing a development fund, participating in investment, and supporting grassroots employment beyond any development and progress group (such as local specialized-trade groups or other levels with capacity).
- Upgrading businesses through establishing chains, networks, and business clusters at the district and higher levels, as well as shaping medium-sized businesses (at this stage, the focus is not solely on national divisions but on environmental capacities).
- Creating and developing medium-sized businesses through various means such as forming cooperative development companies consisting of group members and other economic actors in the target community (at the levels of village, sector, district, province, etc.).
- Conducting joint social and economic actions with public participation to address local community problems.
- Promoting a culture of participation to enhance the level of each of the five livelihood capitals through development and progress groups (these groups, formed with the intervention of the Barakat Foundation, serve as catalysts for progress and development and are responsible for educating and promoting desirable behaviors as well as forming various groups effective in local community growth).

- Forming coalition groups among members of different local communities and directing the path of comprehensive development and progress at local and higher community levels.
- Bottom-up institutionalization and promotion of people's participation in official institutions and decision-making processes.

Utilizing the Problem-Driven Iterative Adaptation (PDIA) approach is a continuous exercise and repetition at both individual and social levels within local development groups. The development groups formed through the Asman Plan are not equivalent to trade unions. Although some group members may internally form specialized trade groups with or without the participation of individuals outside the group, membership in these two groups is not contradictory.

Considering the anticipated mechanisms in the Asman Plan and how it directly intervenes in enhancing the levels of human, social, and financial capital, the Sahab Plan is linked with the Asman Plan in two axes:

- After the implementation of the Sahab Plan and meeting the conditions of the Asman Plan, members of target communities join development groups and pursue empowerment through these groups.
- Individuals who enter the income generation cycle through the Sahab Plan and are engaged in businesses become complementary to the supply chain concerning businesses established through regional development groups. The difference between these chains and those formed solely based on business premises lies in their role in the comprehensive and sustainable development of the local community and the focus on the five capital resources.

Therefore, based on the comprehensive model features of the Barakat Foundation, the Sahab and Asman Plans can be implemented simultaneously in a target area (Barakat Foundation, 2021).

2.1.4.3 Regulations on Poverty Alleviation and Empowerment: Aftab (Sunshine)

Regulation for Poverty Alleviation and Empowerment of Barakat Foundation (Aftab) is implemented as a complementary process alongside the implementation of the Asman Plan or in case of societal suitability in terms of levels of livelihood capital or the effective presence of entrepreneurial individuals as supporters of other business plans or as complements to the supply chain to improve the standard of living at the community level. For this reason, the implementation of the Aftab Plan is contingent upon a portion of the community possessing a certain level of human, social, and financial capabilities. This plan addresses the empowerment of community members by defining forms that the benefits of these forms return to both the members and the community environment in the form of financial, skill-based, and social benefits.

In the realm of business and employment generation, the implementation of the Aftab Plan serves as a means to consolidate resources using two main methods and employing modern technologies to achieve efficient production at scale, enhance the quality and sale of products and services, complete business chains, and develop infrastructure. In the execution of the Aftab Plan, each of the following actions can serve as the basis for activities:

- Establishing business support to ensure the provision of inputs and raw materials, workforce training, improving production processes, marketing, and product sales.
- Establishing formal organizations such as cooperatives with the participation of the local community to create added value through methods like packaging, product processing, branding, marketing, and market development.

Identifying entrepreneurs and facilitating their involvement in completing business supply chains (rural, local, and regional) in such a way that a specific part of a chain (including upstream and downstream businesses, not limited to a specific part of the internal production process that doesn't end with the main product/service) operates as an independent business unit in collaboration with other units (belonging to local community members) in various forms of business relationships or even through crowdsourcing, particularly in the implementation of the Aftab Plan's first mode, is an example of executing the Aftab Plan. In this mode, involvement with entrepreneurs occurs through various methods of support, civil participation, and others, depending on the scale of the business and its impact on the growth or creation of other businesses, empowering local community members, or role-playing in regional development.

The sustainability of businesses is one of the fundamental principles in all plans, which is pursued through various methods including the following:

- Collaboration among established businesses, either in formal or informal groups, to complete the production and sales process or increase bargaining power in accessing raw materials or market environments.
- Securing support from entrepreneurs or investors to provide initial guarantees regarding the provision of raw materials, skill training, familiarity with standards, or benefiting from a designated share of the market (supporters of businesses created through the implementation of the Asman Plan or some businesses resulting from the implementation of the Aftab Plan).

Given the nature and characteristics of the comprehensive model of the Barakat Foundation, in each target area, if conditions are conducive in terms of the composition of livelihood capitals, the possibility of implementing all three main plans: Sahab, Asman, and Aftab, exists both asynchronously (one plan in each community) and synchronously (all three plans in one community). Because each target community can encompass a diverse spectrum of individuals with various capabilities under the dominant culture and environmental capacities. Accordingly, each of the three plans can be implemented based on the level of possession of the five capitals by community members. Additionally, the three plans are interconnected and since they complement each other, they can correspond to the level of community growth. Thus, the complete implementation of the comprehensive Barakat model in each area depends on the intervention strategy and plan considering environmental capacities (characteristics, experiences, past trends, etc.) and intervention policies. Based on the results of environmental assessments and the readiness to implement the comprehensive model in target areas, the Barakat Foundation has planned the implementation strategy of asynchronous models in communities. Depending on the level of capital resources in each target area, especially human, social, and financial resources, each of the three plans is implemented. In all plans, two physical (technical) and environmental resources are considered infrastructure resources, tailored to capacities and essential requirements.

Based on the planned arrangements, it is expected that after creating a conducive environment for relative growth in target communities, each community itself, based on internal dynamics and relying on the capabilities of its members in an evolutionary process, will undertake the implementation of the comprehensive model. Therefore, based on the principle of empowerment and capacity building, long-term presence and creating dependencies of communities on Barakat Foundation actions are never considered. It is worth emphasizing that considering the ultimate goal of development with a Kowsari approach, the comprehensive model of the Barakat Foundation can always be implemented in the path of community progress in each target area. In each period, the status of livelihood capitals in the selected region relative to each other determines the necessary conditions for implementing the appropriate plan and how the comprehensive model is implemented (Barakat Foundation, 2021).

2.1.4.4 Conductor Model of Economic Effort: Mahtab (Moonlight)

The definition, establishment, and sustainability of any business require that its model and plan be tailored to individual and environmental characteristics (including capabilities, capacities, opportunities, weaknesses, and threats). To this end, the Barakat Foundation's Research and Development Committee has compiled a document entitled the "Mahtab Economic Endeavor Model" as a template for designing and disseminating business models. Each "Mahtab" formulated for a specific business includes the following information:

- General introduction of the business and description of the advantages and capacities of the relevant industry.
- Introduction of the methods and processes of production or service provision of that business, emphasizing household, small-scale, or medium-scale dimensions with a focus on community-centric nature.
- Introduction of standards in both process and product (service).
- Description of the profitable scale of activity and its conditions.
- Requirements for implementing the business plan considering environmental conditions such as institutions and raw materials, permits, training, skills, implementation and execution methods, management approach, market access, and sustainability.
- General market conditions and business growth prospects.
- Introduction of the relevant supply chain for the business.

It is mentioned in the menus of the Leader of the Islamic Revolution: "Production does not mean only industrial production. It also includes industry, agriculture, animal husbandry, large, medium, and small industries, and even handicrafts and home crafts, which contribute significantly to the general well-being of local communities. Therefore, they must be planned." Each Mahtab serves as a general guide for understanding, preparing a model, and designing a business plan suitable for the chosen scale. Therefore, a Mahtab does not replace a business plan, and each interested individual must, based on their capacity and circumstances and the operating environment, proceed to develop an appropriate plan. In this regard, the selection of Mahtab titles is based on criteria such as local society needs, feasibility of plan implementation nationwide, market access, and business growth potential (Barakat Foundation, 2021).

2.1.4.5 Summary of the Comprehensive Model for the Progress of Deprived and Rural Areas

In recent decades, despite efforts made toward the development of societies, issues such as poverty, unemployment, resource depletion, and climate and natural disasters have plagued many countries. Rural areas, especially in developing countries, are more intensely affected by these problems. Not only do their natural landscapes face more visible destruction and alterations, but environmental resource indices are also significantly more unstable and fragile than in the past, with inefficient resource utilization. Additionally, various aspects of human life in these areas suffer from constraints and numerous hardships. Therefore, attention to sustainable rural development is crucial and essential in these regions. In Iran, extensive efforts and initiatives have been made in infrastructure development in the years following the glorious victory of the Islamic Revolution, aiming at alleviating poverty and deprivation in various rural areas. Despite all these efforts, according to statistics from the Statistical Center of Iran (1955–2015), we have not been able to prevent the rampant migration of rural residents to cities. Consequently, in recent decades, the ratio of rural to urban population has experienced a downward trend, with the rural population decreasing from about 86% in 1955 to around 62% in 2015. This trend not only exacerbates social issues for urban areas but also, due to rural depopulation, particularly in border regions, leads to the disappearance of villages as economic production entities and as factors in border security preservation.

Although rural-to-urban migration stems from various economic, social, cultural, and natural causes, economic reasons play a predominant role, as many rural residents migrate in pursuit of improving their lives and seeking better employment and income opportunities. This indicates that efforts and policies implemented so far in rural development have either not been effective in creating sustainable employment and increasing income levels for rural residents and inhabitants of deprived or affected areas, or have had very limited impact. Therefore, the Barakat Foundation has set its primary goal as developing sustainable businesses and empowering the economic and social capabilities of deprived communities, considering the relative and competitive advantages of these regions. Sustainable development is considered the intersection point of the economy, culture, and society; nurturing and engaging efficient and skilled human resources will transform a stagnant economy into a dynamic one reliant on knowledge and human capital, while also diligently pursuing environmental protection and ensuring future generations' share of resources.

Rural society is the human and social capital of the country and the fundamental basis for efforts, production, and sustainable development. The Barakat Foundation, through the formulation of the "Conceptual Framework for the Development of Deprived and Rural Areas," based on scientific principles and international development models, as well as focusing on the strengths and weaknesses of rural and deprived areas, seeks to address poverty alleviation in rural areas and create sustainable employment in those communities through a specific and operational plan. Thus, projects such as Asman, Aftab, Sahab, and Mahtab have been designed and formulated (Barakat Foundation, 2021) (Fig. 3).

2.1.5 The Novel Mechanism in Social Entrepreneurship and Rural Development of Iran

Roostatish is a platform for rural product design and sale, which has defined its mission as empowering rural communities and underserved areas. It encourages participation in rural development projects by promoting locally made products. It is also a social entrepreneurship that uses community-oriented collaborative methods to assess the needs, facilitate, manage, produce, and sell rural products and tries to improve the capacities of the local community so that the villagers plan and take actions to influence and change their social environment. To achieve this goal, Roostatish uses mechanisms to improve the socio-economic status of villages to help rural families in their endogenous and sustainable development. In this regard, this company carries out projects that aim to enhance the social participation of the local community. These projects are implemented after a needs assessment and financed through profits from selling rural products. The Roostatish rural activities focus on maximizing benefits for the local communities. Therefore, Roostatish benefits the village in two stages. The first stage involves the purchase of goods produced in the rural area, which are then returned directly to the producer. Second, the implementation of the rural development project seeks collective benefit and attracts the social participation of the local community during the implementation of the project, and increases their motivation to solve the village's problems through the existing capacities in the rural area. This company values the environment and is committed to ecological sustainability, preventing pollution in all its activities and processes. Roostatish has implemented various measures to reduce the environmental impact, as follows:

- The company runs an annual campaign that encourages the reuse of discarded fabrics. Within this initiative, the fabrics undergo meticulous categorization and separation processes before being distributed to collaborating doll makers. By reusing fabric scraps, the company highlights its commitment to responsible waste management practices and reduces its ecological footprint. Moreover, providing fabric scraps to doll makers can lower production costs and encourage creativity. Its offices always segregate garbage into wet and dry waste. This method helps to recycle materials

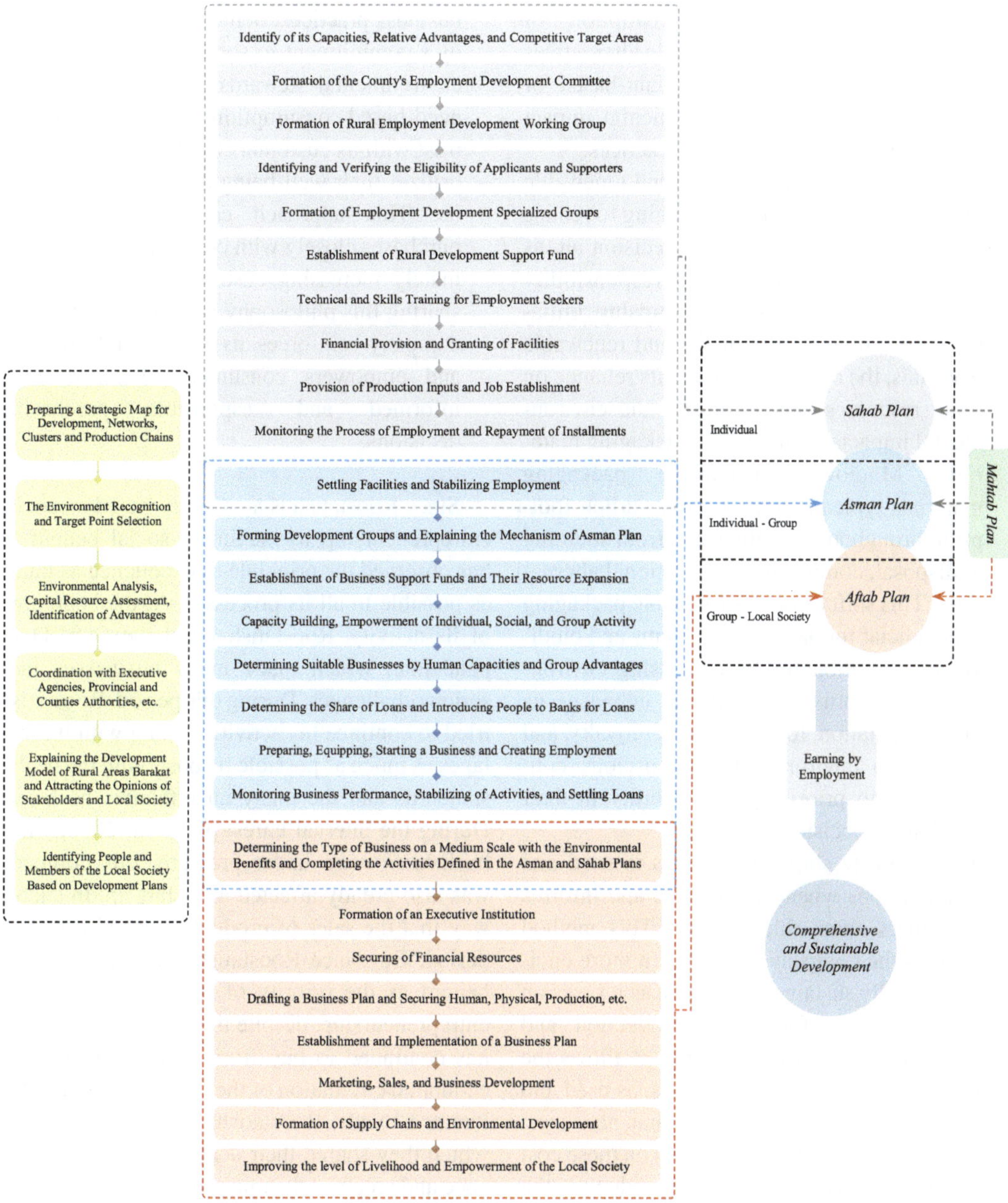

Fig. 3 Components of the execution process for Sahab, Asman, Aftab, and Mahtab plans. *Source* Barakat Foundation, 2021

faster, conserve and prevent resources, and reduce energy consumption.

- The company places great importance on using sustainable raw materials in its product design to minimize environmental impacts. Sustainable raw materials benefits include lower carbon footprints, reduced energy consumption, and minimized waste generation throughout the product lifecycle. Failure to use materials such as leather and forest wood aligns with this objective due to the environmental concerns associated with their production. Similarly, unsustainable logging practices for forest wood can result in deforestation, soil erosion, and disruption of ecosystems. Choosing sustainable alternatives like recycled materials and plant-based or wood options reduces environmental impact and promotes ethical business practices.

- The company uses sustainable and renewable raw materials for product packaging to reduce environmental pollution. This decision aligns with broader corporate social responsibility goals and environmental stewardship initiatives. By utilizing sustainable and renewable materials, the company reduces its reliance on finite resources and minimizes the environmental impacts of traditional packaging materials. Moreover, sustainable packaging materials typically have a lower carbon footprint throughout their lifecycle, from sourcing to disposal, compared to conventional alternatives. This shift towards sustainable packaging is beneficial for natural ecosystems and biodiversity. Additionally, by addressing environmental concerns, the company enhances its brand reputation, fosters customer loyalty, and appeals to environmentally conscious consumers who prioritize sustainability in their purchasing decisions.
- The company ships its products in used and clean cartons whenever possible and informs customers about this practice. This method enables the company to provide a more environmentally sustainable solution as it saves on new resources for carton production and reduces carton waste. Secondly, it allows the company to lower shipping costs, as used and clean cartons are often cheaper than new ones. Therefore, the company can pass on these cost savings to customers, thereby reducing the final prices of products. This approach also enables the company to send a positive message to customers regarding its commitment to environmental responsibility and sustainable values, which may attract and retain customers.
- The company has always emphasized the importance of need-based consumption and avoiding nonoptimal consumption, and it shares this philosophy with its audience. This idea is consistent with ethical and sustainable business practices, demonstrating the company's commitment to social responsibility and environmental stewardship. By emphasizing need-based consumption, the company builds trust with its customers by showing that it prioritizes their well-being over short-term profits. This approach can also help align purchases closely with customers' needs, ultimately increasing customer satisfaction. By sharing this philosophy with its audience, the company reinforces its values and educates and empowers consumers to make more informed and responsible purchasing decisions.

Since Roostatish is an activity based on social entrepreneurship to maximize social benefit, it has observed the principle of excellence as much as possible in all its processes. At the beginning of its activity, Roostatish faced the COVID-19 pandemic, which made many businesses face serious challenges. During this period, Roostatish tried to continue his activities along with the villages as much as possible by observing the health standards and the many uncertainties he faced. During the internal unrest period in 2022, when businesses were under pressure, this company was also greatly affected as a start-up. In such a way that the sales of products reached zero in a period. But since Roostatish has defined social benefit as the priority of its activity as social entrepreneurship, despite the decrease in sales, it has continued to buy from villages so that the economic circulation of the rural community that interacts with Roostatish continues, through which they shared their decisions with the audience and received the minimum opposition in the implementation of these decisions. The company's audience is not only made up of customers but also community members who play a constructive role in building social capital for their villages. As such, the company seeks to foster a close and friendly relationship with its audience.

Roostatish is a registered cooperative company focused on rural development projects. These projects are implemented with the aim of socioeconomic empowerment to increase the resilience of the country's villages. The empowerment and resilience of rural and underprivileged areas form the identity and policy of this company. Roostatish has had three offices, all old houses with local architectural designs and elements. Roostatish has three offices, all old houses with local architectural designs and elements. This company evokes the warm and secure feeling of old houses for the company's employees and customers. Furthermore, the company was able to help develop employee relationships with customers through constructive dialogue and interaction. To increase attention to the development of disadvantaged areas and improve its position in society, Roostatish has reviewed and changed all its processes, relationships, and skills (Figs. 4 and 5).

Statistics show that the company's product sales have significantly increased from \$116,017 in 2019 to \$769,167 in 2022. Additionally, the purchase amount from villages has increased from \$92,773 in 2019 to \$341,488 in 2022. During that period, the price of the dollar experienced an exponential growth of 156% in the country's market.

So far, Roostatish has implemented its projects for 31 villages from 15 standards in 60 self-help groups. It has given employment to 635 people from rural areas of the country. Roostatish bases its projects on the theory of change to define its activities with recognition and needs assessment and to proceed according to a specific framework. In this way, it directs the process of changes in the local community based on the needs and capacities of the community and pursues sensible goals (Fig. 6).

Most of Roostatish's collaborations are with rural women's self-help groups. Almost 90% of the employees of this company are women. The remaining 10% are men who participate in self-help groups of rural women. Roostatish divides its projects into three categories:

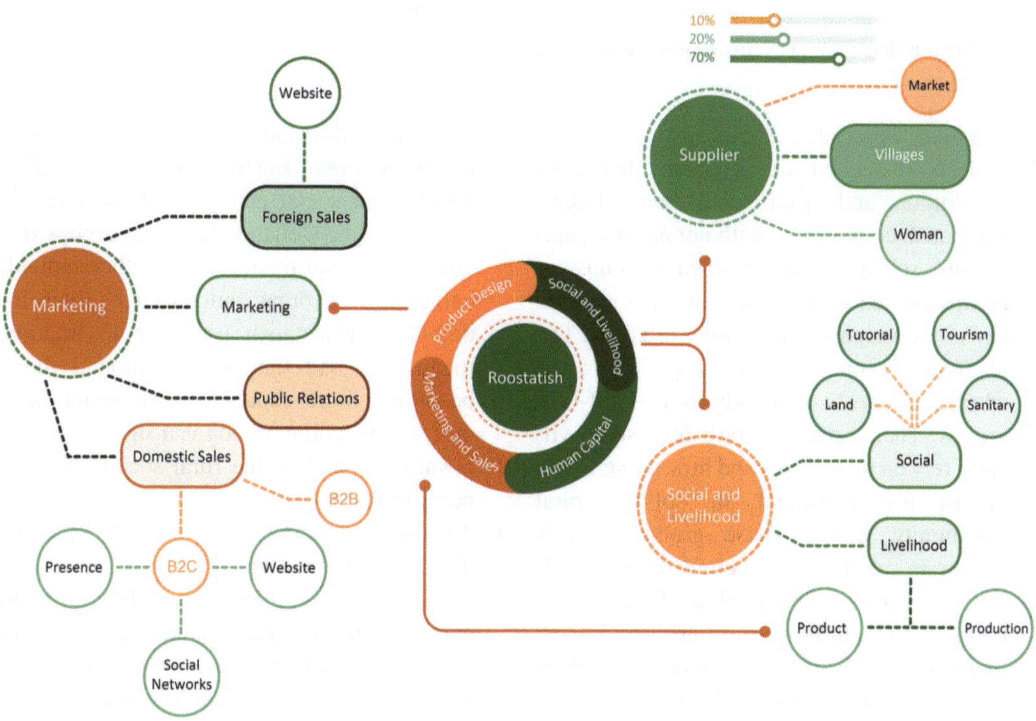

Fig. 4 Roostatish process map. *Source* Roostatish., 2022

Fig. 5 Roostatish process map. *Source* Roostatish., 2022

- The first category is economic projects. These projects are defined in the economic field after recognition and need assessment studies. Roostatish collaborates with nonprofit organizations, self-help rural women's groups, or groups created with the help of other associations in redesigning and selling products. The first type of economic project happens in areas where residents have already formed self-help groups. Their activity shows how some villages rely on themselves and how the self-help groups are beneficial in solving rural community issues. These groups actively engage in solving their socio-economic challenges related to their livelihood. Roostatish collaborates with these rural groups to assist them in sales, redesign, and product development. Other institutions and associations created these groups before they joined Roostatish. Cooperation with existing active groups in these villages has doubled the company's power and promoted the values of collaboration. Roostatish also cooperates with nonprofit organizations such as charities, NGOs, and social entrepreneurs working in various fields. Collaborating with groups that aim to empower them can strengthen individual, group, and organizational capabilities. By doing so, Roostatish was able to improve themselves and the network around them, becoming more capable of interacting with others. This collaboration is a significant step towards expanding the rural social entrepreneurship model.
- The second category is social projects. Roostatish defines the first type of social project through recognition and needs assessment studies in the local community. This type of project does not have an economic aspect. In this category, the entire project is defined and implemented by Roostatish to empower and solve social issues, such as environment, health, and education. In the second type of social project, Roostatish collaborates with

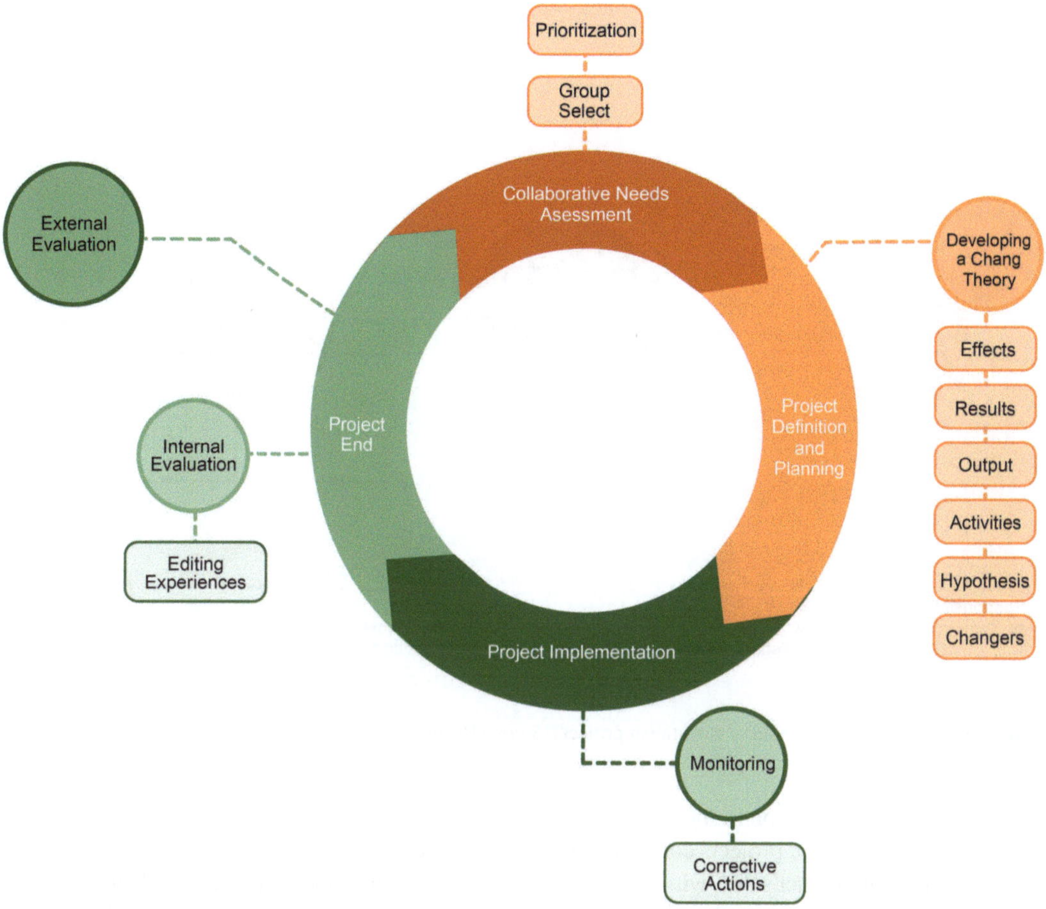

Fig. 6 Implementation path of Roostatish projects. *Source* Roostatish., 2022

companies and organizations to fulfill their social responsibility. The collaboration can cover either part or the entire project.

- The third category is socio-economic projects. In this project, economic and social activities are carried out simultaneously in the villages. Roostatish's collaboration with these villages is not limited to economic activities and livelihood business. Socio-economic projects are defined and implemented based on the theory of change and in a participatory manner to improve the capacities of the target villagers (Fig. 7).

In summary, Roostatish supports the resilience and sustainable rural areas and their ability to adapt to change. They do this through effective mechanisms and initiatives. Firstly, it serves as a platform for developing and selling rural products, aiming to empower rural communities. The company involves the local community in rural development projects through its sales of products. This approach encourages ownership and cooperation in local initiatives. Secondly, Roostatish operates as a social enterprise, employing participatory approaches to address needs, facilitate guidance, and promote rural product production and sales. Through these initiatives, the company empowers local communities to plan and implement actions that have more advantages for their well-being. To achieve its goals, Roostatish employs mechanisms to improve villages' socio-economic status, supporting rural families' livelihoods and fostering self-sufficiency and long-term growth. Based on a needs assessment, these projects aimed to

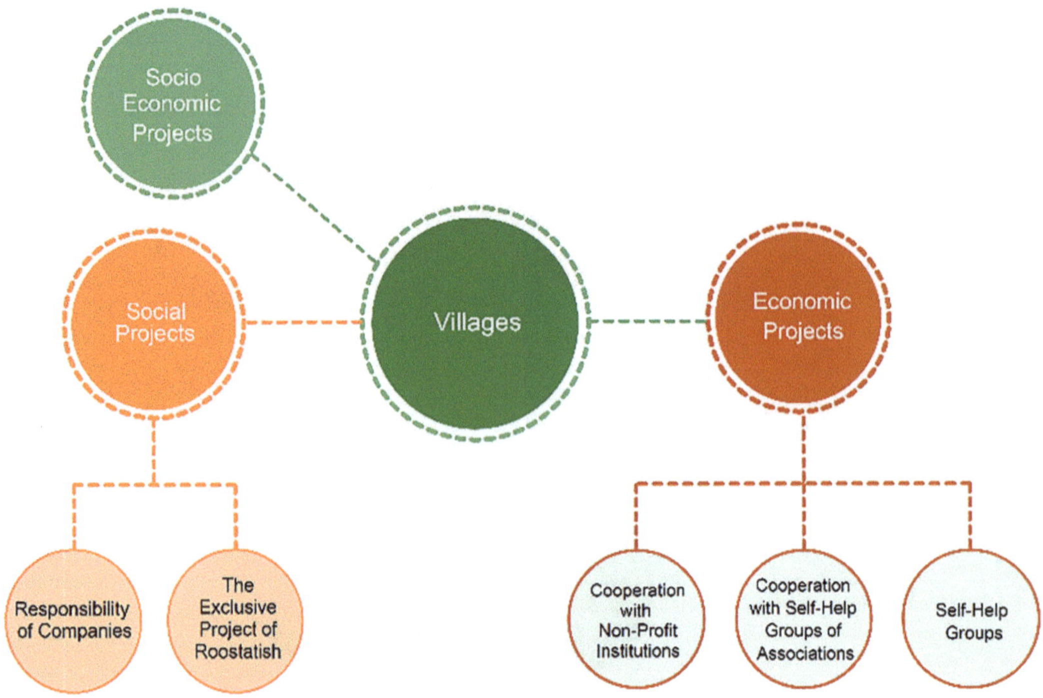

Fig. 7 Flowchart of the division of Roostatish projects. *Source* Roostatish., 2022

increase social participation in the local community. Financing for these projects comes from the proceeds of rural product sales, with profits reinvested back into the villages. By prioritizing the collective benefits of local communities, the company promotes resilience and sustainable development in rural areas. Roostatish also contributes to environmental sustainability by respecting the environment and preventing pollution. It has implemented various environmental measures, such as annual fabric scrap collection, waste separation strategies, sustainable product manufacturing, eco-friendly packaging, and promoting responsible consumption. These initiatives minimize environmental impact and contribute to long-term environmental sustainability. Moreover, the company integrates social entrepreneurship principles into its processes, emphasizing collective decisions, social capital, activity spaces, dialogue-centric relationships, and continuous improvement. By prioritizing dialogue, collaboration, and community involvement, Roostatish builds strong relationships with its audience and fosters mutual development. Overall, Roostatish's comprehensive approach combines socio-economic empowerment, environmental sustainability, and social entrepreneurship, enabling it to lead to rural resilience and ultimately contribute to sustainable rural development.

2.2 Features of Socio-Economic Structure

2.2.1 Rural Settlement in Iran

Before the Aryans of the Aryans to Iran, the southwestern slopes of the Zagros and Northern Khuzestan, which were considered the territory of the Elamites, had a brilliant civilization. At this time, most people lived in sparsely populated and scattered villages and were primarily engaged in agriculture (Mahdavi, 2022). In other words, in the era of the Medes, social formations and various relationships took on a specific form, and villagers were solely dependent on the land according to the laws governing the Medes government, and they played no role in the government. After the Medes and the establishment of the Shahdom in Iran, with the emergence of various social classes and the relative development of urbanization, the majority of people still lived in rural areas. However, since agriculture was the primary source of wealth and economic power in society at that time, and landlords were the most significant taxpayers and had to finance military expenses, rural areas remained constantly under the attention of rulers (Morris, 2019). After the fall of the Medes Empire and during the rule of the Achaemenids, rural agricultural production remained a fundamental source of income for the country. The Achaemenids took measures to revive the land and increase agricultural production. However, the rural community lacked power and did not benefit from prosperity and a better life due to the dominance of the class-based system and the lack of rural participation in administrative and political affairs (Afrakhteh, 2019).

During the Arsacid period, the expansion of trade routes with Western and Eastern Nations, along with urban development, facilitated the expansion of agricultural production (Rizvani, 2018). In this period, Iranian villagers achieved successes through the construction of qanats and irrigation canals (Schultz, 2007). During the Sassanian period, Iranian villages declined, as government officials interfered in the exploitation of natural resources, and harsh administrators executed matters related to the villages. Consequently, rural people were compelled to engage in strenuous agricultural work while also being required to serve in the military (Afrakhteh, 2019). Throughout Iran's history until the present century, it was only during this time that the urban population in Iran experienced significant growth and was able to influence the rural population. This period marks the pinnacle of urbanization in Iran, with one of its most important cities at the time being the city of Madain. During the reign of Ardashir Babakan, according to a comprehensive census, the number of cities was calculated to be 554 points, and the number of villages was estimated to be 6000 points (Ghadiri-Masoum et al., 2019).

During the Seljuk period, commercial crops such as oil and industrial plants, silk cultivation, and beekeeping flourished in the villages of Iran. The Diwan of Ashraf al-Mamalik was established by Khwaja Nazam al-Mulk to plan and manage agricultural affairs, but the prevailing political system and bureaucracy led to the spread of corruption in the villages (Hosseini-Abri, 2004). During this period, Iqta and Feudalism were at their peak. The authoritarian nature of the government's power limited protests and gave the society a kind of cohesive shell that appeared stronger than it was. However, in any case, on the margins and within the general tendencies of society, successive and continuous shifts and stabilizations were taking place (Lambton, 2013). The Mongol invasion led to the collapse of Iran's social and economic life. During the time of Ghazan Khan, with the efforts of his minister Khwaja Rashid al-Din Fazlullah Hamadani, some land reforms were implemented, taxes were reduced, irrigation networks were reconstructed, and rural life prospered (Afrakhteh, 2019). During the Safavid era, particularly during the reign of Shah Abbas, Iranian villages thrived, and agricultural production exceeded the consumption needs of the rural population (Van Der Cruysse, 2017). Afterward, due to the oppression and tyranny of landlords and feudal lords, the villages of Iran turned into ruins.

During the Afsharid era, government properties expanded. In the Qajar period, nomadic tribes gained dominance over rural society. During the reign of Agha Mohammad-Khan, organizations were established for the first time to deal with rural affairs, and agricultural matters and villages were entrusted to the Estate Administration or the Khalsa management. Kadkhoda played a more prominent role in rural affairs. During the first Pahlavi era, more attention was paid to the Khalsa construction, whose income was allocated to the military. The School of Agriculture in Karaj was established to train technical members for agricultural services, which later became the Faculty of Agriculture in Karaj in 1940. During the second Pahlavi period, Dr. Mohammad Mossadegh approved the bill for increasing the share of farmers and the agricultural civil organization in 1951 (Afrakhteh, 2019).

In 1954, the law establishing the National Civil Organization and increasing farmers' shares was approved. Then, in 1955, a bill amending social and rural civil affairs was passed (Mahdavi, 2022). The aforementioned information regarding rural transformations in Iran could only depict the overall situation of Iranian villages with considerable ambiguity. As Kazem Vadiei stated in 1972: "Not much time has passed since the beginning of rural studies and research in Iran. Before the emergence of large cities, the population of Iran was predominantly rural, and urban dwellers constituted a very small percentage of the total population. Nevertheless, except for satellite villages of large cities whose names have occasionally appeared in historical, geographical, and monographic books, it has rarely happened that a village has been studied or several villages have been examined collectively." Therefore, although villages formed the main foundations of power for long-term feudal governments, they have always remained ambiguous and unknown, to the extent that the largest books and dissertations have not contained even minimal information about them (Afrakhteh, 2019).

Sources written under the title of "The Transformations of Iranian Villages Before Land Reforms of the Previous Regime" often address general land ownership systems, exploitation systems, and various types of taxes and tributes. What seems clear is that Iranian villages have been a source of revenue for both the people and the government. Heavy tributes and agricultural land taxes within the framework of farming and leasing systems, which sometimes allocated almost the entire produce to the landlords, pushed villagers into starvation. Villagers had no participation in decision-making systems, so they lacked political power and lived in mud huts lacking necessary strength, as evidenced by reports of heavy earthquake casualties throughout history. In terms of livelihood, rural people lacked food security and faced miserable conditions during droughts and famines. In the nineteenth century, villagers were forced to pre-sell their share of grains and realized during harvest that they did not have enough grain for their winter livelihood (Issawi, 2010) .Rural inhabitants lacked adequate healthcare services. In addition, at best, using communal baths facilitated the spread of various infectious diseases. Despite this, Iranian rural people have become renowned globally for their indigenous knowledge in discovering qanats, traditional irrigation methods, wind, and water mills, and various ways to cope with natural inconsistencies, as well as for nurturing scientists across various fields, indicating their intelligence, resourcefulness, creativity, and diligence in highly unfavorable conditions, this should serve as a model for the youth of the country. Consequently, it is argued that rural households and communities possess two distinct characteristics. On the one hand, excessive poverty and deprivation, and on the other hand, a rich and culturally abundant, sometimes ingenious, heritage. In the 1960s, the Land Reform Law was enacted and implemented in three phases, marking the most significant transformation in rural affairs up to that time. What was known as the hallmark of urbanization in the monarchy's regime ended at the gates of cities, leaving rural areas untouched?

However, when the government's revenue source shifted from land taxes to income derived from oil, extractive industries, and assembly, rural communities regained vitality. Following the Islamic Revolution and the establishment of the Islamic Republic of Iran, rural areas of the country underwent remarkable transformations. The implementation of the Land Reform Law, known as the legal bill land allocation and restoration of land, the performance of seven-member committees (transformation in land ownership system), the establishment of the Jihad of Construction organization, the Housing Foundation of Islamic Revolution, and the implementation of Rural Guide Plan (changes in the physical structure of villages), the implementation of Law on Islamic Councils and Rural Municipalities (changes in rural management), the formation of support organizations such as the Imam Khomeini Relief Committee and the Shahid Rajaee plan (changes in rural livelihoods), and the adoption of supportive procedures and policies (such as purchasing wheat at certain prices, crops insurance scheme), or uncontrolled imports of agricultural products (such as various meat, fruits, silk, tea, and rice), or targeting subsidies (changes in rural agricultural production) are among these transformations and achievements. The reality is that various natural, cultural, economic, political, religious, military, and communication factors intervene in shaping and locational-spatial establishment of rural settlements in the country. In such a structure, any change in these factors can lead to a transformation in the importance and role of that situation. Moreover, the characteristics of the natural environment are relatively stable and observable and examinable in terms of morphology, cultural-environmental factors, in the general sense of material and spiritual culture, act as dynamic and unstable elements that are constantly changing. Therefore, identifying the characteristics of rural areas is accompanied by greater complexity in studies.

Explaining the form and manner of establishment and the existence of rural settlements is not so easily possible and requires detailed and comprehensive investigation. This means that rural settlements have been established in various forms and terms are distributed in diverse areas under the influence of various factors. It is clear that in the past, the country had urban, rural, and tribal settlements. The closer we get to the present time, the number of small rural settlements decreases, while the number of larger rural settlements increases. In recent years, many large villages have been transformed into cities politically and administratively, but the emergence of extensive and rapidly growing settlements around metropolises and large cities, which neither have the authentic characteristics of rural areas nor are compatible with typical urban features, is recognized as an emerging phenomenon. The process of its transformation and management is a subject that, if neglected, can lead to irreparable damages and losses for the future of the country's rural population (Afrakhteh, 2019).

Population distribution in Iranian villages is by no means uniform, and various geographical factors such as topography, climate, soil type, water availability, livelihood, etc., affect its dispersion. Western Iranian villages are more populous than Eastern ones. Moreover, as we move from north to south, the population of villages decreases, and as we move from temperate and humid areas to warm, arid, and desert regions, the distance between settlements increases, and their population decreases. The average population of Iranian villages was over 260 individuals in 1955, which decreased to 230 individuals in 1965 due to the consequences of land reforms and reached 390 individuals in 1990, 338 individuals in 1995, and 348 individuals in 2005.

The trend of changes in the size of village populations has inclined small villages towards becoming densely populated villages and towns. In other words, the continuation of this trend has gradually shifted the balance in favor of large villages, especially since this has been in line with government plans and has increased the possibility of providing welfare facilities to villages. Examining the percentage of the rural population in the country's provinces in the 1955 census shows that the highest percentage of ruralization belongs to Kohgiluyeh and Boyer-Ahmad, Ilam, and Sistan and Baluchestan provinces with 100%, 92%, and 89% respectively, and lowest percentage of ruralization belongs to Tehran, Qom, and Khuzestan provinces with 18%, 40%, and 51% respectively.

Examining the percentage of ruralization in the country's provinces in the 1965 census shows that, similar to the results of the 1955 census, the highest percentage of ruralization belongs to Kohgiluyeh and Boyer-Ahmad, Ilam, and Sistan and Baluchestan provinces with 92%, 91%, and 85% respectively, and lowest percentage of ruralization belongs to Tehran, Qom, and Isfahan provinces with 12%, 33%, and 46% respectively. Continuing to examine the statistics related to the percentage of ruralization in the country's provinces in the 1975 census shows that the highest percentage of ruralization belongs to Kohgiluyeh and Boyer-Ahmad, Ilam, and Kurdistan provinces with 87%, 80%, and 75% respectively, the lowest percentage of ruralization belongs to Tehran, Qom, and Isfahan provinces with 8%, 20%, and 37% respectively. Examining the percentage of ruralization in the country's provinces in the 1985 census shows that the highest percentage of ruralization belongs to Kohgiluyeh and Boyer-Ahmad, Chaharmahal and Bakhtiari, and Golestan provinces with 72%, 63%, and 63% respectively, and the lowest percentage of ruralization belongs to Qom, Tehran, and Yazd provinces with 11%, 13%, and 30% respectively. A notable point regarding the percentage of ruralization in the 1985 census is the increase in the percentage of ruralization in Tehran province due to the occurrence of war and the migration of the urban population to rural areas. Examining the percentage of ruralization in the country's provinces in the 1955 census shows that the highest percentage of ruralization belongs to Kohgiluyeh and Boyer-Ahmad, Hormozgan, and Golestan provinces with 60%, 58%, and 58% respectively, and the lowest percentage of ruralization belongs to Qom, Tehran, and Yazd provinces with 8%, 13%, and 24% respectively. Continuing to examine the statistics related to the percentage of ruralization in the country's provinces in the 2005 census shows that the highest percentage of ruralization belongs to Hormozgan, Kohgiluyeh and Boyer-Ahmad, and North Khorasan provinces with 52%, 52%, and 51% respectively, and the lowest percentage of ruralization belongs to Qom, Tehran, and Isfahan provinces with 8%, 20%, and 37% respectively. Table 3 presents the percentage of ruralization and urbanization in the country by province for different years. According to the results obtained based on the latest census of population and housing conducted in 2015, the highest percentage of ruralization in the country belongs to Sistan and Baluchestan, Golestan, Hormozgan, and North Khorasan provinces, and the lowest percentage of ruralization belongs to Qom, Tehran, Alborz, and Isfahan provinces (Ghadiri-Masoum et al., 2019) (Table 4, Figs. 8, 9 and 10).

The analysis of rural population frequency by province indicates that Razavi Khorasan, Fars, Sistan and Baluchestan, and Mazandaran provinces have the highest population numbers among the country's villages. In addition, Qom, Semnan, Yazd, and Ilam provinces have the lowest rural population numbers.

As shown in Fig. 11, the provinces of Kerman, Sistan and Baluchestan, Fars, and Razavi Khorasan have allocated the highest number of villages to themselves. The provinces of Qom, Alborz, Bushehr, and Chaharmahal, and Bakhtiari have the lowest share of rural areas in the country. To determine the density of rural areas, the number of villages in each province is divided by the total number of villages in the country, and their share is determined. Therefore, the provinces of Gilan, Mazandaran, Kohgiluyeh and Boyer-Ahmad, and Kermanshah have a high density in terms of rural areas (Rezapour et al., 2019) (Fig. 12).

2.2.2 The Population of Iran in Different Census Periods

With the establishment of the General Statistics Office and following the implementation of the 1955 General Population and Housing Census, statistical activities entered a new phase, and various statistical projects in various socio-economic fields were implemented annually. The increasing need of the country's planning institutions for statistics and information and the necessity of close cooperation with the main statistical producer organization led to the separation of the General Statistics Office from the Ministry of Interior in 1964. It was recognized as the Statistical Center of Iran and became affiliated with the Organization of Plan and Budget.

Since 1955, with the stability those were established in different sectors of the country and to exploit more of the rich resources available in the geographical regions of the country, the need for more human power became more noticeable, and the officials realized the importance of the demographics and its quantitative and qualitative aspects. Based on this, the preparation of population statistics began with more seriousness. In the meantime, despite lagging in this area compared to advanced countries of the world that have been dealing with this issue for the past centuries and have had precise information on the population issues of their countries over long periods, efforts were made to reduce this gap by systematically and accurately compiling population statistics and diversifying population topics. Therefore, the working method was such that by preparing the required statistics of different social, economic,

Table 3 Ruralization rate changes by the country's province during 1955–2015

Province name	1955	1965	1975	1985	1995	2005	2010	2015
Total	68.33	62.02	52.97	71.45	69.38	31.54	28.6	25.93
East Azarbaijan	70.28	67.55	58.66	71.48	39.72	33.33	30.8	28.14
West Azerbaijan	79.31	74.47	68.26	48.46	47.32	39.7	31.3	34.58
Ardabil	41.84	83.81	75.44	54.20	51.33	41.6	36	31.82
Isfahan	55.62	46.41	37.06	60.18	25.70	16.68	14.6	11.97
Alborz	–	–	–	44.3	21.6	11.3	9.5	7.36
Ilam	91.98	90.95	80.25	59.00	26.77	30.4	36	31.79
Bushehr	84.77	78.46	63.35	50.00	46.95	34.84	31.8	27.98
Tehran	18.84	12.08	8.48	13.62	13.54	8.65	7.2	6.14
Chaharmahal and Bakhtiari	77.10	70.95	64.43	63.64	54.95	48.44	41.8	35.84
South Khorasan	78.75	70.94	61.86	73.51	42.43	48.67	44	40.98
Razavi Khorasan						31.85	31.9	26.93
North Khorasan						51.62	49.5	43.74
Khuzestan	88.51	48.78	42.56	44.62	37.48	32.78	29	24.45
Zanjan	83.62	82.10	75.09	59.50	52.38	42.01	37.5	32.75
Semnan	69.50	64.36	58.05	41.29	31.71	25.3	23	20.20
Sistan and Baluchestan	89.12	85.65	75.48	59.26	53.88	50.4	51	51.44
Fars	68.09	63.79	57.62	49.61	43.33	38.83	32.4	29.53
Qazvin	77.20	74.83	67.82	54.61	46.14	31.5	26.9	25.25
Qom	40.06	33.34	20.80	11.97	8.84	6.12	4.8	4.82
Kurdistan	87.30	83.48	75.67	60.29	47.58	40.51	34	29.24
Kerman	80.65	75.87	67.40	56.95	47.11	41.47	42.5	41.16
Kermanshah	70.37	65.65	57.13	43.83	38.25	33.21	30.3	24.51
Kohgiluyeh and Boyerahmad	100.00	91.94	87.37	97.72	60.77	52.36	37.4	44.04
Golestan	82.89	76.12	66.37	63.13	58.71	50.83	49	46.64
Gilan	78.95	77.32	70.85	62.32	53.17	43.12	39.7	36.66
Lorestan	80.58	78.42	67.83	53.25	46.35	40.57	38.7	35.44
Mazandaran	79.18	76.09	67.84	60.73	54.10	46.82	45.3	42.22
Markazi	84.59	81.85	72.90	56.00	42.91	30.4	26	23.06
Hormozgan	88.82	84.85	73.30	59.56	58.20	52.82	50	45.18
Hamedan	75.50	74.06	70.11	59.62	51.69	42.42	40.18	36.76
Yazd	64.51	55.37	38.58	30.78	24.85	20.29	17.2	14.64

Source: Statistical Center of Iran detailed the results of the censuses of 1995–2015

and cultural sectors, the process of implementing the country's development plans will be successfully accompanied.

Thus, with the initiation of regular periodic censuses, the country's population information was compiled with greater accuracy and precision, playing a significant role in creating social and economic transformations. Some of the most significant social and economic developments of this period that impacted the country's population distribution and density include the implementation of second and third construction plans, land reforms, expansion of industrial activities, advancement in different sciences and public health, expansion of cities and provision of public services in urban areas, and uncontrolled rural-to-urban migration.

The axes determined for each of the construction plans, while not unrelated to the population of the regions and most designed to improve the economic and social conditions of the country's regions, have not been unaffected in the new formation of these regions. If we look at the review of the second 7-year construction plan (1954–1961), this plan included chapters such as agriculture, communications, industries and mines, and social affairs, focusing on regional development with potential facilities. One of the

Table 4 Rural population, and frequency and density of villages for Iran's provinces in 2015

Province name	Rural population		Frequency of villages		Area (km)	Density
	Quantity	Percent	Quantity	Percent		
East Azarbaijan	1,100,220	5.31	3078	3.14	49.24	62.5
West Azerbaijan	1,129,000	5.45	4518	4.61	40.77	110.82
Ardabil	404,236	1.95	1852	1.89	18.07	102.46
Isfahan	613,073	2.96	3518	3.59	102.99	34.16
Alborz	199,559	0.96	477	0.49	5.2	91.71
Ilam	184,444	0.89	1071	1.09	19.22	55.71
Bushehr	325,495	1.57	851	0.87	21	40.53
Tehran	814,698	3.93	1023	1.04	13.56	75.44
Chaharmahal and Bakhtiari	339,667	1.64	960	0.98	15.37	62.45
South Khorasan	315,070	1.52	3560	3.64	144.39	24.65
Razavi Khorasan	1,733,121	8.36	7360	7.52	116.49	63.18
North Khorasan	377,533	1.82	1298	1.33	28.7	45.23
Khuzestan	1,151,596	5.56	6481	6.62	60.23	107.6
Zanjan	346,283	1.67	1186	1.21	21.94	54.05
Semnan	141,858	0.68	2221	2.27	96.42	23.03
Sistan and Baluchestan	1,427,332	6.89	9632	9.84	161.98	59.46
Fars	1,432,355	6.91	8372	8.55	113.12	74.01
Qazvin	321,610	1.55	1147	1.17	15.64	73.33
Qom	62,317	0.30	356	0.36	11.93	29.84
Kurdistan	468,778	2.26	1884	1.92	28.48	66.15
Kerman	1,302,557	6.28	11,740	11.99	169.34	69.33
Kermanshah	478,444	2.31	3151	3.22	24.57	128.22
Kohgiluyeh and Boyerahmad	314,009	1.51	2230	2.28	14.72	151.54
Golestan	871,546	4.20	1050	1.07	20.67	50.79
Gilan	927,660	4.47	2922	2.98	14.07	207.65
Lorestan	623,896	3.01	3368	3.44	27.02	124.65
Mazandaran	1,386,337	6.69	3625	3.70	23.88	151.78
Markazi	329,690	1.59	1610	1.64	28.43	56.63
Hormozgan	802,512	3.87	2246	2.29	63.95	35.12
Hamedan	639,005	3.08	1209	1.23	19.41	62.27
Yazd	166,724	0.80	3932	4.02	70.93	55.44

Source: Data Analysis of Iran's Statistical Center, 2015

interesting points of this plan was the increase in the share of the agricultural sector compared to the first plan, which was in second place after the communication sector, and the credits allocated in this sector for things such as dam construction studies, irrigation, provision of agricultural machinery and land construction and rural areas of the country were considered.

In addition, the third construction plan for the 5 years (1961–1966) considered important chapters such as communication, agriculture, power, culture, industries, and mines. In this plan, due to the substantial funding allocated for implementing land reforms, special attention was given to the agricultural sector, and the allocated credits for agriculture ranked second only to the communications sector. Thus, it can be observed that the development pattern emphasized in the third construction plan prioritized communications and telecommunications, given the importance of the allocated expenses. Although industrial expenses did not constitute a significant portion, the nature of costs and the project's implementation laid the groundwork for industrialization in Iran according to the desired Western pattern. Among the measures taken in the third construction plan, land reforms were implemented, deemed necessary for the country considering the

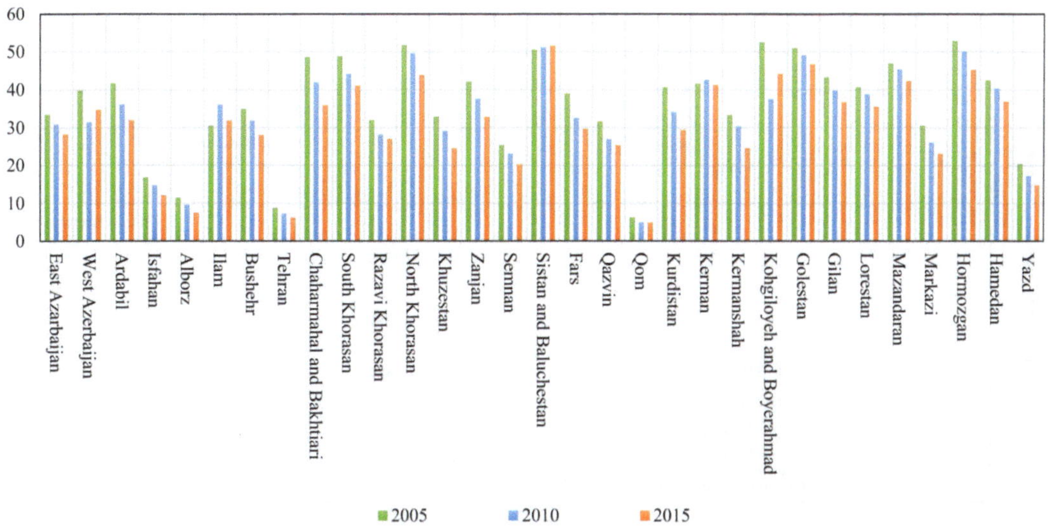

Fig. 8 Ruralization rate changes by the country's province during 2005–2015

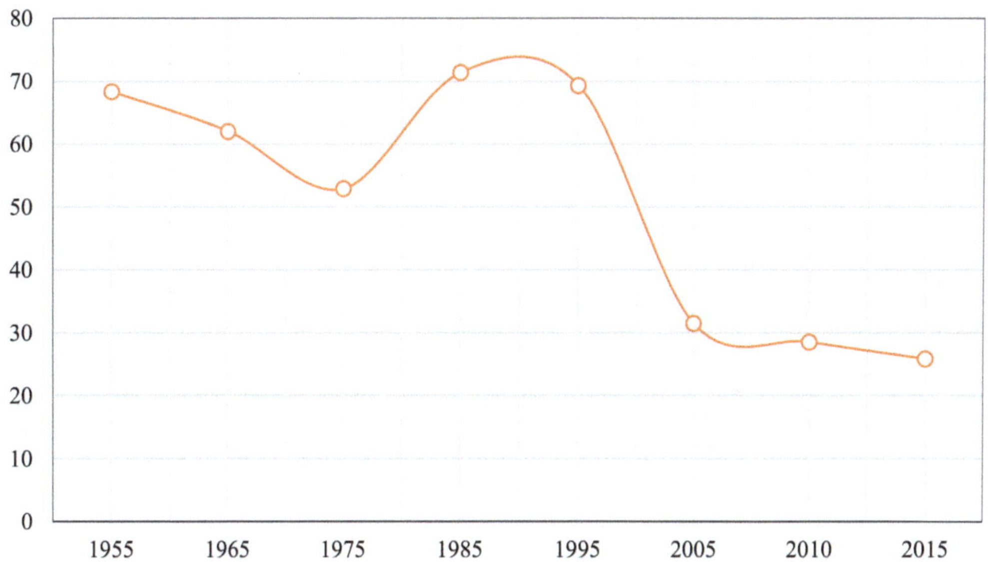

Fig. 9 Country' ruralization changes from 1955 to 2015

temporal and spatial requirements, and executed in three stages. First, the limitation of individual ownership to a six-dong village; second, determining the price at the time of transfer to farmers and paying the amount to the owners; and third, farmers' membership in agricultural cooperatives. Land reforms, which were supposed to create a link between rural areas and rural regions, instead exacerbated rural migration to cities. Although land reforms led to the liberation of farmers, the mechanization of agriculture, increasing cultivation of commercial crops, and limitation of land ownership due to the growing size of families and population growth resulted in rural unemployment. This led to the migration of farmers to cities, especially large cities, and of course, deepened the gap between urban and rural areas.

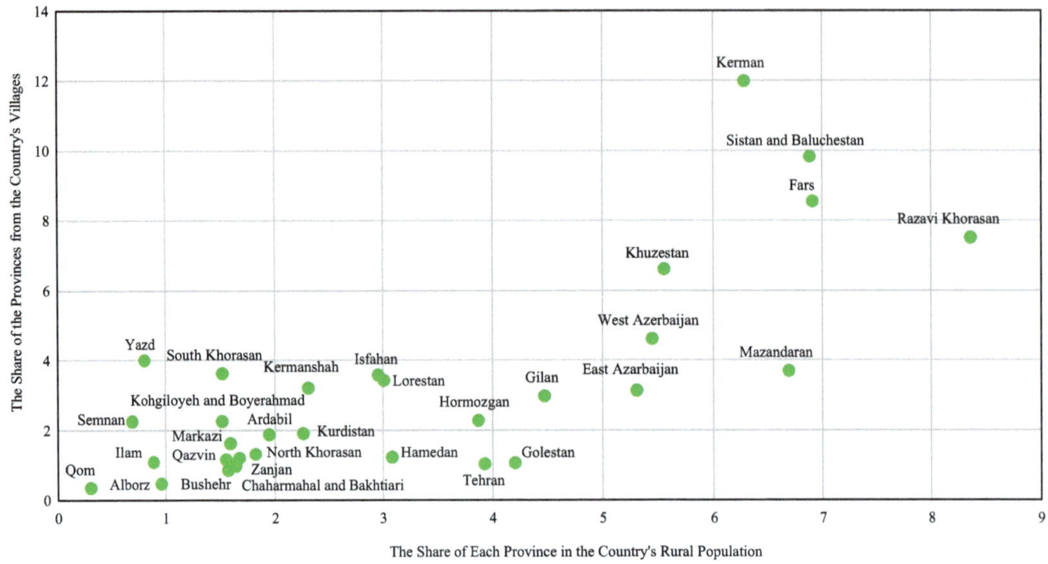

Fig. 10 Ruralization predominance based on the number of villages and rural population in Iran's provinces

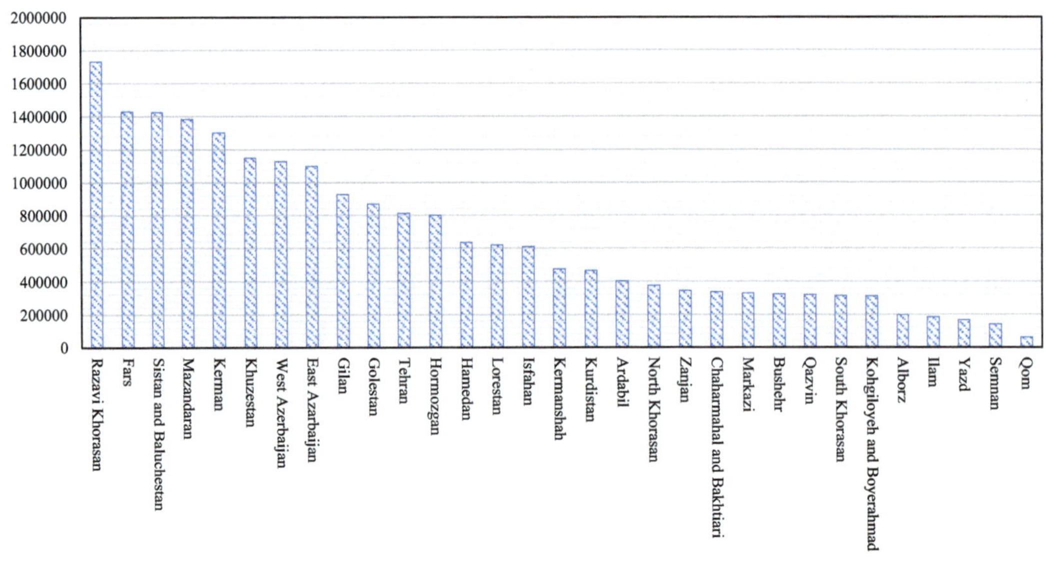

Fig. 11 Share of provinces in the country's rural population

The improper implementation of land reforms led to the abandonment and destruction of villages, the swelling of cities, and pressure on their residential areas to the extent that around most cities, nomads, squatters, slum dwellers, and the like congregated in designated centers, often struggling with poverty and overcrowding. In other words, land reforms facilitated conditions for rapid capitalist development in Iran. The abolition of the old landlord system, the penetration of capital into rural areas, the circulation of money in rural regions, the emergence of a rural bourgeoisie, and the establishment of agricultural banks, mechanized agriculture, and agricultural-industrial companies in rural areas were among the factors that decisively influenced the break-

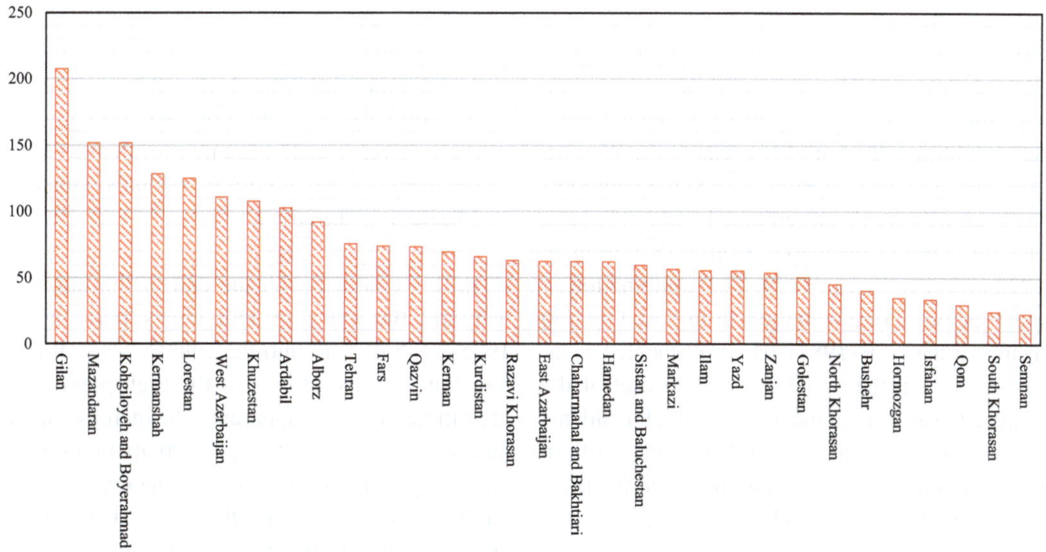

Fig. 12 Share of the country's provinces in the rural areas density

down of previous economic structures and the establishment of new production relations in villages. These events led to a new division of labor that was more compatible with capitalist production relations.

These events led to a new division of labor that was more compatible with the dynamics of the capitalist system. The fourth construction plan, with a budget approximately 2.5 times that of the third construction plan, was devised. Its official aim was to expedite economic growth and enhance national income by increasing production capacity, relying more on industrial development, improving capital efficiency, employing modern methods in all activities, ensuring fair income distribution, reducing reliance on imports, and diversifying the country's export goods. In this plan, the industrial policy of Iran focused on three main sectors: oil and petrochemicals, heavy industries, and consumer goods. Alongside these, there was an emphasis on rapid industrialization and mechanization of agriculture, leading to a gradual decrease in the rural population reliant on traditional agriculture and an increase in the urban population. However, despite the emphasis on industrialization in this plan until 1976, non-oil exports constituted only 2% of the country's total exports, indicating that industrialization under this plan not only failed to lead to independence and self-sufficiency but also increased Iran's dependency on imported consumer goods, resulting in the decline of rural economy productivity and agricultural output in the country's villages (Ghadiri-Masoum et al., 2019).

After the fourth construction plan, the fifth construction plan was implemented during the years 1972–1976. The initiation of this plan coincided with the increase in oil prices and the revenues derived from its sales. Based on this premise, the main objectives of the program placed greater emphasis on industrialization, focusing on capital accumulation in densely populated urban areas, neglecting rural areas, ultimately leading to the formation of population poles in urban areas and the depopulation of villages from active and young populations were critical consequences of this type of planning. Following the victory of the Islamic Revolution, the country's economy revealed its true structure, with the bankruptcy of the banking system due to reckless lending and severe capital flight from the country being the first predictable response to the ailing and dependent economy crisis.

Associated capitalists, feeling insecure, started selling their movable and immovable capitals, and the institutions of consumer goods were closed. On the other hand, due to industries' dependence on Western countries, economic sanctions imposed by them exacerbated the problems. Additionally, with the tricks of the Western countries and the cooperation of some neighboring countries, Iran's oil remained without buyers, and oil revenue decreased from $18 billion in 1978 to $11 billion in 1979. The combination of these factors, along with the war imposed by Iraq against Iran (1979–1987), had multifaceted negative effects on the population. The onset of the imposed war led to the migration of residents from western regions of Iran due to insecurity in other parts of the country. On the other, the presence of young people and the active population on war fronts and related services influenced the age and gender structure of the population. During this period, facilities were provided for youth to form families, and households with larger populations were given priority for benefits and services, with the idea that the continuation of war requires more human power resulting in an increase in population growth to the extent that between 1975 and 1985, the population growth rate reached 4%. On the other hand, the concentration of facilities in urban areas, especially large cities, led to a severe migration of rural residents to cities, which resulted in the urban population growth rate of the country exceeding 7%. The most influential social and economic events on the country's population during this period were:

- The decrease in the age of marriage due to changes in the value system after the victory of the revolution, which, although had many positive aspects, led couples to attain fertility capability at an earlier stage.
- The rapid increase in birth rates and population growth through early marriage and provision of facilities for high-population families such as distribution of goods, housing, allocation of rights, and imposition of restrictions to prevent unwanted births through traditional methods.
- Encouragement of individuals to have more children to meet the human resource needs in various sectors, relying on the traditional assumption that each child can be a useful social and economic factor for the family. Additionally, the notion of "Whoever gives a tooth, gives bread" became prevalent to meet the needs of newly born individuals (Ghadiri-Masoum et al., 2019).

In this context, with the end of the imposed war and the entry of individuals born in the 1975–1985 decade into the country's social arenas, the negative effects of the high population growth rate became apparent. The increase in the number of students, the high participation rate in university entrance exams, the increase in the burden of responsibility, the rising demand for employment, and the need for welfare, medical, health, and housing services were among other factors affected by the rapid population growth rate, requiring a significant portion of the country's credits and revenues to be allocated to them.

The pressures resulting from the sharp increase in population caused even without a clear population control policy, the population growth to experience a downward trend in the following decade, namely 1985–1995, and during this period, it dropped to less than 2%. During this period, the population growth rate decreased to less than 2%. Therefore, the population types during the statistical periods from 1955 to 1995, obtained based on consistent and consecutive planning, indicate that the country's population in 1955, according to the first official census, was 18,954,704 individuals, which compared to previous estimates, reflected the population's dependence on the traditional agricultural economy and reliance on human power. During this period, 68.6% of the country's population lived in rural areas, and 31.4% resided in urban areas. The presence of 42.2% of the population in the age group under 15 years old, the youthfulness of the population, the broad base of the age pyramid, the high burden of responsibility, and the dependence of consumers on the active population indicate significant demographic characteristics.

The second official population census in 1965, with the population reported as 25,788,722 individuals. During this period, approximately 62.1%

of the population lived in rural areas, and 37.9% resided in urban areas. A comparison of the population types for this year with the previous decade indicates an average population growth rate of 3.1%, resulting in 46.1% of the country's population being under 15 years old. Similar to previous years, reliance on human power in economic and social activities, and a low level of life expectancy led families to desire more children.

The third official population census in 1975 coincided with national and international developments that led the country to face a relative decrease in the population growth rate in the decade ending with this period. The population count from the census in that year was 33,708,744 individuals, with 47.1% urban dwellers and 52.9% rural inhabitants. The existence of policies such as urban development, land reforms, changes in oil prices and the resulting increase in revenues, and greater attention to urban areas are among the factors that have contributed to the growth of the urban population. The consequences of these policies, along with some other factors, including adherence to the global policy of population control and luxury and single life, have led to a reduction in the population growth rate in this decade compared to the previous decade, averaging 2.7% per year. However, despite this, the distribution of 44.5% of the society in the age group under 15 years old indicates the youthfulness of the country's population.

The fourth official population census in 1985 was accompanied by significant demographic changes. The Islamic Revolution in 1977 and the subsequent changes in the political system and the governance of Islamic values, the replacement of simplicity with extravagance, the disappearance of manifestations of corruption, and the society's inclination towards salvation, the strengthening of moral and social foundations, and the facilitation of family formation among the youth, as well as granting privileges and incentives to more densely households, and the outbreak of imposed war and the need to provide more human power, are among the factors that have led to changes in the country's population structure during this period. According to the 1985 census, the population of the country was 49,445,010. Generally, three key factors played a role in exacerbating the urban population. Firstly, the results of policies employed by the previous regime regarding inadequate support for rural areas, leading to their migration to cities; secondly, the imposed war by Iraq against Iran, supported by Western countries, resulting in the displacement and migration of citizens and villagers from the western and southwestern regions of the country to other urban areas; and thirdly, incentives such as the distribution of subsidized goods and housing in cities caused the urban population to exceed the rural population, with approximately 54.4% of the country's population being urban dwellers and 45.6% rural residents. Decreased marriage age, lack of birth control or prevention of unintended pregnancies, as well as the encouragement of individuals to have more children during this period, have led to a significant increase in the population growth rate, averaging about 3.9%. This caused 54.4% of people in society to fall under the age group of 15 years. Individuals who were solely consumers and a significant portion of the society's credits, allocated budgets, and household Heads incomes had to be spent on their education in health, treatment, clothing, nutrition, housing, entertainment, and recreation, and its effects became apparent in subsequent periods.

The fifth official population census of the country was held in 1995, while the early years after the population estimate in its previous period coincided with the end of the imposed war and the beginning of the era of reconstruction of the war zones and the resulting devastated industries in other regions. In addition to the reconstruction, which accounted for a large part of the country's construction credits and budgets, the re-approach of society towards consumerism, aesthetics, and the increase in the cost of various living activities in the field of livelihood, housing, education, and health and job provision led to the interruption of family formation and the timely implementation of marriage, which increased the age of marriage among young people. Also, couples who managed to get married tried to reduce the population by controlling their births, so the population of the country, which had the highest growth rate in recent decades in the years before the census, faced a significant decline in population growth rate, reaching less than 2% (1.96%). According to the 1995 census, the population of the country was 60,055,488, of which 61.3% were urban and 38.7% were rural. The social and economic conditions prevailing during this period, which were adopted in part by the stakeholders and planners of government agencies and bodies, designed and the other part based on the self-awareness of the people, caused the rate of people under 15 to be significantly reduced during this period, and for the first time since the beginning of the official population censuses from 1955 onwards, their number has reached less than 40%, i.e., 39.5% of the total population, although this rate still implies the youth of the population in the country, it is promising that if the policies adopted will continue in a correct and calculated way, in the future they will help to reform the age structure of the country's population.

The sixth census was held in 2005, with the trend of the previous period of demographic developments continuing, and the country's population this year reached 70,495,782, of which 68.4% were urban and 31.4% were rural. It is worth noting that the decline in population growth continued as in the previous period, reaching 1.61%, leading to a decline in the population under the age of 15 to less than 30% of the country's population, and raised the alarm for aging and the movement of the population towards aging.

The seventh census period was held in 2015, during which the population of the country reached 79,926,270. The trend of population development continued during this period. During this period, the urbanization process intensified, with 74% of Iran's population being urbanized and 26% being rural. The trend of population growth in Iran from the past to the present shows that the average annual population growth rate in the 1955–1965 decade was equal to 3.1% per year. But in the decade since, between 1965 and 1975, it has been declining to 2.71% a year. The country's population in the 1975s–1985s, has enjoyed strong growth, rising to 3.9% a year. In the second decade of the Islamic Revolution, the policy of population adjustment was achieved by reducing the birth rate and controlling it. In this decade, the population growth rate decreased from 3.9% in the previous decade to 1.96% in the 1985–1995 decade and reached a growth rate of 1.61% in the 1995–2005 decade (Ghadiri-Masoum et al., 2019). The survey of general population and housing censuses and current population statistics shows that Iran faced a sharp increase in fertility in the 1975–1985 decade, with the annual natural growth of the country's population from 2.7% in the 1965–1975 decade to 3.9% in the 1975–1985 decade due to the increase in the number of births. In addition to the urban and rural population, part of the country's population was the nonresident population, and the study of the number of people affiliated with this group during the statistical period 1955–2015 indicates the trend of declining population of this group (Table 5).

Generally, the population of Iran has been consistently increasing over the past 60 years. In a manner that from 1975 to 2015, the country's population has increased by 4.2 times. The population increase has been higher in urban areas compared to rural areas. During the years under discussion, the urban population has increased by 9.9 times and the rural population by 1.6 times. Despite these considerable changes, the population has always maintained its gender balance. In 2015, more than half of the country's population (50.7%) consisted of men and less than half (49.3%) consisted of women (Table 6).

According to the results of the 2015 Population and Housing Census, the population growth rate of Iran has been reported as 1.24%, and during the period between 2010 and 2015, approximately 4,776,601 people were added to the country's population, which averages 955,320 people annually. The results of the 2015 census show that the Hamedan and North Khorasan provinces have negative population growth rates of −0.23% and − 0.11%, respectively, indicating a decrease in the population of these provinces. South Khorasan, Hormozgan, and Alborz provinces, with growth rates of 3.02%, 2.39%, and 2.37%, respectively, have allocated the highest popula-

Table 5 Iran's population changes based on urban, rural, and nonresidential population from 1955 to 2015

Years	Total population	Population of urban areas		Population of rural areas		Nonresident population	
		Quantity	Percent	Quantity	Percent	Quantity	Percent
1955	18,954,704	5,953,563	31.4	13,001,141	68.6	243,797	1.29
1965	25,788,722	9,794,246	38	15,994,476	62	709,799	2.75
1975	33,708,744	15,854,680	47	17,854,046	53	347,812	1.03
1985	49,445,010	26,844,561	54.3	22,349,351	45.2	251,098	0.51
1995	60,055,488	36,817,789	61.3	23,026,293	38.3	211,406	0.35
2005	70,495,782	48,259,964	68.4	22,131,101	31.4	104,717	0.15
2010	75,149,669	53,646,661	71.4	21,446,783	28.5	56,225	0.1
2015	79,926,270	59,146,847	74	20,730,625	25.9	48,798	0.06

Source: Statistical Center of Iran detailed the results of the censuses of 1995–2015

tion growth among the country's provinces to themselves. Of the total 31 provinces in the country, according to the latest national divisions in 2022, only six provinces have had population growth rates exceeding 2%. Additionally, 13 country's provinces have growth rates between 1% and 1.99%, and ten provinces have growth rates below 1%. Furthermore, two provinces of the country have experienced negative growth (Ghadiri-Masoum et al., 2019) (Table 7).

The study of the growth rate of rural population in 2005–2010 shows that the population of rural areas of Kerman, Bushehr, Sistan and Baluchestan, Hormozgan, Mazandaran, Golestan, and North Khorasan provinces has increased. The rural population of other provinces has had a negative growth rate. According to the Geographic Weighted Regression (GWR) function, it was found that there is a correlation between population growth rate and proximity to the sea. Kerman and North Khorasan provinces are exempt from this rule. The reason for the increase in population growth rate in Kerman province is industrial characteristics, exploitation of mines, and the existence of rural industrial areas as well as its countryside. Conversely, North Khorasan, primarily engaged in agricultural pursuits, especially saffron cultivation, has witnessed an increase in its rural population (Rezapour et al., 2019) (Table 8, Fig. 13).

The examination of the population growth rate in rural areas by the province during the period of 2005–2010 indicates that the population of rural areas in provinces such as Qom, Sistan and Baluchestan, South Khorasan, Kerman, Razavi Khorasan, Hormozgan, Kohgiluyeh, and Boyer-Ahmad, and Golestan has shown a positive growth rate. In contrast, the population of rural areas in provinces such as Kermanshah, Isfahan, and Alborz have decreased by more than 2%. During this same period, investigations show that coastal provinces such as Sistan and Baluchestan, Golestan, and Hormozgan have experienced positive growth rates. Although the growth of other coastal provinces has either been moderated or not reported as positive, they are classified among the intermediary villages with low growth rates. The examination of the correlation between the population growth rate in rural areas and climatic factors and access to water indicates a significant relationship between factors such as warm and dry climate, population growth rate, and distance from water sources (sea) with population growth rate. Warm and humid climates, due to the possibility of agriculture beyond the season, are advantageous, and access to water is important for rural areas due to the possibility of benefiting from various activities (Rezapour et al., 2019) (Fig. 14).

Generally, between 2005 and 2015, western regions and parts of the east and center of the country had negative growth rates in rural population, while border areas, the southern and southeastern regions, and some parts of the north of the country experienced positive rural growth rates. Analyzing and studying the number of households in the country from 1955 to 2015 indicates that the number of households has consistently increased over the past 60 years. However, calculations regarding household

Table 6 Iran's population changes by gender in urban and rural areas from 1955 to 2015

Years	Total population			Urban population			Rural population		
	Total	Men	Women	Total	Men	Women	Total	Men	Women
1955	18,954,704	9,644,944	9,309,760	5,953,563	3,070,149	2,883,414	13,001,141	6,574,795	6,426,346
1965	25,788,722	13,355,801	12,432,921	9,794,246	5,096,654	4,697,592	15,994,476	8,259,147	7,735,329
1975	33,708,744	17,356,347	16,352,397	15,854,680	8,291,451	7,563,229	17,854,064	9,064,896	8,789,168
1985	49,445,010	25,280,961	24,164,049	26,844,561	13,769,617	13,074,944	22,600,449	11,511,344	11,089,105
1990	55,837,163	28,768,450	27,068,713	31,836,598	16,435,244	15,401,354	24,000,565	12,333,206	11,667,359
1995	60,055,488	30,514,159	29,540,329	36,817,789	18,805,023	18,012,766	23,237,699	117,101,363	11,527,563
2005	70,495,782	35,866,362	34,629,420	48,259,964	24,576,442	23,683,522	22,235,818	11,289,920	10,945,898
2010	75,149,669	37,905,669	37,244,000	53,646,661	27,023,638	26,623,023	21,446,783	10,853,217	10,593,566
2015	79,926,270	40,498,442	39,427,828	59,146,847	29,841,414	29,305,433	20,730,625	10,630,549	10,100,076

Source: Statistical Center of Iran detailed the results of the censuses of 1995–2015

Table 7 Iran's population growth rates of the provinces from 2010 to 2015

Province name	Total population		Growth rate	Province name	Total population		Growth rate
	2010	2015			2010	2015	
East Azarbaijan	3,724,620	3,909,652	0.92	Fars	4,596,658	4,851,274	1.08
West Azerbaijan	3,080,576	3,265,219	1.17	Qazvin	1,201,565	1,273,761	1.17
Ardabil	1,248,488	1,270,420	0.35	Qom	1,151,672	1,292,283	2.33
Isfahan	4,879,312	5,120,850	0.97	Kurdistan	1,493,645	1,603,011	1.42
Alborz	2,412,512	2,712,400	2.37	Kerman	2,938,988	3,164,718	1.49
Ilam	557,599	580,158	0.8	Kermanshah	1,945,227	1,952,434	0.07
Bushehr	1,032,949	1,163,400	2.41	Kohgiluyeh and Boyerahmad	658,629	713,052	1.6
Tehran	12,183,391	13,267,637	1.72				
Chaharmahal and Bakhtiari	895,263	947,763	1.15	Golestan	1,777,014	1,868,819	1.01
South Khorasan	662,534	768,898	1.43	Gilan	2,480,874	2,530,696	0.4
Razavi Khorasan	5,994,402	6,434,501	−0.11	Lorestan	754,243	1,760,649	0.07
North Khorasan	867,727	863,092	0.98	Mazandaran	3,073,943	3,283,582	1.33
Khuzestan	4,531,720	4,710,509	0.78	Markazi	1,413,959	1,429,475	0.22
Zanjan	1,015,734	1,057,461	0.81	Hormozgan	1,578,183	1,776,415	2.39
Semnan	631,218	702,360	2.16	Hamedan	1,758,268	1,738,234	−0.23
Sistan and Baluchestan	2,534,327	2,775,014	1.83	Yazd	1,074,428	1,138,533	1.17

Source: Statistical Center of Iran detailed the results of the censuses of 2010–2015

growth show that this increase has not followed a specific pattern and has been accompanied by fluctuations and changes in both urban and rural areas. A notable point in these changes is related to the decades after 1985, where it is observed that during the 1985–1995 period, the country's population growth experienced a noticeable decrease. This decrease in the number of households was less pronounced, and in the subsequent decades, namely 1995–2005 and 2005–2015, an increasing trend is evident.

Among the factors influencing the increasing growth of households in the country (despite the decline in population growth) are the shrinking size of households due to the breaking up of households, especially extended families into nuclear families, the younger age structure of the population, and the increase in the number of youths at marriageable ages and establishing their households. It is worth mentioning that since the 1970s, the size of households has decreased in most developed countries. This is due to a noticeable decline in fertility rates, an increase in divorce rates, migration, an increase in life expectancy, and a tendency towards individuals living in old ages (considering that most elderly individuals are women), an increase in the proportion of single-person households, an increase in temporary marriages, and the number of single-parent households. In many families, spouses may also have no desire to have children. It is obvious that the composition and structure of families in each country are influenced by personal choices and preferences, public policies, social and political events, as well as the values, beliefs, and traditions of each country (Fathi & Javaid, 2018) (Tables 9 and 10, Fig. 15).

2.2.3 Instability, Persistence, and Dynamics in Country's Rural Population

In this section, rural areas were studied considering the rural population growth rate. In such a way that the rural population growth rate was classified by rural district. In this classification, a growth rate greater than 0.5 was categorized as population increase, a growth rate between −0.5 and + 0.5 as population stabilization, and a growth rate less than −0.5 as rural population decline. Meanwhile, rural districts with a high density of uninhabited villages were classified as at risk of abandonment. It can be said that areas

Table 8 Number of population and growth rate of rural population in Iran's provinces from 2005 to 2015

Province name	Number of rural population			Rural population growth rate		
	2005	2010	2015	2005–2010	2010–2015	2005–2015
East Azarbaijan	1,200,820	1,145,442	1,100,220	−0.94	−0.8	−0.87
West Azerbaijan	1,148,505	1,148,032	1,129,000	−0.01	−0.33	−0.17
Ardabil	512,195	449,546	404,236	−2.58	−2.1	−2.34
Isfahan	758,890	710,989	613,073	−1.3	−2.92	−2.11
Alborz	234,523	228,142	199,559	−0.55	−2.64	−1.6
Ilam	210,703	199,327	184,444	−1.1	−1.54	−1.32
Bushehr	303,409	325,993	325,495	1.45	−0.03	0.71
Tehran	927,366	877,437	814,698	−1.1	−1.47	−1.29
Chaharmahal and Bakhtiari	414,624	372,746	339,667	−2.11	−1.84	−1.97
South Khorasan	308,305	291,435	315,070	−1.12	1.57	0.22
Razavi Khorasan	1,779,980	1,682,405	1,733,121	−1.12	0.6	−0.27
North Khorasan	414,365	419,068	377,533	0.23	−2.07	−0.93
Khuzestan	1,383,946	1,301,268	1,151,596	−1.22	−2.41	−1.82
Zanjan	405,261	380,925	346,283	−1.23	−1.89	−1.56
Semnan	149,183	144,873	141,858	−0.58	−0.42	−0.5
Sistan and Baluchestan	1,206,547	1,288,160	1,427,332	1.32	2.07	1.69
Fars	1,650,614	1,475,350	1,432,355	−2.22	−0.59	−1.41
Qazvin	365,203	322,787	321,610	−2.44	−0.07	−1.26
Qom	63,639	55,798	62,317	−2.6	2.23	−0.21
Kurdistan	584,337	507,771	468,778	−2.77	−1.59	−2.18
Kerman	1,089,748	1,242,344	1,302,557	2.66	0.95	1.8
Kermanshah	618,718	586,621	478,444	−1.06	−3.99	−2.54
Kohgiluyeh and Boyerahmad	329,849	310,518	314,009	−1.2	0.22	−0.49
Golestan	819,584	870,068	871,546	1.2	0.03	0.62
Gilan	1,109,104	983,701	927,660	−2.37	−1.17	−1.77
Lorestan	691,448	677,416	623,896	−0.41	−1.63	−1.02
Mazandaran	1,368,233	1,391,786	1,386,337	0.34	−0.08	0.13
Markazi	419,184	368,958	329,690	−2.52	−2.23	−2.37
Hormozgan	740,605	786,379	802,512	1.21	0.41	0.81
Hamedan	721,225	716,669	639,005	−0.13	−2.27	−1.2
Yazd	200,988	184,829	166,724	−1.66	−2.04	−1.85

Source: Statistical Center of Iran detailed the results of the censuses of 2005–2015

facing population increase are mainly located in the northern, southern, and southeastern regions and some parts of the country's northeastern parts. Areas with a negative population growth rate are mainly located in the western and some parts of the central and eastern regions.

As illustrated in the Fig. 16, certain sections of the country's border strip are confronted with the risk of rural population evacuation or a negative growth rate. The population stabilization zone predominantly includes a significant portion of Semnan province, while the abandoned or uninhabited zones mainly encompass the provinces of Kerman, Fars, Razavi Khorasan, and Yazd. The risk of village abandonment in the country's central regions is higher than in other areas. Additionally, the rural population growth rate around cities with populations exceeding 500,000 is positive.

Abandoned villages refer to uninhabited villages. The proportion of abandoned villages in each province has been calculated by dividing the number of abandoned villages by the total number of villages. Furthermore, the proportion of abandoned villages in each province relative to the total villages in the country has been calculated by dividing the number of abandoned villages by the total number of villages in the

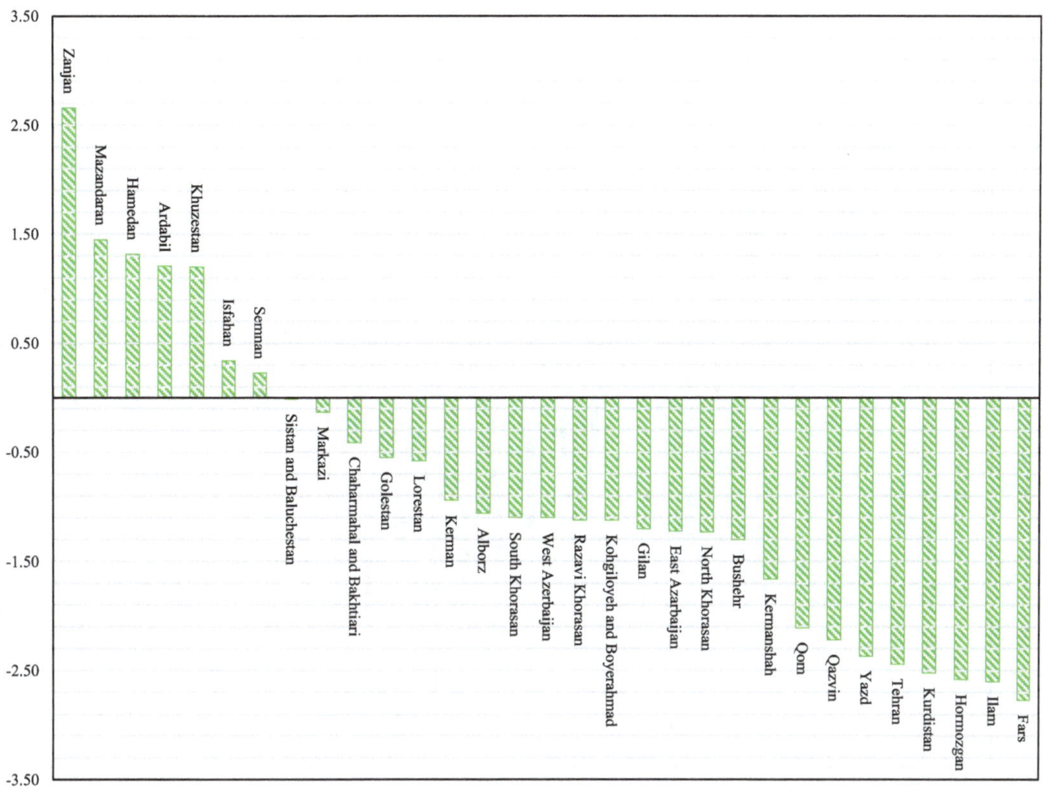

Fig. 13 Growth rate of rural population in Iran's provinces from 2005 to 2010

country (Rezapour et al., 2019). As shown in Table 11, the provinces of Kerman, Fars, Razavi Khorasan, and Yazd have allocated the highest share of abandoned villages from the total country (Figs. 17 and 18).

2.2.4 Consolidation, Transformation, and Annexation of Rural Areas to the City

The villages of Iran have been scenes of profound changes and transformations in recent decades, placing them on an unprecedented and irreversible path. The rural reconstruction path, which refers to deep changes resulting from global and widespread processes centered around economic interests, has placed the country's villages on a path to becoming commodities serving the interests, preferences, and tastes of urban consumers. This has gradually led rural life to lose its intrinsic value, relegating the productive rural economy, which has been both a provider of food security for the people and a source of employment for millions, to the margins. New generations of rural dwellers strive to change the name and identity of their ancestral villages to urban ones. Through this process, they establish a new identity for themselves and smooth the process of commodification of rural spaces, aligning them with the complementary desires and interests of urban society.

Apart from the cultural and social influences on the transformation of villages into cities, policies, laws, and regulations within the country's hierarchical system have had a significant impact on this process. Until before 1955, a historical index was considered the criterion for recognizing a city, and the cities of the country at that time were mainly historical cities and often provincial centers. In other words, despite the enactment of laws such as the Law on the Formation of States and Provinces and the Governors' Regulations enacted in 1906 and the Law on the Divisions of

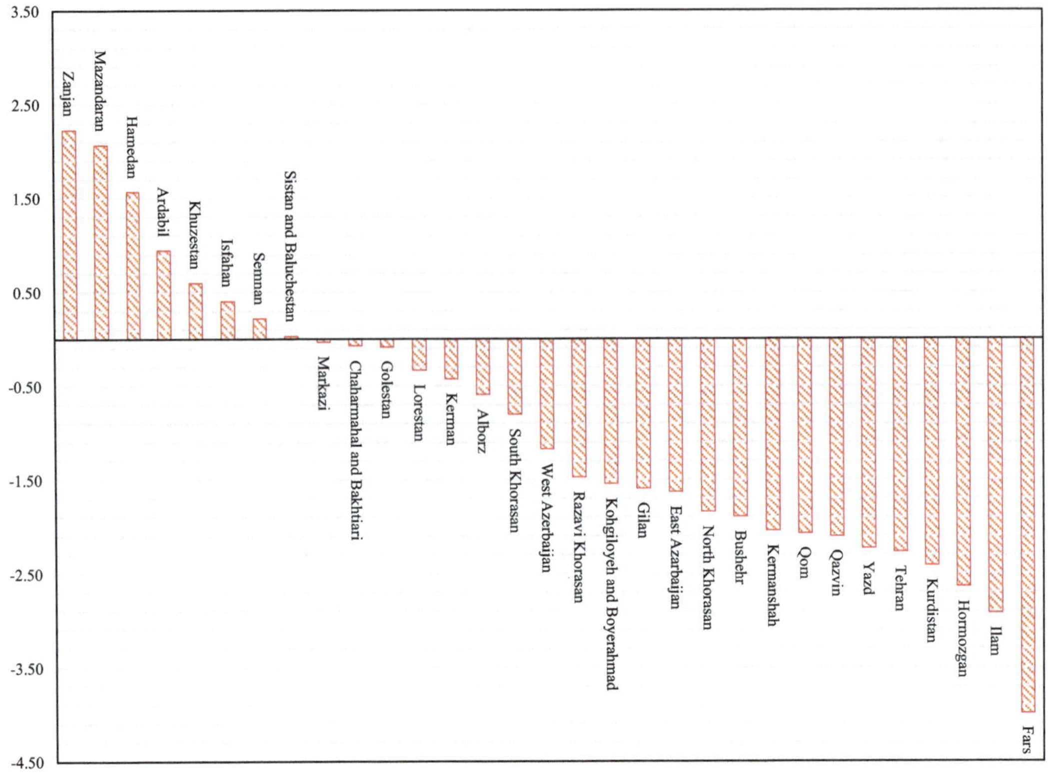

Fig. 14 Growth rate of rural population in Iran's provinces from 2010 to 2015

Table 9 Number of households and family size in Iran by urban and rural areas from 1955 to 2015

	Number of households			Family size		
Years	Total	Urban	Rural	Total	Urban	Rural
1955	3,985,510	1,261,372	2,724,138	4.8	4.8	4.8
1965	5,165,577	1,960,701	3,204,876	5.0	5.0	5.0
1975	6,709,068	3,264,193	3,444,875	5.0	5.0	5.2
1985	9,672,477	5,527,475	4,145,002	5.1	5.1	5.4
1995	12,387,943	7,943,189	4,444,754	4.8	4.8	5.1
2005	17,352,686	12,282,926	5,069,760	4.0	3.9	4.4
2010	21,185,647	15,427,848	5,757,799	3.5	3.5	3.7
2015	24,196,035	18,125,488	6,070,547	3.3	3.3	3.4

Source: Statistical Center of Iran detailed the results of the censuses of 1955–2015

the Country and the Duties of Governors and Prefects in 1936, which governed the country's divisions, no criteria were defined for transforming a village (or hamlet) into a city. From 1955 onwards, for the first time, a population criterion was considered to recognize a city for the convenience of statistical work and distinguishing between urban and rural communities. A population of 5000 was considered as the criterion for urbanization. However, in Material 1 of the Municipalities Law passed in 1954, the criterion for selecting a population point as a city is the consideration of 5000 people. Of course, in the Law of Municipalities approved in 1954, the criterion for establishing a municipality for any place or region was to have a population of 5000 people. This population criterion for determining a place as a city remained until 1362. After the

Table 10 Average annual population growth and household in Iran by urban and rural areas from 1955 to 2015

	Average annual population growth			Average annual household growth		
Annual period	Total	Urban	Rural	Total	Urban	Rural
1955–1965	3.13	5.02	2.13	2.63	4.51	1.64
1965–1975	2.71	4.93	1.11	2.65	5.23	0.72
1975–1985	3.91	5.41	2.39	3.72	5.41	1.78
1985–1995	1.22	1.72	0.60	2.51	3.69	0.7
1995–2005	0.73	1.46	−0.32	3.51	4.46	1.32
2005–2010	1.62	2.74	−0.44	3.89	4.66	2.58
2010–2015	1.29	2.14	−0.67	2.69	3.28	1.06

Source: Statistical Center of Iran detailed the results of the censuses of 1955–2015

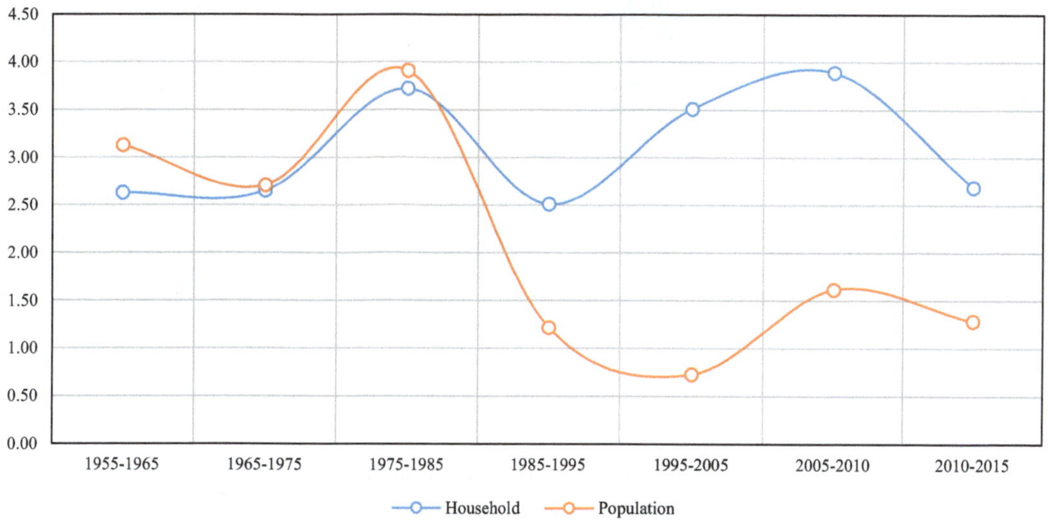

Fig. 15 Average annual population and household growth in Iran's provinces from 1955 to 2015

enactment of the Law on Definitions and Regulations of National Divisions in 1982 and its implementing regulations in 1983, changes were made to the definition of a city to control and prevent the expansion trend of villages from turning into cities.

According to Material 4 of this law, the population criterion for a city doubled, and having a minimum population of 10,000 was considered the criterion for being a city. This law was amended in 2009, and according to the fifth annex to Material 4, the growth of central village regions with any population and villages meeting the conditions, if they had a population of 3500, would be recognized as cities. The reduction of the population criterion for a city from 10,000 in 1982 to 3500 in 2009 accelerated the transformation of villages into cities.

What has occurred as a result of these changes and transformations in the definition of cities and villages in Iran is the continuous expansion of urban areas and the abandonment of some rural areas. The trend of increasing urban areas over the past six decades (1955–2015) has always been upward, and the main reason for this is the transformation of villages into cities. The number of urban settlements in Iran, which according to censuses of the years 1955, 1965, 1975, and 1985 were 199, 272, 273, and 496 urban points respectively, significantly increased during the censuses of the years 1995, 2005, and 2010, reaching 612, 1013, and 1242 urban points respectively. On the

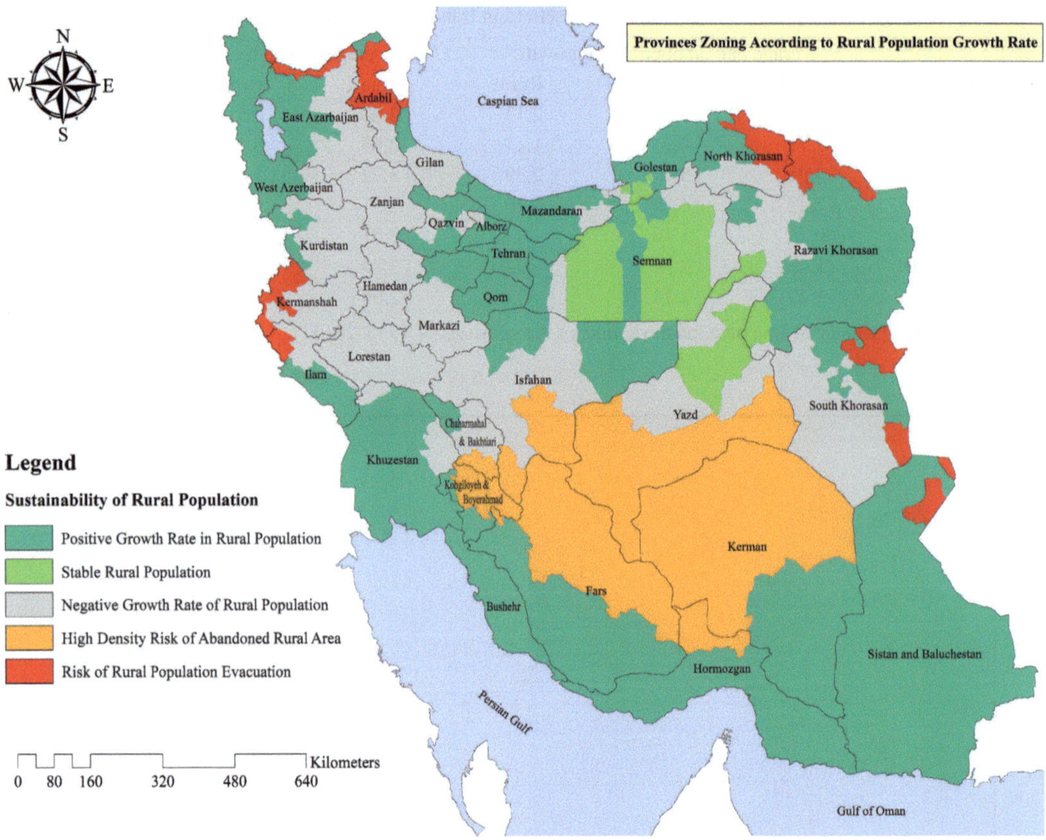

Fig. 16 Iran's provinces zoning according to rural population growth rate. Source: Rezapour et al., 2019

other hand, the population of rural areas increased from 13 million in 1955 to about 21 million in 2015, while the population of urban areas increased from 5.9 million in 1955 to about 60 million in 2015. Therefore, the rural population's share of the total population of the country increased from 7.68% in 1955 to about 9.25% in 2015. The Statistical Center of Iran's population projection for the country over the next 15 years also indicates that by 2035, this share will decrease to less than 19%. The United Nations report (2019) on the country's population estimate also indicates that the negative growth of the rural population will continue, and with an urbanization rate of 85% and ruralization rate of 15% in 2050, Iran will become the tenth country in terms of the rate of reduction of the rural population during the period 2050–2018 (Khorasani & Babaei, 2022a, 2022b). Table 12 and Fig. 19 examine the growth rates of urban and rural populations as well as the total country in the recent century.

The first discussion in examining rural-urban relationships is the process of rural transformation into cities and the emergence of cities in the country. From 1955 to 2015, both the urban population and the number of cities has consistently increased. Meanwhile, the population of rural areas has been declining since the 1995 census. In such a way in the process of urbanization of human settlements in the country, the number of urban points has increased more than six times, from over 200 urban points to over 1200 points by 2015 (Table 13).

From 1955 to 2015, nearly 634 rural points have been converted into cities. Fars province ranks first with the transformation of 54 rural points into cities. Isfahan and Khuzestan provinces rank second and third respectively, with 49 and 48 points in the following rankings. On the

Table 11 The number and share of abandoned villages in the according to the Iran's provinces

Province name	Rural population	Number of villages	Number of abandoned villages	Share of the province	National share
East Azarbaijan	1,100,220	3078	368	11.96	0.38
West Azerbaijan	1,129,000	4518	1499	33.18	1.53
Ardabil	404,236	1852	152	8.21	0.16
Isfahan	613,073	3518	1827	51.93	1.87
Alborz	199,559	477	146	30.61	0.15
Ilam	184,444	1071	399	37.25	0.41
Bushehr	325,495	851	223	26.2	0.23
Tehran	814,698	1023	293	28.64	0.3
Chaharmahal and Bakhtiari	339,667	960	179	18.65	0.18
South Khorasan	315,070	3560	1784	50.11	1.82
Razavi Khorasan	1,733,121	7360	4085	55.5	4.17
North Khorasan	377,533	1298	459	35.36	0.47
Khuzestan	1,151,596	6481	2445	37.73	2.5
Zanjan	346,283	1186	262	22.09	0.27
Semnan	141,858	2221	1676	75.46	1.71
Sistan and Baluchestan	1,427,332	9632	2761	28.66	2.82
Fars	1,432,355	8372	4229	50.51	4.32
Qazvin	321,610	1147	278	24.24	0.28
Qom	62,317	356	133	37.36	0.14
Kurdistan	468,778	1884	207	10.99	0.21
Kerman	1,302,557	11,740	6123	52.16	6.25
Kermanshah	478,444	3151	554	17.58	0.57
Kohgiluyeh and Boyerahmad	314,009	2230	595	26.68	0.61
Golestan	871,546	1050	47	4.48	0.05
Gilan	927,660	2922	362	12.39	0.37
Lorestan	623,896	3368	420	12.47	0.43
Mazandaran	1,386,337	3625	712	19.64	0.73
Markazi	329,690	1610	444	27.58	0.45
Hormozgan	802,512	2246	551	24.53	0.56
Hamedan	639,005	1209	134	11.08	0.14
Yazd	166,724	3932	2812	71.52	2.87

Source: Rezapour et al., 2019

other hand, the analysis of the density of rural-urban settlements shows that the northern region of the country, including Tehran, Alborz, Mazandaran, and to some extent Gilan, has the highest density of settlement in unit area. Also, examining the density of rural-urban settlements in the southwest region indicates a significant correlation between the distance from the capital and the distance from the central cities of the provinces with the formation of such rural-urban settlements. This means that with the distance from Tehran and the provincial capitals, the number of small villages transformed into cities increases. This issue can be considered as an administrative-management strategy to strengthen decentralization and regional development in deprived regions, as well as a policy to fill the urban hierarchy gap in provinces, counties, and newly formed districts.

In general, the mentioned statistics indicate that the spatial development policies of the country, especially the transformation or annexation of villages to cities, have been in a way that rural areas have turned into repellent areas and urban areas into attracting areas for population. What influences the positive or negative effects of the

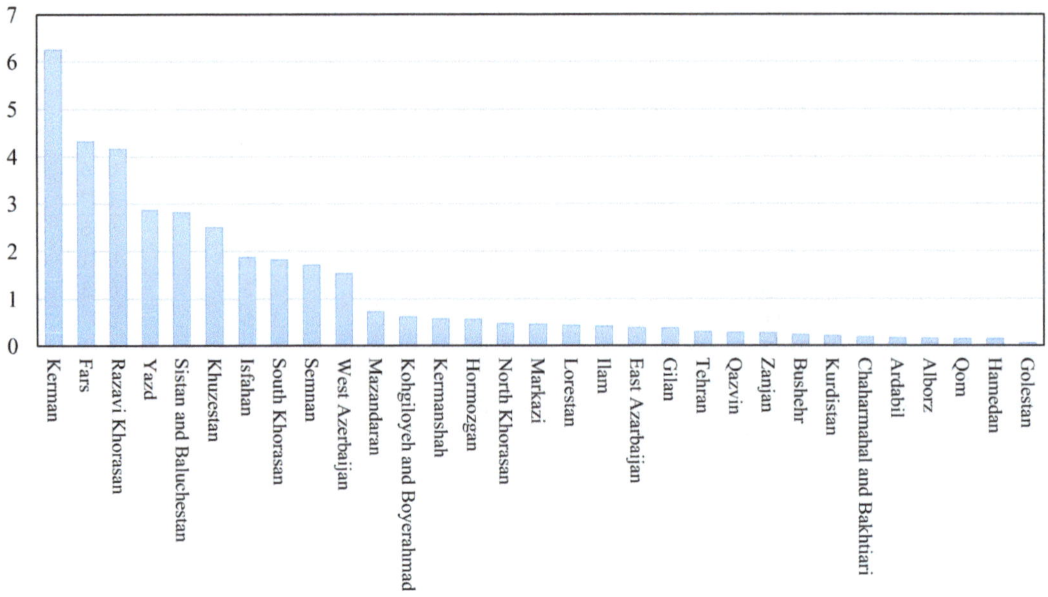

Fig. 17 Comparison of the share of abandoned villages in each province from the entire country (percentage)

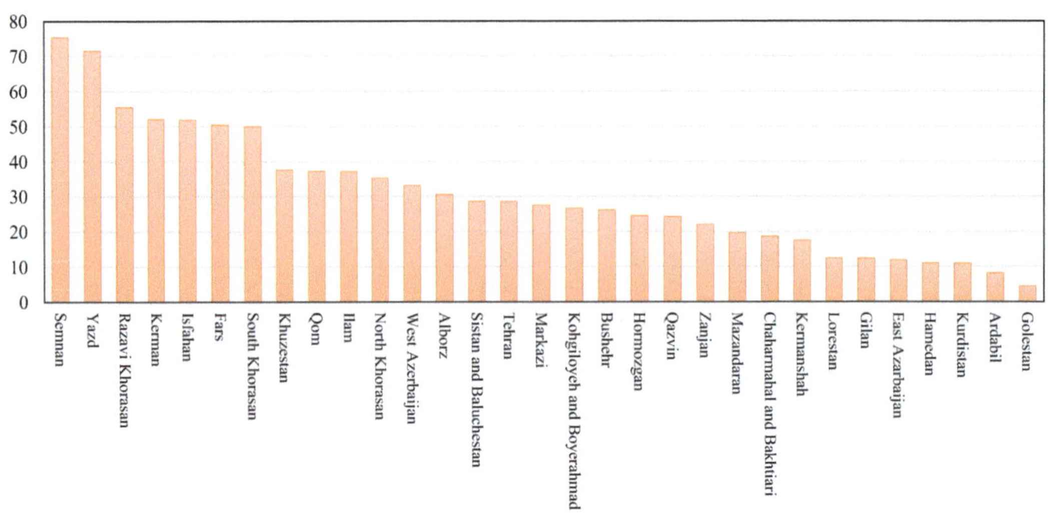

Fig. 18 Comparison of the share of abandoned villages in each province out of all villages in that province (percentage)

phenomenon of rural conversion or annexation to the city are the approach and expectations of the government, reflected in policies and legislation. The current approach governing the country in transforming a rural settlement into a city is a dual approach. In the sense that human settlements are often viewed in two forms: either as urban or rural, and decision-makers seek to name a place as a city or a village, without balanced and mutually reinforcing network connections between them. To achieve this, certain inflexible and definitive criteria such as population size, area, and similar factors are determined, while criteria such as demographic structure, cultural-social characteristics, and dominant economic structure are neglected.

Table 12 Growth rate of urban, rural, and total national population in various census periods

	Number of population (thousands)			Population share (percent)		Population growth (percent)	
Years	Total	Urban	Rural	Urban	Rural	Urban	Rural
1915	9707	2718	6989	28.00	72.00	–	–
1925	11,185	3210	7975	28.70	71.30	18/10	14/10
1935	12,833	3773	9060	29.40	70.60	17/53	13/60
1945	16,237	4936	11,301	30.40	69.60	30/82	24/73
1955	18,955	5954	13,001	31.41	68.59	20/62	15/04
1965	25,789	9795	15,994	37.98	62.02	64/51	23/02
1975	33,709	15,855	17,854	47.03	52.97	61/86	11/62
1985	49,445	26,845	22,600	54.29	45.71	69/31	26/58
1995	60,055	36,818	23,237	61.31	38.69	37/15	2/81
2005	70,368	48,245	22,123	68.56	31.44	31/03	−4/79
2015	79,926	59,146	20,730	74.00	25.94	22/59	−6/29

Source: Statistical Center of Iran

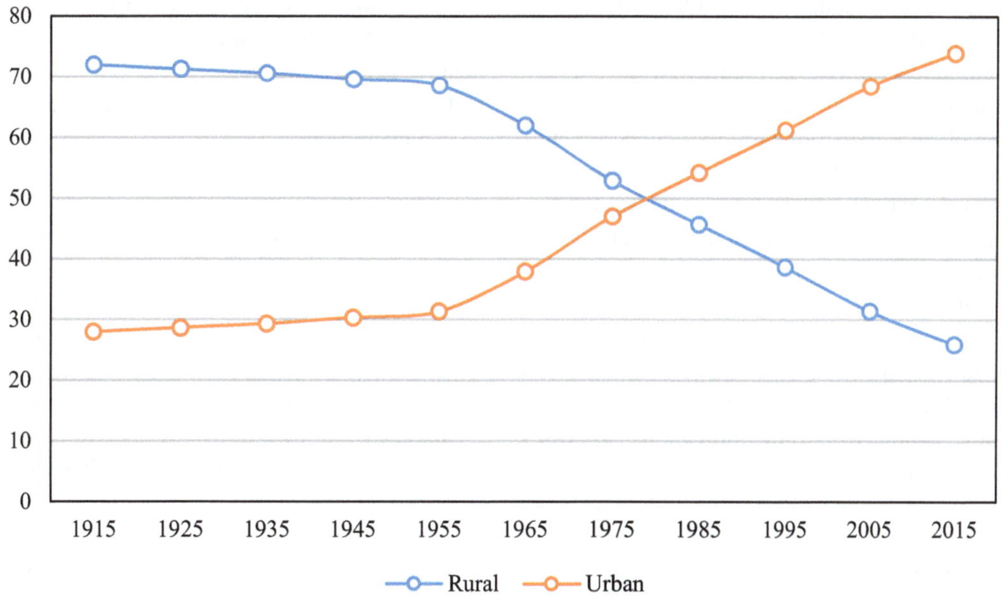

Fig. 19 Percentage of urban and rural population from 1915 to 2015

Therefore, the prevailing approach stems from neglecting the diverse dimensions, and features of rural settlements. So much so that the current criteria for rural-to-urban conversion in the country, rather than being based on scientific and logical foundations, are influenced by political factors and demands arising from prioritizing cities in the country's planning system. This has led to the formation of cities with incomplete and dysfunctional functions, which have not only disrupted the hierarchy system of the country's settlements but also imposed a double financial burden on the public sector. Previous studies by domestic researchers indicate that a significant portion of cities resulting from the nonprofessional conversion of villages have not only failed

Table 13 Trend of changes in the geographical elements of the country from 1955 to 2015

Years	Province	County	District	Rural district	City
1955	11	101	383	–	201
1965	13	145	454	–	271
1975	23	161	462	–	373
1985	24	195	501	1583	496
1990	24	229	617	2093	520
1995	26	265	707	2227	678
2005	30	336	889	2400	1015
2010	31	397	982	2497	1139
2015	31	430	1059	2590	1244

Source: Ministry of the Interior, Administrative Divisions Office

to achieve their primary goals, such as stabilizing the population and reducing migration to major cities but also lost their rural influence (Shahraki, 2013), it has provided the ground for their direct communication with the metropolis.

In any case, with the continuation of the process of rural transformation into cities, the agricultural and productive function of villages is seriously endangered, and due to the stark difference in added value between the agricultural sector and other economic sectors, the groundwork and motivation for converting agricultural land use are created, resulting in accelerated legal and illegal conversion of agricultural lands into more profitable uses, and we will witness a purely economic approach to rural indigenous processes. Also, the transformation of these villages into cities creates swollen population glands in some areas and abandoned and underutilized lands in other areas, resulting in security threats and the loss of nonoperational defense requirements in some areas, abandonment of mass water and soil resources, reduced production in the agricultural sector, transformation of rural buffer zones into marginal settlements devoid of urban and rural identity. The transformation of villages into cities can have positive or negative effects (Khorasani & Babaei, 2022a, 2022b). Below, possible positive and developmental effects as well as negative and non-developmental effects of rural transformation into cities in Iran have been mentioned. These effects have also been mentioned to some extent in the results of the studies and research under study and are agreed upon by experts in the field of rural planning and development.

2.2.4.1 The Economic Effects of the Transformation or Annexation of the Village to a City in Iran

Positive Effects:

- Increasing the activity of monetary and financial institutions.
- Creating new economic and social opportunities.
- Creating more job opportunities in the service sector.
- Getting out of the crisis (economic, identity, etc.) in villages without productive functions.
- Reduction of production cost as a result of faster and cheaper access to inputs in the local market.
- Increasing the possibility of attracting private-sector funds.
- Activities of agriculture diversification and updating due to the effects of the market and urban technology.
- Development of the market for agricultural products at the district level.
- Growth and development of industries in various scales.

Negative Effects:

- Transformation of land into tradable economic commodities.
- An irrational and sudden increase in land and housing prices.
- Reduction in agricultural sector activities.
- Creation of consumer markets for imported goods and expansion of consumerism and decline in productivity.

- Change in land use from agricultural to service and industrial activities.
- Abandonment or destruction of agricultural lands.
- Economic recession in the agricultural sector due to restrictions stemming from urban laws.
- Involvement of settlement economy in land and housing speculation; non-productivity of rural economy.
- Growth of pseudo-nonproductive businesses such as middlemen.
- Disappearance of many traditional and advantageous rural occupations.
- Decline in motivation, especially among youth, to engage in agriculture.
- Increase (leap) in operational and service budgets.
- Increased cost of living compared to before for residents, such as payment of municipal fees and taxes.
- Incomplete economic functioning based on urban economic standards.

2.2.4.2 The Sociocultural Effects of the Transformation or Annexation of the Village to a City in Iran

Positive Effects:

- Increasing life expectancy.
- Increase the quality of life.
- Increasing the self-confidences.
- Creating a sense of belonging to the urban society, especially among the youth group.
- Access to new cultural facilities such as library.
- Achieving a more favorable position in political circles and macro-decision-making.
- Reducing migration from the villages to big cities due to the diversification of job opportunities.

Negative Effects:

- Prevalence and intensification of consumerism.
- Loss of historical and cultural attractions in some villages.
- Creation or increase of unprecedented sociocultural problems (divorce, theft, etc.)
- Creating cultural duality due to the change in the physical and cultural context of the settlement.
- Disintegration of social cohesion.
- Dominance and dominance of urban lifestyle.
- Increase in the level of residents' expectations from the performance of the local management system.
- Increase in dissatisfaction and frustration among the people.
- Increasing society's expectations and demands regarding the solution problems by the government.
- The gradual loss or oblivion of cultural authenticity (folklore, clothing, music, ceremonies, etc.) in villages.
- Gradual reduction of people's participation in various public affairs.
- Creating motivation and mentality in other villages to become a city.
- Confusion among villagers regarding the rural or urban identity of the settlement.
- Bureaucratization and the ever-increasing expansion of bureaucracy.
- Threatening food security and increasing dependence on foreign countries.

2.2.4.3 The Environmental-Infrastructure Effects of the Transformation or Annexation of the Village to a City in Iran

Positive Effects:

- Reducing production activities based on excessive use of available natural resources.
- Improving per capita and physical access to all kinds of services.
- Accelerating the process of improving rural infrastructure.

Negative Effects:

- Spatial imbalance in the hierarchical system of settlements at the national level.
- Creating a distance between the hierarchies of settlements at the level of rural areas.
- Intensification of land grabbing and changing the use of agricultural lands and natural resources.

- The unsteady growth of the city due to the lack of design by the population and predetermined urban services.
- Destruction of rural housing livelihood functions in new urban patterns.

As it is evident, Iranian rural development experts and elites have addressed the economic, social, cultural, and environmental-infrastructural benefits of rural transformation into cities and believe that one cannot ignore the positive events that follow a planned and logical transformation of villages into cities. However, it is worth mentioning that this same group of experts considered recent experiences in the country regarding the transformation of rural areas into cities, mixed with political manipulation, as the source of numerous damages. Analyzing these positive and negative effects, it can be concluded that firstly, the number and variety of negative effects resulting from transformation outweigh the positive effects. Secondly, the main and fundamental mission of small towns is to attract the population of small and scattered villages to improve their quality of life and well-being and balance urban and settlement hierarchies in the country. While numerous case studies have shown that small towns resulting from the process of transforming villages into cities have largely failed in this mission. The major success of small towns has been in providing more and better services to rural people, while this service provision comes at the cost of loss of production and activity in the agricultural sector, land use conversion or destruction of agricultural lands, reduction of social capital, increased land prices, and its alienation from the local community by urban capitalists, and more unpleasantly, the creation of a perception of improving welfare and quality of life conditional on transforming villages into cities among rural people.

2.2.5 Rural-Urban Migrations

Until the year 1920, most internal migrations in Iran constituted the displacement of ethnic groups, tribes, and nomads, as well as migration from city to city and sometimes from village to village. The existence of neighborhoods, districts, and relatively old caravanserais with names of residents from other cities and surnames associated with distant and nearby cities might be signs of a history of migration. Abdolghaffar Najm Al-Dowlah, a teacher at Darolfonoon, in his 1884 report on the census of the caliphate, refers to a noticeable proportion of Isfahan and Azerbaijani populations residing in Tehran. The first general population census of Iran in 1955 indicates that out of the total population of 18,954,704, 2,081,082 individuals or 10.97% lived at a location other than their birthplace at the time of the census. The 1955 census covered political boundaries comprising 13 provinces. Of the total migrants in that decade, over 44% were absorbed into the Markazi province. Additionally, migration balance in Kurdistan and West Azerbaijan provinces, due to military forces and their border importance, was positive, while other provinces of the country acted as migrant senders. The relative weight of migrants in the population of provinces, which is an index of population readiness and migration capacity, has been such that the Markazi province attracted 80% of Gilan province migrants, 85% of Mazandaran province migrants, and 60% of East Azerbaijan province migrants. An important point to note during this decade is the difference in the age composition of migrants compared to the total population, with children under 15 years old constituting over 42% of the country's population, whereas this ratio drops to 22% for migrants, indicating migration of individuals in working and active age groups compared to other age compositions.

In the 1965 census, the number of the country's provinces increased from 13 to 21, and the number of migrants increased from 2 million in 1955 to 3.2 million in 1965 (13%) of the total population, despite this, the provinces of Khuzestan and Mazandaran had the highest percentage of migrants at 34% and 20.16% respectively. These indices were the lowest in Chaharmahal and Bakhtiari, Hamedan, and Ilam provinces. In other words, the most important focal point and pole of migrant reception was the Markazi province, which due to the current location of Tehran province within its jurisdiction, attracted over 67% of the country's total migrants.

Another noteworthy point is the change in the role of Mazandaran province in the country's migration system, which transformed from being a migrant-sending province in the 1955s to a migrant-receiving province in 1965. In 1975, according to the Iran official statistical data, out of the population of 33 million, over 25 million were native residents (76.8%), and 7.6 million (22.6%) had changed their place of residence at least once. Paying attention to the above figures indicates a 157% increase in migrants during this decade compared to the previous decade. Two Markazi provinces with 35% and Khuzestan with 23.4% have acted as the two major poles of migrant reception in the country, while other provinces have been mentioned as migrant senders. Migration statistics of the country's provinces show that East Azerbaijan, Hamedan, Gilan, Khorasan, and Isfahan provinces have been the most migrant-sending provinces.

In the general census of population and housing in 1985, the trend of demographic changes appears differently compared to previous periods. Along with the very rapid population growth, the number of immigrants also reaches 11 million (about 22% of the total population). Since 1976, due to the opening of planning offices at the provincial level, limited population centers have been formed at the national level and in some provincial centers, and internal migrants settled within the provinces. This means that provincial centers are attractive at their level and repellent at the national level, sending and receiving migrants. As in past decades, Tehran province has the highest proportion of immigrants with 38.9%, while Chaharmahal and Bakhtiari province ranks lowest. The wartime conditions prevailing in this decade and the influx of Afghan migrants into the country, especially in Khorasan province, have increased the proportion of immigrant individuals. Similar conditions can be considered somewhat true for Khuzestan province, as this province has been one of the most immigrant-receiving provinces in recent decades.

The pattern of internal migration has also been dissimilar so in Tehran province, only 9.3% of immigrants have come from within the province's counties. Meanwhile, this figure reaches 78% in Kurdistan province. A study of migration trends in 1995 indicates that the total number of immigrants was 8,718,770, with the pace slower than the preceding period. The dominant pattern of migration has been inter-provincial, mainly occurring in urban areas, with 29% of immigrants residing in urban areas of other provinces, 21.2% in urban areas within the same province, and 14.1% residing internally within counties (totaling 64.3%). These migrations, representing a new type of internal and urban-to-urban migration, are significant because these migrants differ from rural migrants in terms of aspirations, needs, desires, and other factors.

Factors such as family migrations, marriages, temporary and seasonal jobs, intensive exploitation of natural resources, mines, etc., have led to a type of migration labeled as village-to-village, which includes about 10% of the total migrants. Between 1995 and 2005, a total of 12,148,148 individuals relocated within the country. Of this number, 11,783,772 (97%) moved within the country, and 260,495 (2.14%) moved out of the country, while 103,881 individuals (0.86%) did not declare their previous place of residence. Of the total migrants who changed their place of residence, 8,999,709 individuals (74.1%) settled in urban areas, and the remaining migrants (3,148,439 individuals) moved to rural areas. Based on this, out of the total 11,783,772 individuals who changed their place of residence within the country, 5,385,665 individuals (54%) migrated from city to city, 2,330,054 individuals (20%) from village to city, 1,064,042 individuals (9%) from city to village, and 2,004,012 individuals (17%) from city to village.

Tehran and Qom provinces have the highest and lowest proportions, respectively, of urban-to-urban migrations, with 70% and 29%. In rural-to-urban migrations, Kohgiluyeh and Boyer-Ahmad provinces have the lowest proportion with 34%, while Mazandaran and Khuzestan provinces have the lowest proportion with 15%. In rural-to-rural migrations, Kohgiluyeh and Boyer-Ahmad provinces have the highest proportion with 23%, and Qom province has the lowest with 1%. Similarly, in urban-to-rural migrations, Mazandaran and Bushehr provinces have the highest proportion with 27%, while Qom province has the lowest with 5%. Out of the total individuals who changed their place of residence between 1995 and 2005, 4,774,041 individuals (41%) moved between provinces, and 7,009,131 individuals (59%) moved within the provinces they resided in. Of the total migrations that occurred within provinces, 3,898,042 individuals (56%) were from the census city, and 3,111,689 individuals (44%) were from other cities within the province. Nearly one-third of these migrations (30.1%) were to Tehran province. After Tehran province, Razavi Khorasan, Isfahan, Gilan, and Mazandaran provinces are ranked in terms of the number of migrants. Regarding migrants leaving provinces during this period, Tehran province ranks first with 17.25%, followed by East Azerbaijan, Khuzestan, Razavi Khorasan, and Isfahan provinces. Net migration of provinces has been calculated based on the number of immigrants entering and leaving each province. Accordingly, Tehran province had the highest net migration with 615,027 individuals, while East Azerbaijan province had the lowest with −143,733 individuals from 1995 to 2005. Considering the difference in the population shares of provinces in the total population of the country, it can be argued that the net migration type is not a very suitable indicator for explaining the migration situation in provinces.

In the analysis of migration in Iran and its causes and factors, a broad theoretical framework has been developed, which mainly revolves around the urban attractions and rural repulsions, ultimately relying on the economic benefits of urban areas. Migration, in its modern sense and current dimensions, slowly began during the first Pahlavi era. This period coincided with the growth and expansion of capitalist relations and modernization, concurrent with the increasing role of oil in Iran's political economy. During this period, government policies on forced settlement of nomads had significant impacts. In the second Pahlavi era, simultaneous with the country's rapid industrial developments, land reforms strengthened population displacement trends. The liberation or constriction of labor from land, the prevalence of cash transactions among villagers, the expansion of consumer markets, and new consumption patterns in villages all significantly influenced the trend of rural-to-urban migration. Overall, oil revenues led to Iran's social and economic progress, subsequently accelerating the urbanization process. At a time when the country stood at the threshold of major civilizations, the industrial sector's share of total non-oil GDP was 20%, while services accounted for 56% (Ghadiri-Masoum et al., 2019) (Table 14).

Examining demographic data over the past few decades reveals interesting realities regarding the migration trend. The raw figures show that the country's migration figure increased from 1955 to 2005, from 5.82 million people in 1975 to 12.148 million in 2005. The fact is that these three decades marked the peak of the country's demographic changes, accompanied by the disruption of urban-rural population balances and the intensification of urban population density. During this period, the average annual migration rate more than doubled from 1975 to 2005. From 2005 to 2010, the migration rate was 5.534 million people, showing a somewhat decreased trend compared to the previous period, reaching 1.106 million people annually. Therefore, the period

Table 14 General characteristics of the country's immigration from 1975 to 2010

Annual period	1975–1985	1985–1995	1995–2005	2005–2010
Total immigration	5,820,625	8,718,770	12,148,148	5,534,666
Migration to urban areas	2,022,902	9,132,675	8,999,709	4,302,086
Migration to rural areas	1,777,723	2596–95	3,148,439	1,232,580
Ratio of annual immigration to the total population	1.4	1.59	1.86	1.52
Ratio of urban migration to total migration	69.4	70.2	74.08	77.73
Ratio of rural migration to total migration	30.5	29.78	25.92	24.27

Source: Statistical Center of Iran

Table 15 Immigration situation in Iran from 1995 to 2010

Years	1955	1965	1975	1985	1995	2005	2010	2015
Total population	18,955	25,789	33,709	49,445	60,055	70,496	75,149	79,926
Number of immigrants	2100	3378	7600	1090	8719	1314	5534	4300
Ratio of immigrants to total population	11	13.1	15.5	21.6	14.5	17.2	7.3	5.3

Source: Statistical Center of Iran

from 1995 to 2005 could be considered the peak of migration in the country, and the last 5 years have seen a trend toward stabilization and a relative decrease in migration. This trend is largely attributed to the relative stabilization of the urban-rural population and is expected to continue slowly. Overall, a noteworthy point in this regard is the declining trend in total migration, which is expected to continue in the coming years, aside from unpredictable events, the slowing pace of structural economic changes in the country, which significantly affects the distribution and weight of influential population centers, is considered the main factor in this trend (Table 15).

During the mentioned decades in Iran, the intensification of rural-urban migrations was born out of economic and social changes following land reforms and the collapse of the feudal system in the forties, urbanization expansion, opening up rural economy gateways, and the widening gap between urban and rural areas. The analysis of migration share trends shows that while urban migrations initially exceeded rural migrations during statistical periods, there has been a significant gradual increase. Urban migration accounted for over 69% of total migration from 1955 to 1985, increasing to 77% in 2010. In contrast, the share of rural migrations has decreased, with rural migration accounting for 30.5% from 1975 to 1985, decreasing to 22% in 2010 (Ghadiri-Masoum et al., 2019).

This decline can be attributed to two main factors: a significant reduction in the proportion of the country's rural population and the trend of stabilizing the urban-rural population in recent years. Analyzing the migration situation over the past 50 years shows that the highest proportion of migrants to the total population occurred in the 10-year periods corresponding to the statistical period of 1995–2005, with 17.2% of the total population, and the lowest was in the year 1955, during which only 11% of the total population were migrants. According to statistics, it can be said that the trend of migrations in Iran over the past 50 years has experienced growth, with major factors being the increase in spatial disparities and inequalities. The examination of migration factors in Iran in the year 2010 indicates that family follows, with 46.12%, is the most significant reason, while completing education, with 0.79%, is identified as the least common cause of migration (Ghadiri-Masoum et al., 2019) (Fig. 20).

The examination of migration trends during the period 2005–2010 demonstrates that approximately 7.2% of the total population of the country has been relocated internally (domestic migrations). Among these, the provinces of Tehran, Razavi Khorasan, Isfahan, and Fars respectively experienced the highest levels of migration absorption. These provinces have accounted for about 38% of all internal migrations. Furthermore, the reduction in rural-urban migration from 19.77% in the period 1995–2005 to 13% in the period 2005–2010 indicates the decrease in the rural population's migration to urban areas due to the decline in the rural population (Rezapour et al., 2019).

Considering Table 16, it can be observed that 15% of internal migrations during the period of 2005–2010 were from urban areas to rural areas. As shown in Table 17, in seven provinces including Mazandaran, Tehran, Bushehr, Gilan, Golestan, Razavi Khorasan, and Kermanshah, the number of migrants from urban to rural areas has been higher than migrants from rural to urban areas. The reasons for urban-to-rural migration can be examined in three cases:

- Migration to Rural Areas with Suitable Climates: The presence of suitable climatic conditions, access to inexpensive land, and the

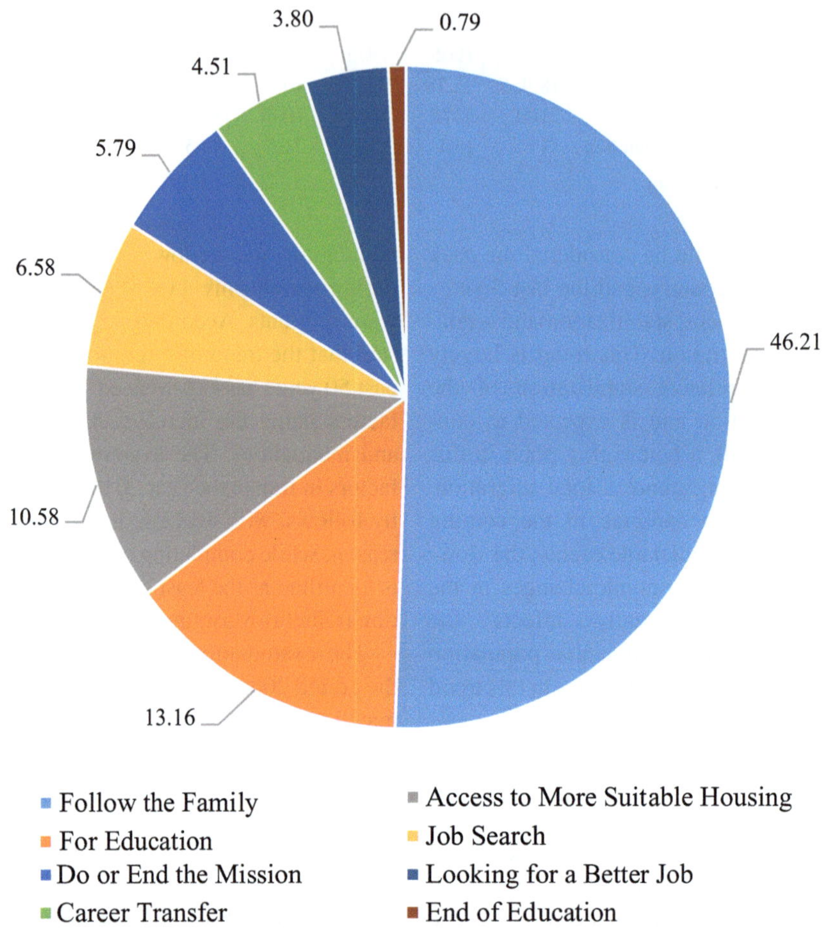

Fig. 20 Examination of the factors of country migration in 2010

Table 16 Internal migrants frequency in Iran according to the origin and destination of migration from 1975 to 2010

	1975–1985		1985–1995		1995–2005		2005–2010	
Type of migration	Quantity	Percent	Quantity	Percent	Quantity	Percent	Quantity	Percent
Total	4,816,702	100	8,556,813	100	11,783,774	100	5,038,545	100
City to city	1,912,240	39.70	4,062,171	47.47	6,385,665	54,19	3,259,040	64.68
City to village	678,717	14.09	1,540,694	18.01	2,004,012	17,01	755,546	15.00
Village to city	1,535,527	31.88	1,889,905	22.09	2,330,054	19,77	655,251	13.00
Village to village	690,218	14.33	1,064,043	12.44	1,064,043	9,03	368,708	7.32

Source: Statistical Center of Iran

short distance between the city and the village have led to migration. Rural migration in the Northern provinces of Iran includes this category.

- Migration to Villages around Metropolises: These villages are essentially satellite cities located a short distance from major cities. Provinces like Tehran and Razavi Khorasan enjoy such a situation. These villages have largely lost their rural functions and have mainly become dormitory areas for urban low-income residents. These villages experi-

Table 17 Distribution of internal migration between the country's provinces by origin and destination in 2010

Destination provinces	Total	Village to village	Village to city	City to city	City to village
Whole country	5,038,545	368,708	655,251	3,259,040	755,546
East Azarbaijan	192,808	12,462	29,947	118,817	31,582
West Azerbaijan	163,495	18,570	31,692	84,954	28,279
Ardabil	66,457	3957	13,342	39,736	9422
Isfahan	342,773	13,423	33,218	259,729	36,403
Alborz	279,415	4851	17,388	235,660	21,516
Ilam	39,451	2088	4820	24,319	8224
Bushehr	108,311	7750	13,909	59,027	27,625
Tehran	871,069	19,601	56,621	715,525	79,322
Chaharmahal and Bakhtiari	49,424	5225	12,160	26,395	5644
South Khorasan	79,717	8966	19,044	34,797	16,910
Razavi Khorasan	373,068	30,730	59,764	211,534	71,040
North Khorasan	76,160	9031	18,358	31,981	16,790
Khuzestan	221,021	15,822	21,805	153,333	30,061
Zanjan	69,529	7251	17,727	31,525	13,026
Semnan	65,782	2966	7301	47,091	8424
Sistan and Baluchestan	88,914	9352	10,356	57,373	11,833
Fars	324,784	23,263	58,084	202,565	40,872
Qazvin	80,478	5393	13,491	48,685	12,909
Qom	58,223	640	5987	49,300	2296
Kurdistan	124,895	14,415	30,115	58,912	21,453
Kerman	152,836	12,579	21,825	98,628	19,804
Kermanshah	105,852	6989	14,100	60,725	24,038
Kohgiluyeh and Boyerahmad	61,388	11,613	16,261	23,706	9808
Golestan	143,035	27,300	18,810	66,065	30,860
Gilan	215,052	27,681	22,574	120,127	44,670
Lorestan	63,333	5128	8518	37,481	12,206
Mazandaran	219,942	31,780	18,063	112,541	57,558
Markazi	102,026	4203	16,929	66,845	14,049
Hormozgan	108,716	13,238	15,506	65,920	14,052
Hamedan	99,386	8663	16,734	52,396	21,593
Yazd	91,205	3778	10,802	63,348	13,277

Source: Statistical Center of Iran

enced the highest rural population growth in the 2015 census.

- Migration in Border Provinces: Kermanshah province is a prominent example of this type of migration. Rural areas are essentially barracks, military zones, and military residential complexes. The reason for migration to these areas is to find work, find better jobs, and perform military service duty. It can be said that the northern regions of the country have a higher capacity for migration compared to other regions of the country (Rezapour et al., 2019) (Fig. 21).

The examination of the extent of women's and men's participation in recent migration processes in Iran indicates that the share of women in internal migration flows increased from 47% in the 5 years from 2005 to 2010 to approximately 49% in the 5 years from 2010 to 2015. Therefore, women's presence in domestic migrations is on the rise. In the past 5 years from 2010 to 2015, out of a total of 4,309,988 individuals who changed their place of residence within the country, 83% have an urban origin and 17% have a rural origin. The gender ratio of migrants with rural origin is 124, while it is 102 for migrants

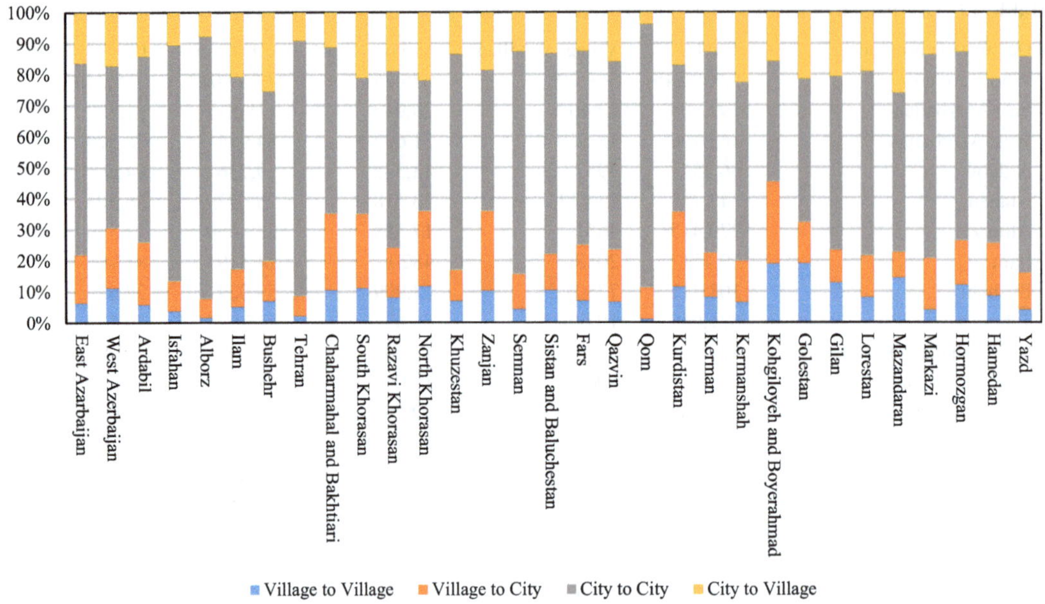

Fig. 21 Distribution of internal migration between the country's provinces by origin and destination in 2010

Table 18 Migrants entered over the past 5 years categorized by gender, urban and rural areas

Type of residence	Total	Man	Woman	Gender ratio
Total residents	4,300,988	2,210,387	2,090,601	105.7
Residents in urban areas	3,564,735	1,803,306	1,761,429	102.4
Residents in rural areas	736,253	407,081	329,172	123.7

Source: Statistical Center of Iran

with urban origin. Consequently, the presence of women in internal migrations with urban origin is more significant and noticeable compared to rural migrants (Table 18).

During the period from 2010 to 2015, 4,309,988 individuals changed their place of residence within the country or entered Iran from abroad. Based on the latest movement, the previous place of residence was 18.27% within the county of enumeration, 30.76% in other counties of the same province of enumeration, and 47.85% in counties of other provinces. A total of 2.85% had a previous residence outside the country, and the residence of the rest was not stated. Comparing the latest previous residence of migrants with the place where they were enumerated shows that, during the 5 years before the November 2015 census, 14.55% migrated from rural to urban areas, 68.28% from urban to urban areas, 5.04% from rural to rural areas, and 12.33% from urban to rural areas (Fig. 22).

In the recent period, approximately 51% of internal migrations were intra-provincial, and 49% were inter-provincial. The share of women in intra-provincial migrations (52%) exceeded that of inter-provincial migrations (48%). This finding is consistent with Ravenstein's migration laws, which state that women migrate more over short geographical distances while men have greater dominance over long geographical distances (Fathi & Javaid, 2018).

2.2.6 Demographic Window of Opportunities

The demographic window refers to a period during which the dependency ratio on a scale unprecedented in history decreases. The beginning and end of this short period are almost evident and

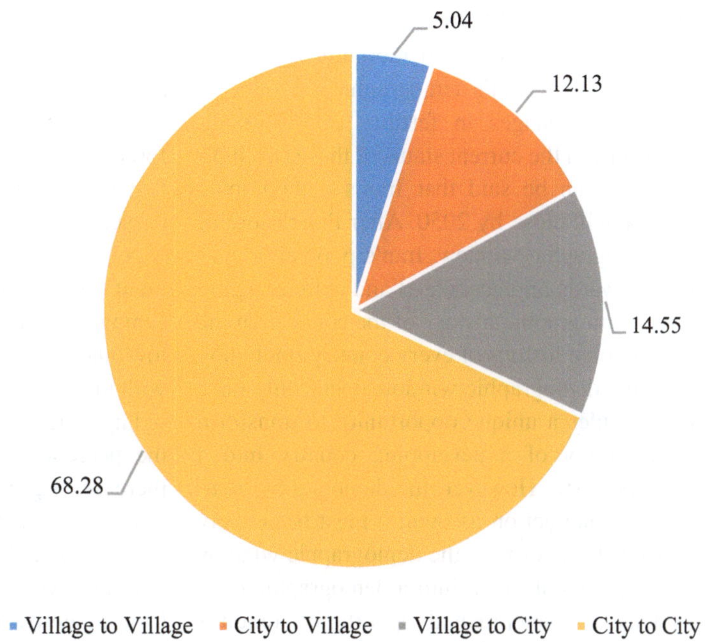

Fig. 22 Relative distribution of internal migrants by migration flow from 2010 to 2015

will last for a few decades. In the demographic window, the population situated in the working age reaches its highest level in the demographic history of a country, providing an opportunity to increase per capita production (Bongaarts, 2009).

In the first transition phase, a phase in which fertility is approximately constant and mortality is decreasing, mainly due to the reduction in child mortality, the percentage of children under 15 years old in the population increases. Consequently, the population unprecedentedly moves towards youth, and with the relative stability in the percentage of the elderly population (65 years and older) in this phase, the pressure on the population increases due to the rising proportion of children in the working-age population, and a significant portion of resources is allocated to childcare, healthcare, education, and other expenses related to childhood. This situation is not conducive to economic development. Most developing countries, including Iran, have been in such a situation until recently.

Examining the trend of the demographic window in Iran indicates that the beginning of the demographic window in the country can be attributed to the early years of World War I. The population of Iran started its first demographic transition phase between the two World Wars, and with relative stability in fertility levels and a decrease in mortality rates (which mainly resulted from the reduction in child mortality), its age dependency ratio gradually increased and reached approximately 0.9 (90 individuals not in working age for every 100 individuals in working age), nearly parallel to the 1995 census. Thereafter, due to the long-term resistance of fertility to change and its persistence at a very high level, the age dependency ratio remained very high, at 0.9, for nearly four decades. This level of age dependency with such a long period may be unparalleled in the experience of developing countries worldwide. Iran's long-term fertility resistance to change was finally broken around 1988, and the fertility rate declined at a pace perhaps unprecedented in the history of fertility changes worldwide (Saraie, 2022).

Following the very prompt decline in fertility, the age dependency ratio also decreased rapidly due to the quick reduction in the proportion of the population under 15 years old and relative stability in the population aged 65 and older, reaching 43 individuals per 100 persons in the working-age population on the eve of the 2010 (Saraie, 2022). Iran's population entered the demographic window in the late 1960s, considering the deviation of fertility rates from mortality rates. The sustainability of Iran's demographic window will depend on changes in fertility and mortality trends. Given the current status of these two indicators, it can be said that Iran's demographic window will close by 2050. After the closure of the demographic window, Iran's population will move towards unprecedented and extreme aging in the demographic history of the country. In the demographic history of every country (including Iran), the demographic window opens only once and provides a unique opportunity to transform the economy of a developing country into a developed one. However, the demographic window does not act on its own; it must be actively planned. Transforming the demographic window from a potential threat into a demographic dividend as a developmental opportunity requires readiness, policy-making, planning, and proper implementation of plans (Ghadiri-Masoum et al., 2019).

The population of Iran, after exiting the demographic window, will move towards extreme aging, and considering the relative stability in the percentage of children (under 15 years old) in the population, gradually reduce the potential working-age population (15–64 years old) and increase the population aged 65 and older. Consequently, the pressure exerted by the elderly (especially those lacking savings) on the population of working age creates conditions that may have negative consequences for the country's development. If this historical opportunity is properly utilized, it can transform a developing country into a developed one; however, failing to do so could turn this phenomenon into a threat to the country's future, leading to much more complex problems that could set any country back for years to come (Saraie, 2022). In a period where the population window is open, policy-making specifically must concentrate on labor supply and demand. In terms of supply, adopted policies should aim to form a skilled and motivated workforce. In other words, these policies should emphasize improving the nutrition, health, and well-being of the workforce (especially skilled labor) and providing effective training. In terms of demand, policies should help create favorable political, social, and economic conditions to attract and employ the large generations resulting from past high fertility levels entering the workforce in waves (Ghadiri-Masoum et al., 2019).

On the other hand, in populations that have passed through their demographic transition and experienced decades of fertility either at replacement level or below replacement level, the percentage of children (under 15 years old) remains low and relatively constant. This occurs when with increased life expectancy, larger waves enter old age and live longer in these ages. As a result, the percentage of elderly in these populations increases significantly. With a relatively stable ratio of children, the increase in the ratio of the elderly coincides with a decrease in the population ratio within working ages (15–64 years old). Consequently, the dependency of the elderly (especially those without financial savings) on the working-age population increases, and a considerable portion of society's financial resources will be allocated to the care, health, and other costs of the elderly. Such a situation cannot be a long-term incentive for development, especially if accompanied by the dependence of the elderly and national poverty.

European countries have already reached this stage, and other countries including Iran are moving towards this situation, ultimately reaching this stage. Between these two conditions, there comes a period where a relatively rapid decline in fertility rates and slow changes in fertility rates lead to a decrease in the proportion of children in the population, thus shifting the age structure away from childhood. In other words, while the proportion of the elderly witnesses minor changes during this period, the proportion of children in the population sharply declines. Consequently, the share of the potential working-age population (15–64 years old) from the total population becomes significantly larger than ever before. In this situation, as the proportion of the population outside the working age (children and elderly) relative to the potential working-age population reaches a minimum, highly favorable demographic conditions for potential economic growth will be provided. Population experts refer to this condition as the demographic window, while economists term it as a demographic dividend, recognizing the opportunity it presents for economic development (Saraie, 2022).

This situation, which occurs in the demographic history of any nation usually only once, has a short lifespan, typically lasting a few decades (about four decades), and if properly planned, it will greatly contribute to the country's development. Otherwise, instead of timely guidance and exploitation toward achieving economic and social development, it becomes a major obstacle in the path of development and may lead to the decline of a nation.

The economic miracle of East Asian countries is an example of the timely and proper utilization of this situation. In fact, about one-third of the extraordinary economic successes of these regions have been attributed to paying attention to the demographic window making timely use of this exceptional opportunity and properly guiding it. At present, those born in the 1980s and early 1990s constitute a significant portion of the active population, and the share of the active population reached 71% in 2015. Two-thirds of the country's population constitutes the active population. The demographic window usually lasts about 30 years in countries, starting from 2005 in the country and will continue until 2035 to 2045. Given the fertility trend in the country, it can be expected that over the next 30 years, 30% of Iran's population will become elderly (Ghadiri-Masoum et al., 2019) (Fig. 23).

The relative distribution of the population aged 65 and over indicates that the country's population is currently young, but in the coming years, it will tend towards aging. The aging of the population in the next 35 years will occur through the entry of a considerable number of the current young population (aged 20–29) into the 65 and over age groups. In other words, just as the current young population has currently led to the "Youth Bulge" phenomenon in the country, in the next 35 years, by transitioning from the adult age group to the 65 and over age groups, they will contribute to the emergence of the "Aging Population" phenomenon. It is worth mentioning that due to women's higher life expectancy compared to men, the feminization of the elderly population will also take shape.

Examining the country's population by major age groups reveals significant changes in the age structure of the population. The population aged 0–14 in 1975 was over 40%, while this number decreased to 25% in 2005. During this period, the ratio of the population aged 15–64 has increased

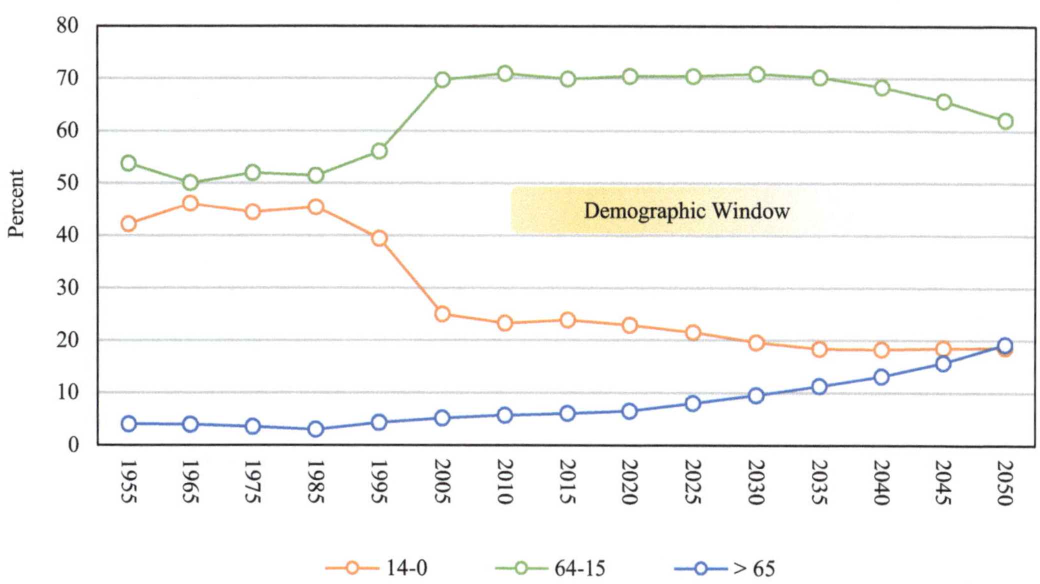

Fig. 23 Changes in Iran's age groups and population window from 1955 to 2050

by about 1.3 times, from about 40% to about 70%. Additionally, the elderly population (65 and over) has increased from about 4% to over 5%. The result of all these statistics and figures indicates that the country's population is gradually moving towards aging (Fathi & Javaid, 2018) (Table 19, Fig. 24).

2.2.7 Unemployment Situation in Iran

Based on the results of the general population and housing censuses from 1965 to 2015, the number of unemployed individuals and their rates in the total population of the country has undergone considerable changes. By studying Table 20, it can be concluded that the volume of unemployed men in 1965 was 7.1 times the volume of unemployed women. In 2015, the volume of unemployed men was 2.3 times the volume of unemployed women. This indicates that although a significant percentage of women aged 10 and older in the country are homemakers, in recent years, with the increase in the level of education and literacy among women, there has been a tendency towards participation in activities outside the home.

The noticeable decrease in the unemployment rate in 1995, especially for men, can be attributed to fundamental changes in social, economic, cultural, and political conditions after the end of the imposed war and the improvement of the conditions and the status of the labor market in the country. In 1985, the unemployment rate increased for both genders, but the intensity of the increase was much higher for women than men. This situation is related to the socio-economic conditions of the imposed war era, labor market pressures, and reduced job opportunities, especially for women. In 2005 and 2010, the unemployment rate increased for both genders, especially for women. This could be due to the entry of the population aged 10 and older into

Table 19 Population share according to gender and major age groups from 1955 to 2015

	Man			Woman		
Years	14–0	64–15	+65	14–0	64–15	+65
1955	42.2	53.6	4.2	42.1	54.1	3.9
1965	46.4	49.6	4.0	45.7	50.5	3.7
1975	45.0	47.5	3.6	44.1	52.5	3.4
1985	45.5	51.5	3.0	45.5	51.5	3.0
1995	39.7	55.8	4.5	39.3	56.6	4.1
2005	25.3	69.4	5.4	24.9	70.1	5.0
2010	23.7	70.6	5.7	23.0	71.1	5.8
2015	24.3	69.7	5.9	23.7	70.0	6.3

Source: Statistical Center of Iran

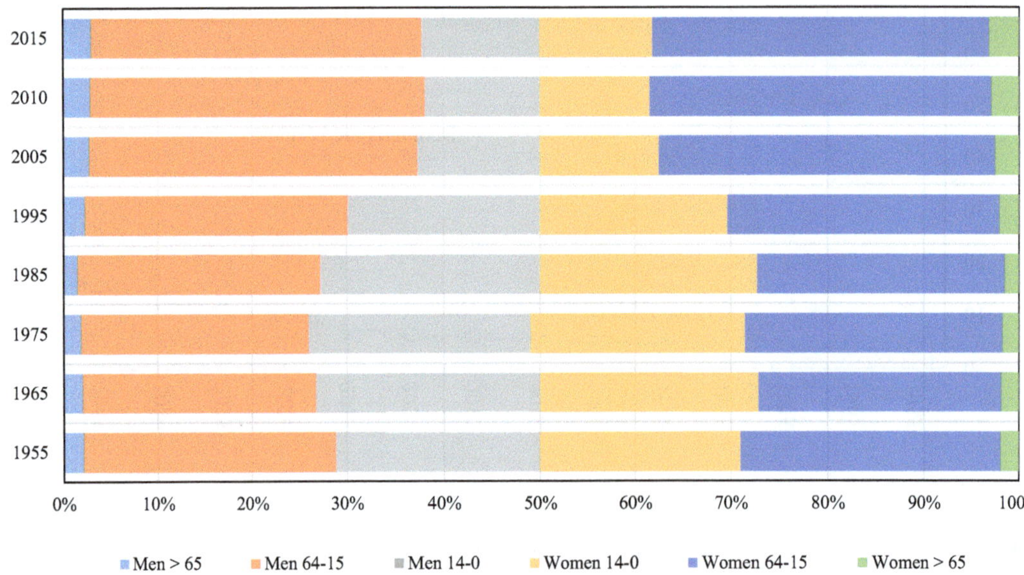

Fig. 24 Population share according to gender and major age groups from 1955 to 2015

Table 20 Unemployment situation in the provinces of Iran from 1965 to 2015

Years	Area	Unemployed population		Unemployment rate		Average annual unemployment growth	
		Men	Women	Men	Women	Men	Women
1965	Total	635,881	89,855	9.3	8.7	–	–
	Urban	146,386	11,681	6	3.7	–	–
	Rural	489,495	78,174	11.2	10.9	–	–
1975	Total	759,650	236,986	9.1	16.4	1.8	10.2
	Urban	193,637	29,291	5	6	2.8	9.6
	Rural	566,013	207,695	12.6	21.6	1.5	10.3
1985	Total	1,486,138	332,602	12.9	25.4	6.9	3.4
	Urban	856,507	216,467	13.6	29.2	16	22.1
	Rural	629,631	116,135	12	20.5	1.1	−5.6
1995	Total	1,184,086	271,565	8.5	13.3	−2.2	−2
	Urban	713,496	141,594	8.4	12.5	−1.8	−4.2
	Rural	470,590	129,971	8.6	14.4	−2.9	1.1
2005	Total	2,142,001	846,105	10.8	23.4	6.1	12
	Urban	1,317,749	565,937	9.8	22.5	6.3	14.9
	Rural	824,252	280,168	12.9	25.5	5.8	8
2010	Total	2,696,240	863,782	13.1	24.2	4.7	0.4
	Urban	1,939,690	745,562	13.5	26.7	8.0	5.7
	Rural	754,900	117,813	12.2	15.1	−1.7	−15.9
2015	Total	2,278,572	979,643	6.10	6.22	−3.3	5.2
	Urban	1,718,090	803,192	0.11	7.23	−4.2	5.1
	Rural	559,554	176,029	5.9	8.18	−8.5	4.8

Source: Statistical Center of Iran

the labor market, with a significant share of the population, and the economic boom in the country.

In 2015, the unemployment rate for women slightly decreased, one of the reasons being changes in the age structure of the population and a decrease in the young population. There is an imbalance between the supply and demand of labor in the Iranian labor market. Unemployment in the country is not solely due to demographic conditions and recent decades' developments but is partly due to the country's economic situation and its overall structure. This imbalance and the weak adaptation of skills and expertise of large sections of the country's workforce to the fundamental needs and requirements of sectors of economic-social growth have resulted in this situation (Fathi & Javaid, 2018) (Fig. 25).

2.2.8 Employment Situation in Iran

The employment rate of men during the years 1965 to 2015 has not followed a specific pattern. This is while this ratio has decreased for women until 1985 and then increased. The decrease in the employment rate in 1985 can be justified by the fact that the Iranian labor market underwent fundamental changes and transformations after the Islamic Revolution, especially during the years of the imposed war, with a recession in production and economic siege.

Just as in normal economic conditions and even in the slow trend towards development, women are considered as a reserve and potential workforce, in times of crisis and economic inflation accompanied by recession, employment and investment opportunities are limited to men. In such conditions, which are accompanied by the

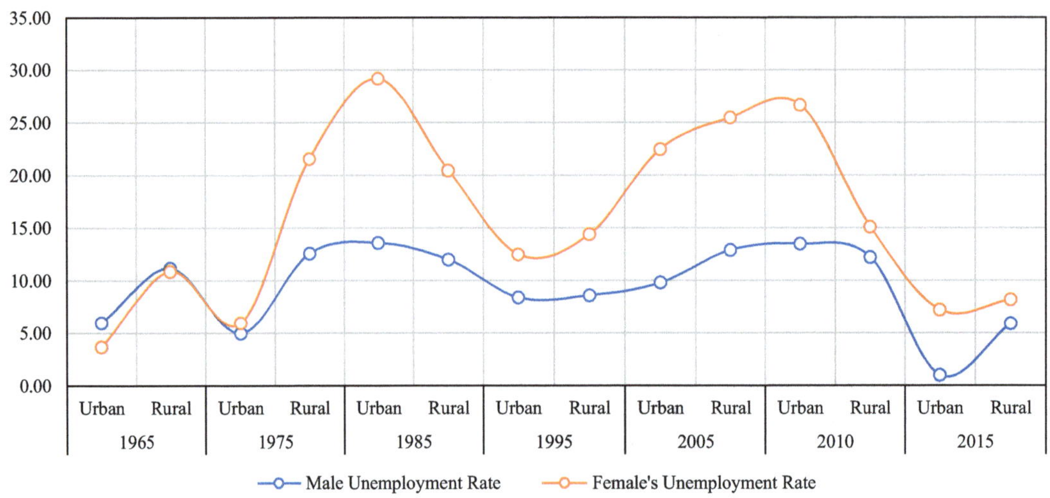

Fig. 25 Unemployment rate in the provinces of Iran from 1965 to 2015

emergence of unemployment and underemployment, providing equal employment conditions for women is very difficult, and social and economic conditions play a very influential role in women's employment.

The recent increase in the number of employed women can be justified by urbanization, increasing levels of education, and the increasing tendency of young women to participate in the labor market. The employment rate increased for both genders in 2005 and remained almost stable in 2010 and 2015. Although the annual growth rate of the population of employed women has been significant, there is still a stark difference between the employment of women and men. In 2015, the volume of employed men was 5.7 times the volume of employed women (Fathi & Javaid, 2018). Table 21 illustrates the changes in the employment rate by gender (Fig. 26).

2.2.9 Economic Participation Situation in Iran

One of the important indicators in examining the characteristics of human resources in any country is the level of economic participation of the population of that country. The discussed index is obtained by dividing the active population by the population aged 10 and above. In order to more accurately address the issue, the rate of economic participation based on the age index of the population is also utilized. This rate is obtained by multiplying the ratio of the active population (employed and unemployed) in a specific age group to the population of that age group by 100. Over the years under discussion, the volume of the active male population has always been higher than the volume of the active female population. In 1965, the volume of the male population was 6.6 times the volume of the female population, whereas in 2005, the volume of the male population became approximately 5.5 times the volume of the female population.

Although these figures indicate a more favorable situation for women, it is evident that job opportunities for men are still significantly greater than those for women. The rate of economic participation of men decreased from 1965 to 1995, then increased in the years 2005, and again increased in the years 2010 and 2015. The increase in the rate of economic participation of men in 2005 may be due to the younger age structure of the country's population. Its decrease in the years 2010 and 2015 was due to changes in the age structure of the population. Changes in the rate of women's overall activity have not been uniform and have always fluctuated. This rate sharply decreased in 1985 and then increased again. This indicates a greater tendency of women to enter the labor market and increased job opportunities for women compared to the years follow-

Table 21 Employment situation in the provinces of Iran from 1965 to 2015

Years	Area	Working population		Employment rate		Employment ratio		Average annual growth rate of employment	
		Men	Women	Men	Women	Men	Women	Men	Women
1965	Total	6,172,187	943,600	90.7	91.3	70.2	11.5	–	–
	Urban	2,302,912	307,382	94.0	96.3	65.1	9.6	–	–
	Rural	3,869,275	636,218	88.8	89.1	73.6	12.7	–	–
1975	Total	7,587,400	1,212,020	90.9	83.6	64.3	10.8	2.1	2.5
	Urban	3,652,629	460,007	95.0	94.0	60.7	8.5	4.7	4.1
	Rural	3,934,771	752,013	87.4	78.4	68.1	13.0	0.2	1.7
1985	Total	10,026,230	975,305	87.1	74.6	59.5	6.1	2.8	−2.1
	Urban	5,428,176	524,846	86.4	70.8	57.7	5.9	4.0	1.3
	Rural	4,598,054	450,459	88.0	79.5	61.9	6.3	1.6	−5.0
1995	Total	12,806,170	1,765,402	91.5	86.7	55.6	7.9	2.5	6.1
	Urban	7,808,032	991,391	91.6	87.5	53.6	7.1	3.7	6.6
	Rural	4,998,138	774,011	91.4	85.6	59.1	9.2	0.8	5.6
2005	Total	17,694,521	2,781,874	89.3	76.9	58.6	9.5	3.3	4.7
	Urban	12,093,727	1,953,165	90.2	77.5	57.9	9.7	4.5	7.0
	Rural	5,570,512	819,466	87.1	74.5	59.9	9.1	1.1	0.6
2015	Total	17,835,317	2,711,557	86.9	75.8	56.1	8.6	0.2	−0.5
	Urban	12,405,194	2,044,619	86.5	73.3	54.2	9.0	0.5	0.9
	Rural	5,414,721	663,161	87.8	84.9	60.9	7.6	−0.6	−4.1
2015	Total	19,311,460	3,359,528	89.4	77.4	57.5	10.2	1.6	4.4
	Urban	13,952,405	2,591,853	89.0	76.3	55.9	10.5	2.4	4.9
	Rural	5,342,659	761,705	90.5	81.2	62.1	9.3	−0.3	2.8

Source: Statistical Center of Iran

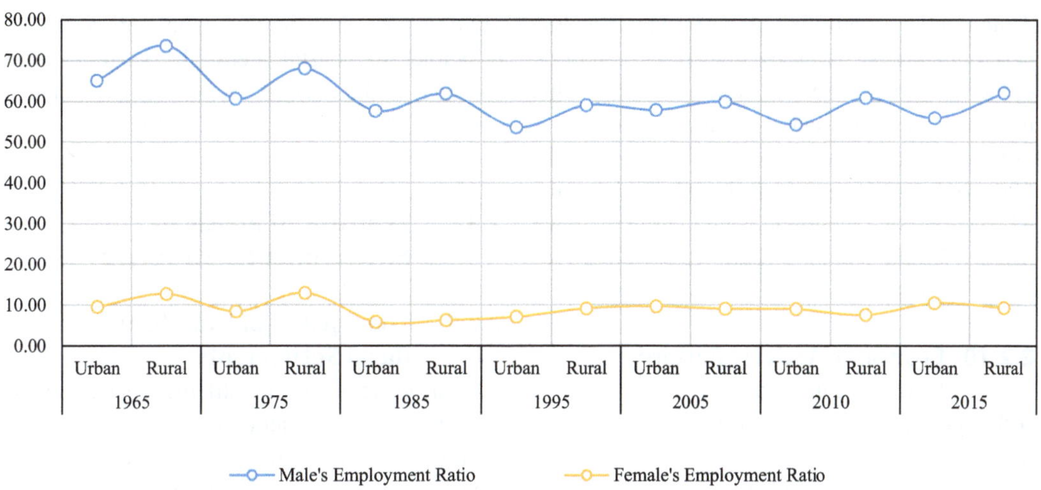

Fig. 26 Employment rate in the provinces of Iran from 1965 to 2015

ing the Islamic Revolution and the war (Fathi & Javaid, 2018).

The noticeable decrease in the unemployment rate in 1995, especially for men, can be attributed to fundamental changes in social, economic, cultural, and political conditions after the end of the imposed war and the improvement of the conditions and the status of the labor market in the country. In 1985, the unemployment rate increased for both genders, but the intensity of the increase was much higher for women than men. This situation is related to the socio-economic conditions of the imposed war era, labor market pressures, and reduced job opportunities, especially for women. In 2005 and 2010, the unemployment rate increased for both genders, especially for women. This could be due to the entry of the population aged 10 and older into the labor market, with a significant share of the population, and the economic boom in the country. In 2015, the unemployment rate for women slightly decreased, one of the reasons being changes in the age structure of the population and a decrease in the young population.

There is an imbalance between the supply and demand of labor in the Iranian labor market. Unemployment in the country is not solely due to demographic conditions and recent decades' developments but is partly due to the country's economic situation and its overall structure. This imbalance and the weak adaptation of skills and expertise of large sections of the country's workforce to the fundamental needs and requirements of sectors of economic-social growth have resulted in this situation (Fathi & Javaid, 2018) (Table 22, Fig. 27).

2.2.10 Economic Activity Pattern Situation in Iran

Although the primary function of villages is settlement, in addition to this fundamental function, villages can be classified into three main functions based on their types of activities: agricultural function, industrial function (rural production), and service function. In order to identify the predominant function of villages in each province, the Location Quotient (LQ) index has been utilized. Sections with a coefficient above 1.25 have been identified as the predominant performance of the province's villages. In provinces where a coefficient above 1.25 does not exist, a coefficient above 1 is considered the predominant performance (Table 23).

The examination of the spatial distribution of rural activities in the country, categorized by counties and economic sectors, indicates that villages in the western, eastern, and southern regions of the country are predominantly classified as agricultural villages. Due to their climatic and topographic conditions influenced by the Zagros mountain range and its folds, villages in the western region engage in agricultural activities. Villages in the southern region, due to their capacity to produce tropical products such as dates and bananas, as well as off-season products, are advantageous. Similarly, villages in the eastern region of the country (Khorasan) engage in agricultural activities due to their high capacity for saffron production.

Analysis of the distribution of industrial sector employees indicates that access to major arteries and the northwest-southeast axis has led to an increase in the share of rural workers in the industrial sector around this axis. Furthermore, the distribution of rural workers in the service sector is also consistent with the aforementioned principle, In coastal areas, the share of rural workers in the service sector is higher than in other sectors (Fathi & Javaid, 2018).

2.3 Features of Physical-Environmental Structure

2.3.1 The Pattern of Establishing Rural Settlements

For various reasons, the settlement patterns identification and the hierarchical classification of rural settlements in Iran are not straightforward. Constant changes in administrative divisions, transformations in rural areas, data limitations, scarce research, and the volatility of factors affecting rural settlements hinder the understanding and classification of these settlements in Iran. Nevertheless, environmental, social, economic, and political indicators play a significant role as

Table 22 Economic participation of Iran's provinces from 1965 to 2015

Years	Area	Population of 10 years and older		Economically active population		Economic participation rate	
		Men	Women	Men	Women	Men	Women
1965	Total	8,793,948	8,205,868	6,808,068	1,033,445	77.4	12.6
	Urban	3,537,257	3,209,182	2,449,298	319,063	69.2	9.9
	Rural	5,256,691	4,996,686	4,358,770	714,392	82.9	14.3
1975	Total	11,796,414	11,206,085	8,347,050	1,449,006	70.8	12.9
	Urban	6,017,794	5,410,183	3,846,266	489,298	63.9	9.0
	Rural	5,778,620	5,795,902	4,500,784	959,708	77.9	16.6
1985	Total	16,841,407	16,032,870	11,512,368	1,307,907	68.4	8.2
	Urban	9,411,616	8,869,177	6,284,683	741,313	66.8	8.4
	Rural	7,429,791	7,163,693	5,227,685	566,594	70.4	7.9
1995	Total	23,021,747	22,379,336	13,990,256	2,036,967	60.8	9.1
	Urban	14,560,788	13,943,245	8,521,528	1,132,985	58.5	8.1
	Rural	8,460,959	8,436,091	5,468,728	903,982	64.6	10.7
2005	Total	30,202,231	29,237,460	19,806,240	3,618,736	65.6	12.4
	Urban	20,905,108	20,190,566	13,411,476	2,519,102	64.2	12.5
	Rural	9,297,123	9,046,894	6,394,764	1,099,634	68.8	12.2
2010	Total	31,804,774	31,435,818	20,531,557	3,575,339	64.6	11.4
	Urban	22,897,505	22,696,976	14,344,884	2,790,181	62.6	12.3
	Rural	8,884,225	8,716,782	6,169,621	780,974	69.4	9.0
2015	Total	33,562,612	32,859,377	21,590,032	4,339,171	64.3	13.2
	Urban	24,942,055	24,656,474	15,670,495	3,395,045	62.8	13.8
	Rural	8,598,106	8,184,312	5,902,213	937,734	68.6	11.5

Source: Statistical Center of Iran

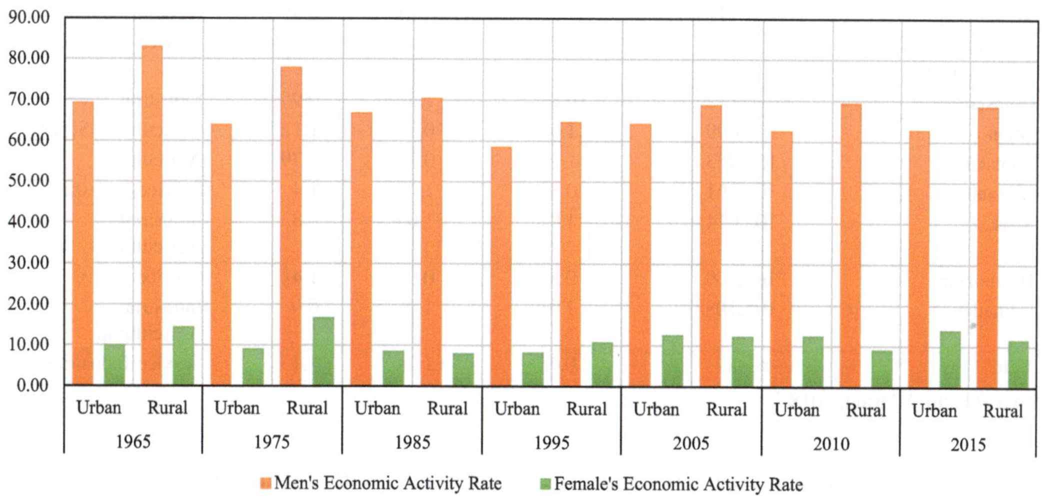

Fig. 27 Economic participation rate of cities and villages in Iran from 1965 to 2015

macro-indicators in determining rural areas (Saidi, 2009). Moreover, natural factors, particularly topography, water resources, and slope, have had a crucial impact on the location of rural settlements in Iran (Azizpour & Shamsi, 2014). Additionally, the diversity of sociocultural and economic-political systems across the country has led to the spatial distribution of rural settlements in the form of large and small, scattered villages. Rural settlements in the interior regions

Table 23 Location quotient of rural workers in major economic sectors by province

Province name	Number of employees	Agriculture LQ	Industry LQ	Services LQ	Dominant performance
East Azarbaijan	355,304	1	1.18	0.79	Industrial
West Azerbaijan	293,550	1.15	0.88	0.79	Agriculture
Ardabil	126,199	1.21	0.83	0.7	Agriculture
Isfahan	203,727	0.73	1.32	1.24	Industrial-Service
Alborz	48,874	0.58	1.24	1.69	Service-Industrial
Ilam	47,479	1.26	0.56	0.91	Agriculture
Bushehr	86,630	0.46	1.4	1.79	Service-Industrial
Tehran	219,000	0.28	1.7	1.84	Service-Industrial
Chaharmahal and Bakhtiari	83,876	0.79	1.51	0.9	Industrial
South Khorasan	91,838	1.19	0.81	0.79	Agriculture
Razavi Khorasan	486,625	1.08	1.09	0.7	Industrial-Agricultural
North Khorasan	116,306	1.07	1.22	0.58	Industrial-Agricultural
Khuzestan	206,729	1.09	0.82	0.99	Agriculture
Zanjan	112,045	1.15	1.12	0.52	Industrial-Agricultural
Semnan	37,964	0.85	1.01	1.34	Service-Industrial
Sistan and Baluchestan	155,664	1.24	0.55	0.96	Agriculture
Fars	351,797	0.97	0.93	1.16	Service
Qazvin	87,798	1.04	0.97	0.96	Agriculture
Qom	14,466	0.98	1.09	0.94	Industrial
Kurdistan	139,977	1.23	0.82	0.68	Agriculture
Kerman	284,660	1.29	0.65	0.72	Agriculture
Kermanshah	129,416	1.33	0.44	0.88	Agriculture
Kohgiluyeh and Boyerahmad	60,041	0.92	1	1.19	Service-industrial
Golestan	226,112	0.87	1.3	0.95	Industrial
Gilan	311,699	1.08	0.74	1.12	Agriculture- Service
Lorestan	154,463	1.22	0.74	0.79	Agriculture
Mazandaran	409,561	0.82	1.01	1.4	Service-Industrial
Markazi	103,014	1.07	1.03	0.8	Industrial-Agricultural
Hormozgan	162,109	0.88	0.9	1.4	Service
Hamedan	195,108	1.01	1.06	0.9	Industrial-Agricultural
Yazd	48,887	0.78	1.34	1.1	Service-Industrial

Source: Fathi & Javaid, 2018

of Iran are dependent on groundwater resources. In these areas, the construction of deep and semi-deep wells and the qanat use (traditional water management system of Iran) have played a vital role in the village's formation. In places with harsh natural conditions, the lack of settlements is evident.

In a country, the natural structure and geographical features undoubtedly play a significant role in the rural settlement formation of its various locations. The natural state of rural settlements in Iran is directly related to environmental factors such as topography, water resources, and vegetation cover. Therefore, environmental factors are the basis for the initial classification of the location of rural settlements in Iran. The fundamental step in regional zoning is dividing the country and rural settlements based on primary water basins (Saidi, 2009). Overall, natural factors have played a significant role in the rural areas establishment throughout Iran. These factors include topography, climate, and soil and water resources, which have all impacted the formation and development of these settlements. As a result, there is a system to categorize villages based on natural factors influencing their formation. This classification is as follows: plain, coastal, foothill, valley, and mountainous villages. The rural settlement location indicates many characteristics, including population, type of activity, access to resources, and potential for growth and development. Based on the climate factor, the country's climatic diversity is evident in the physical, economic, and social structures of the country's villages. Considering Iran's location in the arid and semiarid region of the world, water availability plays a significant role in the village formation and survival in the country. Also, elevation is a critical factor in the formation and distribution of settlements in Iran, as it affects soil and climate diversity (Afrakhteh, 2019). According to the official statistics of the 2006 census, out of the total villages in the country, there are 56,448 plain villages, 1321 mountainous villages, 18,080 plain-mountainous forest villages, 1843 plain-forest villages, and 5555 mountainous-forest villages. Its classification system does not encompass 2580 rural settlements.

This classification shows the diversity and dispersion of villages in different natural locations. It is undeniable that villages are affected by various hazards and incidents due to the diversity of natural environments and the spatial distribution of villages in the country. The location of villages in the path of faults and floods, climatic hazards, landslides and geotechnical hazards, droughts, outbreaks of human and animal diseases, and other predictable and unpredictable hazards and crises increase this vulnerability. If rural structures and communities lack awareness and resilience, hazards and incidents will cause increased damage in rural settlements. As is now the case, in incidents such as earthquakes, the damage in villages is more than in urban settlements. An examination of the natural factors influencing the formation of rural spaces and their spatial distribution reveals that 50% of the country's villages are in the foothills, 17% in the mountains, 13% in the plains, and the remaining villages are in another environment. In an alternative classification, 73% of the country's villages are in uneven areas, while the remaining 7% are in flat area (Ahmadian & Mohammadi, 2012) (Tables 24 and 25).

2.3.2 Climatic and Environmental Diversity

Iran is a large country covering 1,648,000 km^2 located in a warm zone between 25 and 40 degrees north latitude. In terms of altitude, it is considered a high plateau. Only a small percentage of Iran's land is less than 475 meters above sea level. The Caspian Sea to the north, the Persian Gulf to the south, and two long and high mountain ranges, the Alborz in the north and the Zagros in the west are some of the country's most prominent geographical features. Iran has a

Table 24 Ecological typology of country's rural settlements

Type	Mountainous	Desert	Coastal	Plain	Foothills	Wetland	Rivers	Island	Total
Frequency	18,982	5432	3719	14,742	55,941	1434	12,101	405	112,756
Percent	16.8	4.8	3.3	13.1	49.6	1.3	10.7	0.4	100

Table 25 Distribution of the country's villages based on natural conditions

Situation	Plain	Mountainous	Plain- mountainous	Mountainous-forest	Plain-forest	Other	Total
Frequency	56,448	41,321	18,171	5555	1843	2571	125,909
Percent	44.8	32.8	14.4	4.4	1.46	2.04	100

Source Riahi & Ghadermarzi, 2018

unique climatic situation that is influenced by various natural factors, resulting in geographical diversity throughout the country. The differences in climatic factors and elements are considered fundamental characteristics of Iran. The most commonly used climatic classification divides the country into four major climate zones. These include the temperate and humid climate on the southern Caspian coast, the cold climate in the western mountains, the hot and dry climate in the central plateau, and the hot and humid climate on the country's southern coast.

Dr. Mohammad Hossain Ganji (the late professor of the Faculty of Geography in Tehran University and the father of modern geography in Iran) proposed a classification based on the Köppen climate classification for Iran. According to this classification, 2% of the country's area has a temperate and humid climate, which includes three provinces. 31% of the area or 13 provinces have a cold climate, while 63% of the area or 13 provinces have a hot and dry climate. Finally, 4% of the area or two provinces have a hot and humid climate (Riahi & Ghadermarzi, 2018).

2.3.3 Typology of Rural Settlements

Based on scientific investigations, there are significant disparities in development indicators. These existing gaps and inequalities largely stem from a lack of recognition of the resources and talents in different regions concerning economic development and inadequate planning in geographical areas. The existing inequalities result in a hierarchical spectrum formation of urban and rural settlements, where the highest level comprises prosperous settlements, and conversely, at the lowest level, weaker settlements are formed, or in other words, deprived settlements emerge. In 2016, out of the total rural residential units in the country, 98.7% had electricity, 62.8% had landline telephones, 80.9% had minimum plumbing facilities, and 44.8% had minimum gas piping facilities (Mir-Bagheri et al., 2016). These facilities and amenities are essential for fire prevention in rural areas. Overall, the natural features and environmental factors have played a significant role in the formation and distribution of rural settlements in Iran. Despite numerous challenges at the borders, the country faces considerable spatial heterogeneity.

The Alborz Mountains range has separated the humid, low-lying northern region from the dry, elevated central plateau. It has resulted in two entirely different climates on either side of the Alborz mountain range. Considering these climates, various types of rural settlements emerge (Riahi & Ghadermarzi, 2018). In terms of typology, there are several differences between rural settlements located in the northern and southern parts of the Alborz Mountains:

- The Northern Alborz side typically experiences a humid climate due to its proximity to the Caspian Sea and receives higher precipitation. The terrain is often characterized by lush greenery and fertile valleys. In contrast, the Southern Alborz side generally has a drier climate, with less precipitation and more arid conditions. The terrain tends to be more rugged, with rocky landscapes and sparse vegetation.
- Rural settlements in the Northern Alborz often engage in agriculture, benefiting from the fertile soil and ample water supply. Farming activities, such as rice cultivation and tea production, are prevalent in this area. Agricultural practices in the Southern Alborz may be more limited due to the arid conditions. Residents may rely on irrigation systems and focus on cultivating drought-resistant crops such as wheat and barley.
- Settlements in the Northern Alborz may be clustered in valleys or along rivers, taking advantage of the available water resources. The villages in this area are connected by winding roads that traverse the mountainous terrain. Settlements in the Southern Alborz may be more dispersed, with villages located at higher elevations to avoid water scarcity and take advantage of grazing lands for livestock.
- Traditional architectural styles in the Northern Alborz often reflect the local climate and materials, with houses constructed using wood and incorporating features such as steeply pitched roofs to shed rainwater. Infrastructure

such as roads and utilities is more developed in some areas. Architecture in Southern Alborz may be influenced by the availability of stone and other durable materials, with houses designed to withstand harsher climatic conditions. Infrastructure is less developed, with fewer paved roads and access to services.

Overall, the northern and southern regions differ in climate, geography, agricultural practices, settlement patterns, and architecture despite the Alborz Mountains shaping both areas, resulting in distinct rural settlement typologies.

In the northeastern part of the country, particularly in the mountainous regions of Northern Khorasan, the geographical features, including the mountain ranges, act as significant determinants in shaping the landscape. These mountain ranges, which are regarded as extensions of the Alborz Mountain range, play a crucial role in the formation of valleys and springs. The process begins with the natural geological formations and the movement of tectonic plates, which create uplifts and depressions in the landscape. As a result, valleys are formed between the mountain ranges, where water from precipitation and melting snow collects, forming streams and rivers. Additionally, the geological structure of the region may include porous rocks or fault lines that allow groundwater to emerge as springs. These valleys and springs provide essential water sources for both human settlements and agriculture in regions of Northern Khorasan. The availability of water promotes the establishment of rural settlements in these areas, as communities historically settled near water sources for sustenance and irrigation purposes. Furthermore, the diverse topography offers a range of environmental conditions suitable for various agricultural practices and livelihoods.

The Zagros Mountains, spanning much of the western region of Iran, have played a crucial role in the formation of rural settlements in several ways (Riahi & Ghadermarzi, 2018):

- First, the Zagros Mountains influence the distribution of water resources through their geological features. These mountains capture precipitation, which then feeds rivers and streams that originate from their slopes. These water sources are vital for agriculture, providing irrigation for crops and supporting rural livelihoods in Iran's western and center.
- Second, the foothills of the Zagros Mountains offer relatively flat and fertile lands suitable for agriculture and settlement. The gradual slopes and valleys at the base of the mountains provide ideal conditions for farming activities, allowing communities to establish villages and agricultural and industrial towns. The availability of arable land and water from the mountain streams facilitates the development of rural settlements in these areas.
- Third, the Zagros Mountains act as natural barriers, shaping winding routes for transportation and trade networks. Historically, passes through the mountains served as important trade routes connecting different regions. As a result, settlements emerged along these routes, serving as trade hubs and centers of economic activity.
- Fourth, numerous folds and the presence of many salt domes have turned Zagros into an oil-rich region, which has attracted human settlement and economic activities.
- Fifth, Zagros oak forests play a valuable role in groundwater nutrition, agricultural sustainability, soil protection, and flood control. However, climate changes, including the decrease in rainfall, have dried up a significant part of these valuable resources (Heshmati et al., 2017).
- Sixth, dams serve as vital water management infrastructure, regulating the flow of rivers originating from the Zagros Mountains. They control the release of water downstream, which ensures a water supply for irrigation, domestic use, and industrial activities in rural settlements. The reservoirs created by dams in the Zagros Mountains provide a reliable water source for agricultural activities in rural areas. Many dams built on the Zagros mountain range also have hydropower generation capabilities. The energy produced from these dams contributes to the electrification of rural settlements, businesses, and infrastructure develop-

ment in remote areas in the western and central parts of Iran. The Zagros Mountains dams serve as recreational areas, attracting tourists and visitors. The reservoirs created by these dams offer opportunities for activities such as boating, fishing, and camping, which can contribute to the local economy of rural settlements through tourism revenue. Dams help mitigate the risk of floods by regulating the water flow during periods of heavy rainfall or snowmelt. By storing excess water in reservoirs, dams can prevent downstream flooding, protecting rural settlements and agricultural lands from potential damage. In recent years, due to increased catastrophic flooding in upstream areas of the rivers leading to the dams in the Zagros Mountains and to prevent additional pressure on these dams, the water discharge has increased several times compared to normal conditions. This has led to the destruction and submergence of downstream rural areas, and the displacement of residents to other areas has occurred. For example, For example, the devastating floods in 2018, following severe and widespread rainfall in the basins of Khuzestan province, resulted in the inundation of rural settlements, roads, railways, and agricultural lands, causing extensive damage to all sectors, including oil and energy, transportation, and commerce, industry, and agriculture. After investigating different aspects of the floods in Khuzestan province, it has become apparent that the primary issue that made them worse was the flood management officials' lack of trust and confidence in the predictions and simulations. This has caused a delay in implementing timely flood control measures and utilizing the maximum capacity of reservoir dams to control and mitigate floods (Rajabizadeh et al., 2019).

In a vast portion of the country's central regions, single mountains have been influential in the formation and continuity of rural settlements. In these areas, villages are developed distinctly like dense spots. Along the southern coasts of the country, although not to the extent of the northern coasts, the primary potential exists for forming villages.

Overall, the distribution of rural settlements in Iran is heavily influenced by topography and slope, surface and underground water resources, agricultural and horticultural capabilities, livestock breeding, and the exploitation of dependent natural resources (Afrakhteh, 2019).

In addition to natural factors, human factors such as population growth, migration, tourism potential, nonagricultural employment, lifestyle changes, and urbanization expansion have significantly impacted the evolution of population dynamics and the formation and expansion of rural settlements. These issues have led to demographic changes and transformations in the rural settlement system in recent decades (Ahmadian & Mohammadi, 2012). Approximately half of the country's villages have populations of less than 50 people. Larger villages have experienced a growth trend and population concentration. Currently, despite natural population growth in rural areas, the proportion of the rural population has decreased. It should be noted that in recent decades, migration from villages to cities has been a major driver of changes in the settlement system in the country. Villages that have remained in proximity to urban areas while maintaining mutual relations with cities have transformed into urban neighborhoods. The residents of these areas, due to their lack of sociocultural and economic-physical cohesion with the city, despite their physical-biological cohesion, have a different experience compared to urban dwellers. These conditions are evident in many densely populated poor countries.

Entrepreneurship attention can prevent rural migration to cities and provide employment, welfare, and progress opportunities for them in the same villages where they reside (Ghadermarzi, 2015). Furthermore, it is crucial to implement suitable measures to promote resilience and ensure sustainability in rural areas. These include protecting rural environments, reducing natural hazards and disasters, safeguarding and expanding the sustainability foundations of settlements, removing and mitigating factors that hinder sustainability, eliminating barriers to effective management alongside promoting participation to support efficient spatial management at local and regional levels, and enhancing the efficiency and spatial role of settlements (Saidi & Hosseini, 2009).

2.3.4 Environmental Hazards in Rural Settlements

Iran, situated within the seismic-prone Alpine-Himalayan orogenic belt, has endured a significant history of devastating earthquakes, notably exemplified by the Sar-Pol-Zahab, Kermanshah, Iran Earthquake of 2017, measuring a magnitude of 7.3, as one of its most recent and destructive occurrences. Seismologists, drawing from recorded data and seismo-tectonic analyses, anticipate the likelihood of substantial earthquakes across various regions of the country. Moreover, Iran faces a spectrum of natural hazards, including flash floods, droughts, and landslides. The floods in March and April of 2019 in provinces such as Golestan, Mazandaran, Lorestan, and Khuzestan underscored the nation's vulnerability to extreme environmental events. Additionally, Iran grapples with a series of unprecedented climate-related challenges, exemplified by phenomena such as the desiccation of vital water bodies like Lake Urmia, Zayandeh-Rud River, and Wetlands of Gavkhoni, Horulazim, Shadegan, Bakhtegan as well as frequent dust storms and prolonged droughts. These multifaceted challenges, coupled with ongoing political conflicts and surrounding country's wars, have the potential to precipitate profound socio-economic and demographic transformations, giving rise to emergent issues, notably large-scale and protracted migration (Fekete et al., 2020).

In Iran, 93% of the territories are prone to earthquakes, and 50% of the country's area is susceptible to floods and droughts. More of these threats pertain to rural settlements, highlighting the necessity for serious attention to rural crisis management. The crises of rural settlements are classified into four main groups based on their origin and diversity. The first group comprises crises with natural origins, such as earthquakes, landslides, droughts, water and soil pollution, depletion of groundwater, and weather-related crises like severe cold winds, excessive heat, or torrential rain and floods. The second group encompasses crises with biological origins, such as plant pests and the spread of diseases among livestock. Socially originated crises form the third group, including anomalies, crimes, cultural changes, drastic demographic shifts, migrations, and addiction. The fourth group consists of crises with economic origins, such as extreme poverty, inflation, unemployment, inequality, and fluctuations in supply and demand markets.

Given that villages constitute a continuous set of environmental and human factors that have shaped within a defined geographical area, and residential units, farms, orchards, or the substrate of any economic activity, rural public spaces, and communication networks are common to all of them, the occurrence of multiple crises and simultaneously in rural areas is not perceived as an extraordinary phenomenon. If these factors and sets are considered in classifications such as rural typology and its categorization concerning population size, settlement type, diversity, and dispersion of villages relative to the expanses of Iran's nature, various complex forms and types of rural settlements emerge. It not only delineates the common characteristics of each form but also indicates the difficulty and complexity of rural evaluations in Iran.

Villagers and their settlements face various risks that threaten their safety and livelihood. In other words, rural settlements, like other habitats, are exposed to a range of risks and damages, some of which are predictable and preventable, while others are unpredictable. However, environmental threats are more prominent in rural settlements due to the heightened influence of causative factors compared to urban areas. Moreover, if these threats are considered alongside the natural structure of different regions in Iran, a wide variety of environmental threats in rural areas becomes apparent. The factors shaping the risks and damages of rural regions vary in each rural basin. Hence, environmental challenges in rural areas differ, and understanding these challenges involves the frequency of each factor shaping and classifying them based on two main variables, such as factors shaping risks and damages. There are significant differences in safety issues between rural and urban areas due to various factors. Firstly, in urban areas, fire incidents in numerous dimensions and forms are the most significant safety concern, whereas, in rural areas, this seems to encompass only a minor portion of safety issues. Secondly, in urban areas, the systematic and coherent registration of safety-related data in firefighting organizations contributes significantly to planning and decision-making regarding future safety measures. However, in rural areas, there is no systematic recording of environmental hazards and safety issues. Thirdly, the structure and approach to dealing with safety and relief in urban areas fall under the responsibilities of established institutions such as firefighting organizations. Whereas in rural areas, emphasis is placed on local cooperation and voluntary assistance based on neighborly relations.

Nowadays, significant changes in the global perspective towards risks have been observed in recent years, shifting from an overall focus on vulnerability reduction to increasing resilience against transformative disasters. However, there has been little attention to elucidating dimensions and enhancing resilience among Iran's rural settlements against environmental and socio-economic hazards simultaneously. The mixture of these risks increases the possibility of irreparable disasters for the country's rural societies in the absence of management systems to promote resilience and risk reduction (Saleh et al., 2022).

Iran's rural areas are not only damaged when faced with various hazards, but the lack of resilience and the level of awareness and perception of the society regarding the degree, type, and nature of the hazards also cause damage. In the country's tumultuous history, rural communities have employed various methods to mitigate unforeseen risks in the face of natural, socio-economic, and agricultural hazards. These include preventing avoidable risks through strategies of resilience and dealing, preventing risks through improving infrastructure and taking measures, and taking events for granted and implementing compensatory mechanisms.

In 2021, the Rural Development Studies Center was established within the Natural Disasters Research Institute of the Ministry of Science, Research, and Technology with the primary objective of assessing rural crises and preparing programs to mitigate damages resulting from hazards in these settlements. Additionally, it was formed to plan, develop, and implement rural projects and establish a network among organizations and influential bodies in rural areas of the country, especially in deprived regions. The main objective of the Rural Development Studies Center is to undertake development projects with appropriate intellectual and scientific support to enhance planning and management systems in rural settlements. The main activities of this group include:

- Preparation of Rural Constellations Sustainable Development Plan (RCSD).
- Conducting studies on organizing space and rural settlements.
- Preparation of scientific and specialized reports on the social-economic and environmental dimensions of rural settlements for identification of challenges and potentials.
- Implementation of rural community empowerment plans.
- Identification of obstacles and infrastructural problems in rural settlements and providing executable solutions to overcome these challenges.

2.3.4.1 Frequency and Diversity of Hazards in Rural Settlements

The various risks threaten the rural populations of the country. Poverty, deprivation, inequality within rural areas, unemployment, rural migrations, village evacuations, and the instability of rural settlement establishment systems are among the most significant problems and sometimes hazards in the country's rural regions. Additionally, the lack of employment opportunities, the absence of appropriate mechanisms for participation, the unclear duties of responsible rural development and planning authorities, and the disparity between urban and rural areas are other challenges facing the country's villages. Although one of the fundamental objectives of rural development is to reduce unemployment and increase employment opportunities, food security is also a critical issue in Iran, which highlights the need for more attention to the agricultural sector. Also, the proper disposal of waste in rural areas is crucial because it can lead to diseases, contamination, and environmental degradation. Waste can have negative impacts on the environment and society. Therefore, it is essential to handle waste responsibly to prevent such negative impacts. Furthermore, these predominantly human-related risks can lead to poverty or the destruction of the country's vegetative cover. This arises from four factors, namely unfavorable natural conditions, direct utilization of rangelands, direct occupation of rangelands for the development of other sectors, and indirect utilization of rangelands such as overgrazing, wildfires, earthquakes, floods, subsidence, landslides, and lightning in rural areas. The occurrence of these hazards depends on environmental conditions, and the extent of their impact on rural communities depends on the readiness of rural populations and other factors.

Each village's identification of hazards and their frequency and types play a pivotal role in enhancing community safety and preparedness and reducing vulnerability. According to Iran's official data during the period 2005–2008, floods with 10,650 occurrences, fires with 10,561 occurrences, and earthquakes with 3937 occurrences have accounted for the highest number of disaster events, respectively. Analysis of the statistics reveals that accidents and disasters affect most villages in the country's provinces, necessitating comprehensive safety measures. Each province has varying hazard numbers, with the highest fires in Kerman and South Khorasan and the lowest in Qom, Tehran, and Alborz. Floods occur more frequently in West Azerbaijan, East Azerbaijan, and Razavi Khorasan provinces, while Qom, Alborz, and Tehran provinces experience fewer floods in villages. As for earthquakes, the highest number of occurrences is in villages in Kerman and Lorestan provinces, while the lowest amount is in villages in Ilam, Alborz, Qom, Sistan-and-Baluchestan, Tehran, and Hormozgan provinces (Riahi & Ghadermarzi, 2018) (Table 26).

2.3.4.2 Human Casualties Caused by Hazards in Rural Settlements

Sustainable rural development is unattainable without ensuring rural inhabitants' physical and financial security. Individuals within the community are perceived as its most valuable assets. Surveys during the period 2006–2008 indicate that rural residents across all country provinces are susceptible to various incidents and hazards. Notably, the provinces of Kerman, Fars, and Razavi Khorasan incur the highest human casualties. Conversely, rural inhabitants of Qom and Ilam provinces experience the lowest human losses during incidents and hazards. The frequency and severity of incidents and hazards in rural areas directly correlate with the safety and security of the people residing in those regions. Thus, analyzing these statistics can provide valuable insights into the effectiveness of safety measures and emergency response mechanisms in Iran's rural communities (Table 27).

2.3.5 The Necessity of Resilience in Safety Development in Rural Areas

The inclusion of Iran among the top ten disaster-prone countries, experiencing 30 of the recognized natural hazards and accounting for 6% of human casualties resulting from world's natural disasters, despite comprising only 1% of the world's population, underscores the undeniable necessity to focus on the safety development and crisis management (Torkian et al., 2022). According to official statistics from the 2006 census, there were 56,448 plains villages, 41,321 mountain villages, 18,080 plains-mountain villages, 1843 plains-forest villages, and 5555 mountain-forest villages in the country, with 2580 settlements not classified in these categories. Examination of official data from villages with over 100 households reveals their distribution as follows: 15,088 villages in plains regions, 11,130 villages in foothill areas, 4218 villages in intermountain regions, 2670 villages in coastal areas, and 2612 villages in mountainous regions (Riahi & Ghadermarzi, 2018).

Iran, a nation frequently beset by natural calamities such as floods and earthquakes, has

Table 26 The frequency of hazard events in rural settlements separate by Iran's provinces during the period 2006–2008

Province	Fire	Flood	Subsidence	Earthquake	Avalanche	Lightning	Other	Total
East Azarbaijan	649	1114	107	10	9	80	1774	3743
West Azerbaijan	459	1704	57	157	22	25	838	3262
Ardabil	506	294	30	114	2	34	1205	2185
Isfahan	323	253	103	21	27	8	741	1476
Alborz	25	17	28	0	8	2	1245	1325
Ilam	79	55	10	0	0	13	239	396
Bushehr	129	126	60	7	4	26	286	638
Tehran	19	20	14	2	4	1	1459	1519
Chaharmahal and Bakhtiari	111	102	13	45	3	14	280	568
South Khorasan	368	188	65	29	0	9	801	1460
Razavi Khorasan	1140	990	260	245	93	32	1658	4418
North Khorasan	327	793	49	152	4	70	664	2059
Khuzestan	501	651	26	16	0	77	1076	2347
Zanjan	234	178	52	77	0	40	557	1138
Semnan	118	144	19	33	4	8	309	635
Sistan and Baluchestan	257	343	65	1	15	26	458	1165
Fars	394	285	134	36	1	40	944	1834
Qazvin	317	153	32	47	8	15	670	1242
Qom	36	12	9	1	0	0	63	121
Kurdistan	281	271	32	13	37	48	590	1272
Kerman	1389	433	176	1086	2	90	2190	5366
Kermanshah	356	366	53	141	13	114	832	1875
Kahkiloyeh and Boyerahmad	92	52	7	12	0	16	166	338
Golestan	295	660	59	101	0	10	524	1649
Gilan	621	126	43	176	113	30	2790	3899
Lorestan	308	520	63	823	15	35	527	2291
Mazandaran	408	431	67	444	0	7	808	2165
Markazi	326	69	42	19	0	9	496	961
Hormozgan	158	105	36	3	0	30	434	766
Hamedan	196	111	52	83	3	20	638	1103
Yazd	139	84	65	43	2	4	415	752
Total	10,561	10,650	1828	3937	389	933	25,677	53,968

Source Riahi & Ghadermarzi, 2018

endured approximately 247 major disasters from 1900 to 2020, according to the International Emergency Events Database. These events have claimed roughly 163,000 lives, affected over 55 million people, and inflicted significant loss of life and property. The 2003 Bam earthquake is considered a poignant example, decimating over 70% of the region's infrastructure, resulting in more than 26,000 fatalities out of a population of 150,000, and rendering over 75,000 people homeless.

Despite the recurrent nature of these hazards, little attention has been paid to assessing the socio-economic and environmental resilience of Iran's communities, though efforts are primarily confined to case studies of micro-communities or specific disaster types like floods. Apart from the study by Javadpoor et al., 2021, there has been a conspicuous absence of comprehensive initiatives aimed at identifying indicators for assessing rural resilience in Iran without restricting the scope to a particular disaster or community type.

Table 27 Human casualties caused by hazards in rural settlements separate by Iran's provinces during the period 2006–2008

	The number of dead		The number of injured		Total	
Province	Frequency	Percentage	Frequency	Percentage	Frequency	Percentage
East Azarbaijan	372	2.46	555	2.52	927	2.50
West Azerbaijan	277	1.83	304	1.38	581	1.57
Ardabil	272	1.80	340	1.55	612	1.65
Isfahan	623	4.12	1200	5.45	1823	4.91
Alborz	31	0.21	26	0.12	57	0.15
Ilam	56	0.37	93	0.42	149	0.40
Bushehr	192	1.27	341	1.55	533	1.44
Tehran	86	0.57	97	0.44	183	0.49
Chaharmahal and Bakhtiari	163	1.08	176	0.80	339	0.91
South Khorasan	409	2.71	716	3.25	1125	3.03
Razavi Khorasan	1020	6.75	1879	8.54	2899	7.81
North Khorasan	321	2.12	527	2.40	848	2.28
Khuzestan	388	2.57	1066	4.85	1454	3.92
Zanjan	226	1.50	205	0.93	431	1.16
Semnan	66	0.44	186	0.85	252	0.68
Sistan and Baluchestan	382	2.53	914	4.15	1296	3.49
Fars	1131	7.48	1797	8.17	2928	7.89
Qazvin	314	2.08	728	3.31	1042	2.81
Qom	21	0.14	65	0.30	86	0.23
Kurdistan	206	1.36	187	0.85	393	1.06
Kerman	5437	35.98	5077	23.08	10,514	28.33
Kermanshah	306	2.02	517	2.35	823	2.22
Kahkiloyeh and Boyerahmad	114	0.75	105	0.48	219	0.59
Golestan	304	2.01	430	1.95	734	1.98
Gilan	614	4.06	1440	6.55	2054	5.53
Lorestan	190	1.26	507	2.30	697	1.88
Mazandaran	702	4.65	1161	5.28	1863	5.02
Markazi	258	1.71	349	1.59	607	1.64
Hormozgan	115	0.76	204	0.93	319	0.86
Hamedan	295	1.95	417	1.90	712	1.92
Yazd	221	1.46	391	1.78	612	1.65
Total	15,112	100	22,000	100	37,112	100

Source Riahi & Ghadermarzi, 2018

In response to this void, the mentioned study endeavored to develop and evaluate a framework for resilience indicators, resulting in the creation and testing of the Baseline Resilience Indicators for Communities (BRIC). Rooted in the theoretical framework of the Disaster Resilience of Place (DROP) model, the BRIC framework adopts a localized approach to understanding community resilience. The DROP model, crafted to illuminate the interplay between inherent resilience and vulnerability, facilitates a comparative assessment of disaster resilience at local and community levels. While the DROP model defines resilience as a dynamic process contingent on antecedent conditions, severity of incidents, inter-event intervals, and external influences, the BRIC framework focuses on assessing and illustrating the resilience of inherent or pre-event within communities.

The study undertook the categorization of 35 indicators across five dimensions of resilience. These dimensions encompass social, economic, community capital, infrastructure and housing, and environmental factors. This classification was implemented upon the BRIC framework, initially presented by Cutter and Derakhshan

(2020). The study's objectives encompass the identification and validation of comprehensive resilience indicators in Iran, along with estimating and mapping resilience levels across the country's diverse provinces. The authors believed the study is significant as its findings can illuminate the strengths and weaknesses inherent in various provinces. They also thought that the study results could be beneficial in shaping resilience-building plans and policies at the spatial levels. Additionally, they believe that findings can serve as a basis for conducting follow-up studies over time to track and better understand changes.

The study by Javadpoor et al., 2021 revealed that provinces located in the Country's northern and close to the Caspian Sea had higher resilience values. In contrast, provinces in Iran's northwest exhibit moderate resilience levels. Ardabil is the province with the highest resilience value, while Tehran, Alborz, and Sistan and Baluchestan have the lowest resilience. Eight provinces, primarily along the Caspian Sea coastline, exhibit relatively high resilience levels. Across Iran, nine provinces display lower levels of resilience. Also, 42% of provinces have average resilience levels (Fig. 28).

In the following, the dimensions of resilience have been investigated:

- In terms of social resilience, none of the provinces in Iran have high values, with only Sistan and Baluchestan and Alborz showing the lowest levels. Although provinces with relatively high social resilience lack a distinct distribution pattern, those located centrally in the country tend to fall within this category. The provinces located in the northwestern and

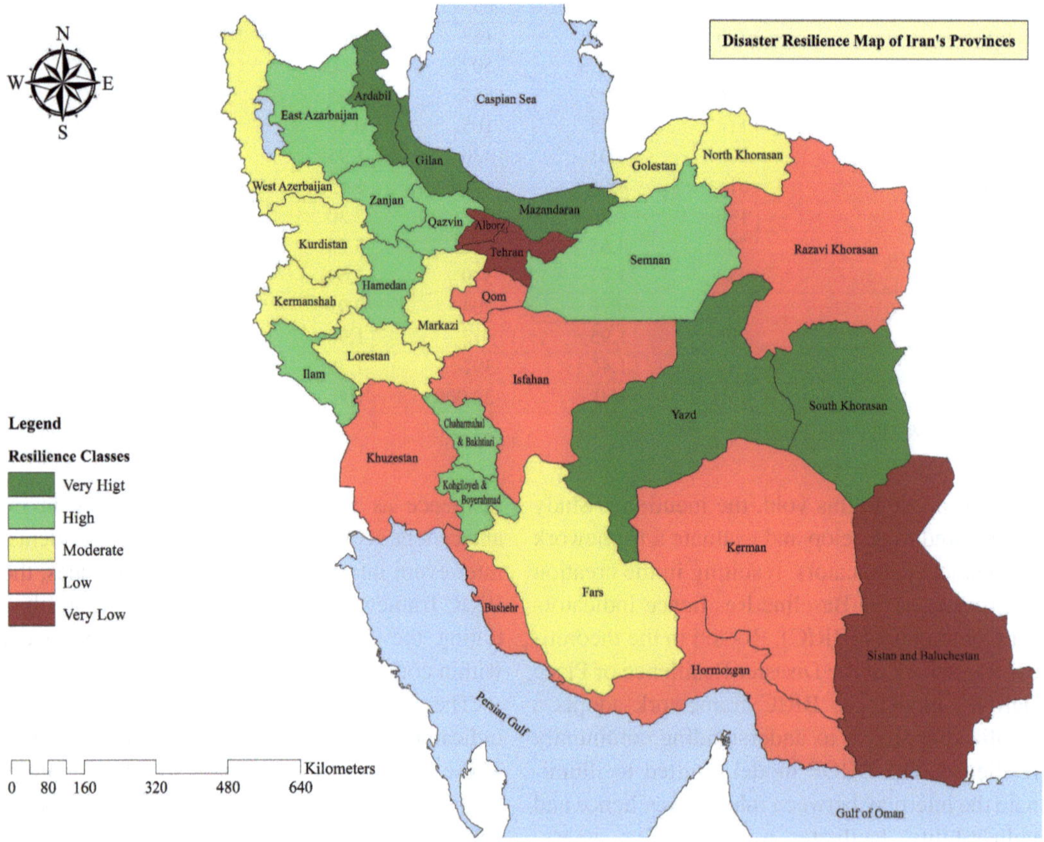

Fig. 28 The overall situation of disaster resilience in Iran's provinces. *Source* Javadpoor et al., 2021

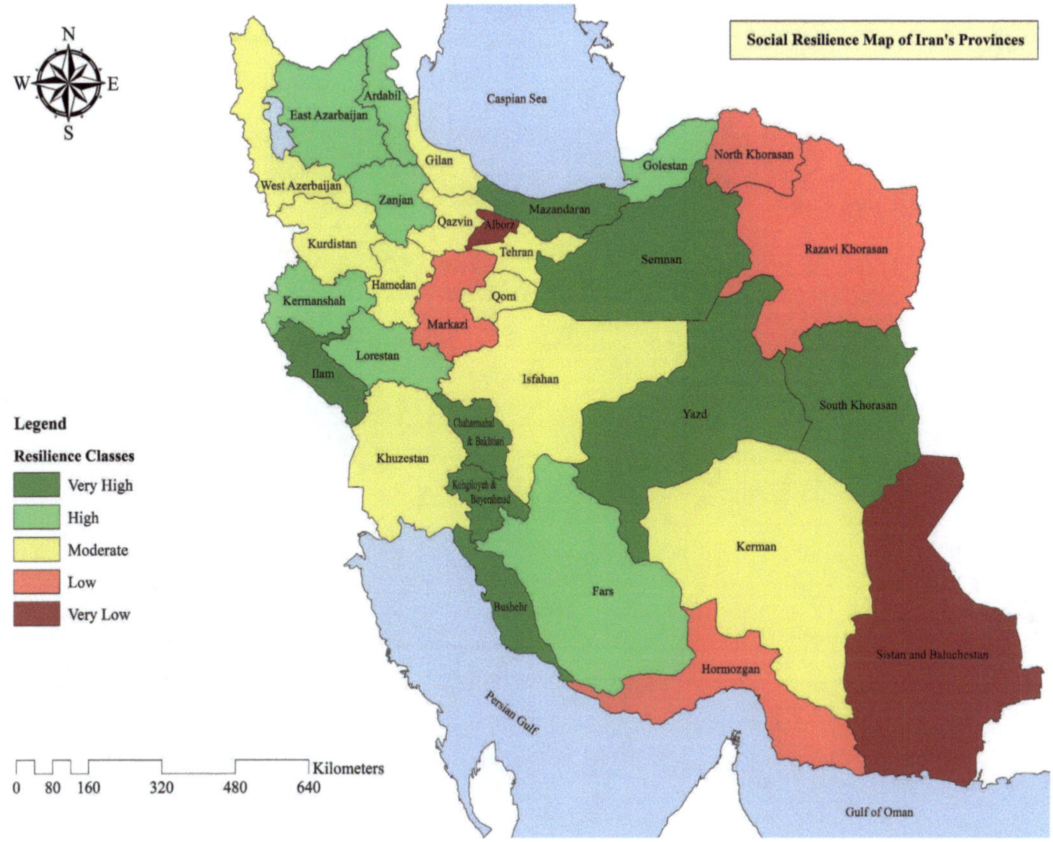

Fig. 29 Economic resilience situation in Iran's provinces. *Source* Javadpoor et al., 2021

southwestern regions have a lower level of resilience compared to others. However, Alborz and Markazi provinces, which are located centrally, also demonstrate a low and relatively low level of resilience, respectively, despite being surrounded by provinces with moderate to relatively high social resilience. The western and northwestern regions, on the other hand, have the highest concentration of provinces with moderate social resilience levels (Fig. 29).

- Regarding Economic resilience, none of the provinces demonstrate high value, with Sistan and Baluchestan exhibiting the lowest economic resilience, particularly noteworthy given its border with Pakistan. Western provinces generally display relatively lower economic resilience, while those with higher resilience are primarily situated in the northern part of the country (Fig. 30).

- Concerning community capital resilience, only South Khorasan province boasts high resilience, whereas Tehran and Alborz exhibit the lowest performance levels. Provinces show random distribution regarding community capital resilience scores, although there is a notable concentration of provinces with moderate resilience levels in the northwest and south, with better performance observed in provinces bordering the Caspian Sea (Fig. 31).

- Regarding resilience of infrastructure and housing, Tehran, Yazd, and Ardabil provinces rank highest, contrasting with Sistan and Baluchestan at the bottom. Among the

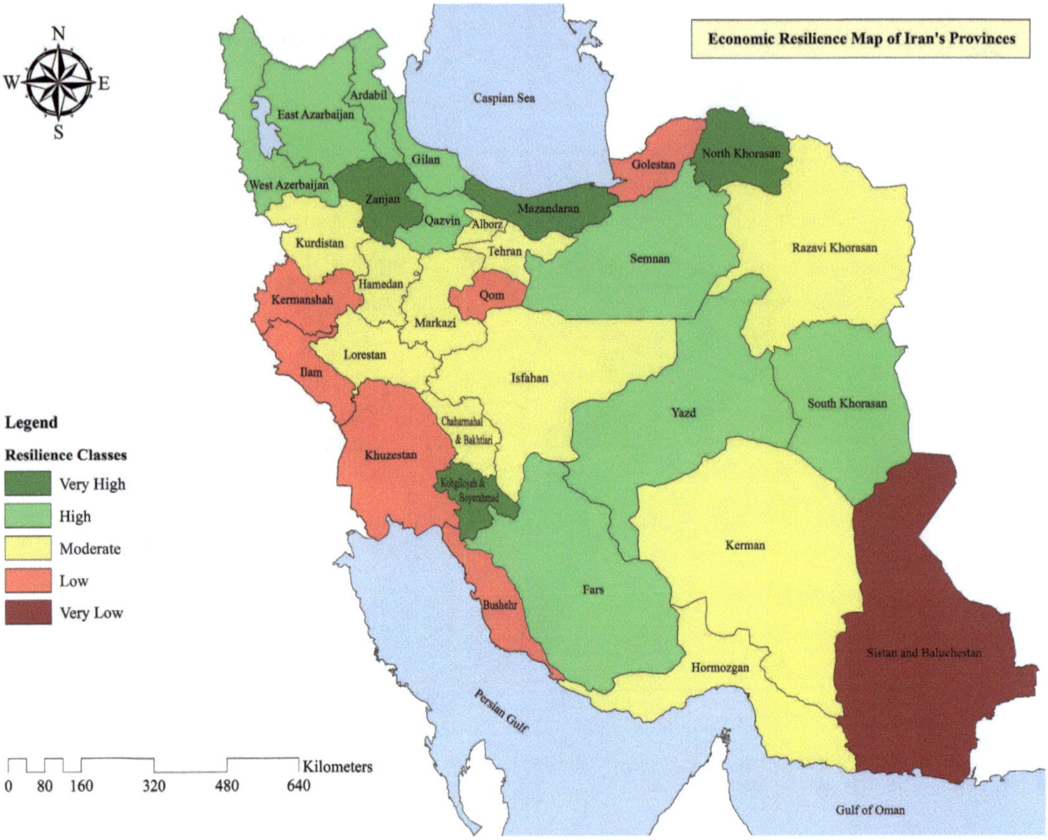

Fig. 30 Economic resilience situation in Iran's provinces. *Source* Javadpoor et al., 2021

explored dimensions, the proportion of provinces with performance below moderate levels is highest for this aspect, approximately 42% (Fig. 32).

- Environmental resilience displays a spatial correlation among provinces, with clusters identified based on performance. A decreasing pattern of resilience performance is observed from the northwestern mountainous region to the mainly desert areas in the southern and southeastern parts. Ardabil province stands out with the highest resilience, while Tehran ranks lowest, akin to community capital resilience. Among the dimensions explored, environmental resilience records the highest proportion of provinces with high and relatively high performance, approximately 42% (Fig. 33).

In general, findings indicate that the western and northern regions experience better conditions in contrast to their southeastern and eastern counterparts. Among the various dimensions of resilience, environmental resilience exhibits a unique spatial distribution wherein a gradual decline in environmental resilience is evident from the northwest to the southeast of the nation. The changes in the landscape and the increase in severe weather conditions towards the southeast have contributed to this geographical trend.

The low resilience values observed in Tehran, the capital of Iran, and Alborz, its neighboring province, formerly part of Tehran province until 2009, can be attributed to imbalanced growth

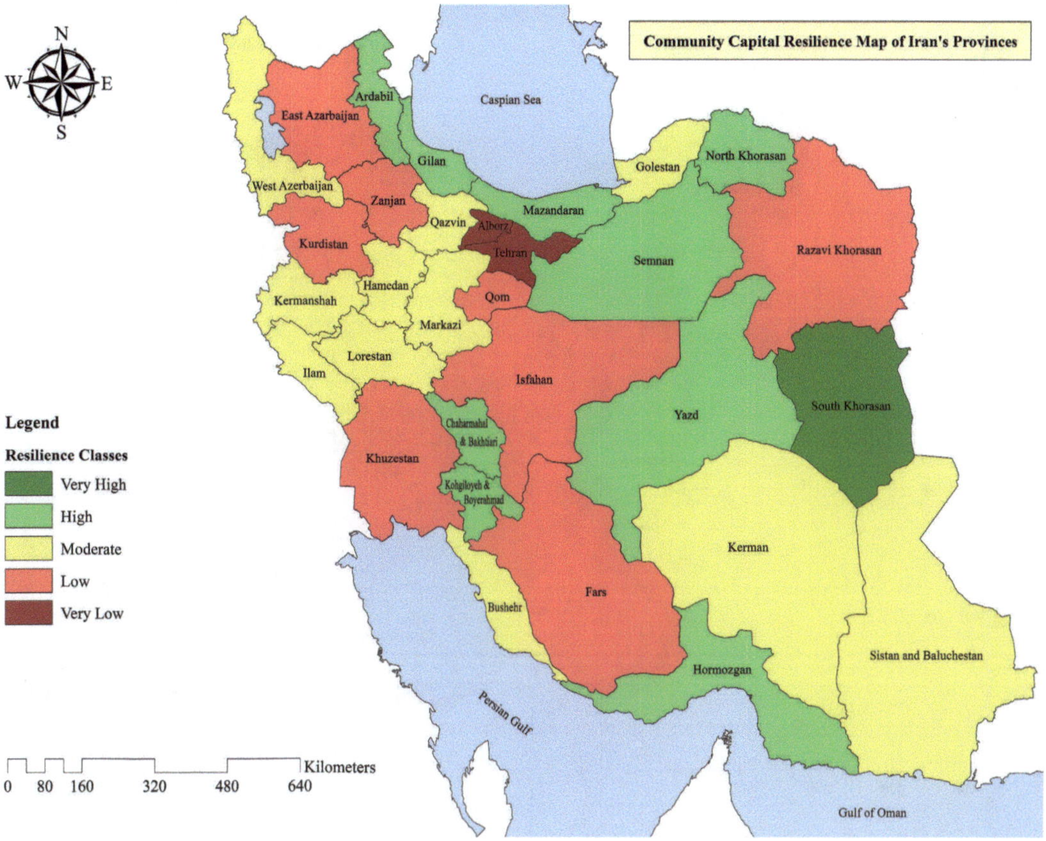

Fig. 31 Community capital resilience situation in Iran's provinces. *Source* Javadpoor et al., 2021

policies, prioritizing physical and economic expansion at the expense of social and environmental aspects. This underscores the necessity for more equitable approaches in future development strategies. Additionally, certain peripheral provinces in both the eastern and western regions exhibit limited resilience capacities due to a confluence of economic, political, and environmental factors, demanding further investigation to comprehend the underlying causes and propose tailored strategies for bolstering resilience. Assessing the resilience of Iranian provinces can help planners and government officials identify their susceptibility to disasters of each region, as well as specific strengths and weaknesses. This information can then be used to create targeted interventions that will fortify resilience (Javadpoor et al., 2021).

3 Methodology

3.1 Study Area

Iran lies between 25° and 40° N latitude and 44°–63° E longitude, with an area of 1,873,959 km^2 (Sadeghi & Hazbavi, 2022). The country is situated in a unique geographical position, acting as a bridge among Africa, Europe, and Asia, connecting India and Russia in the north–south corridor, and playing a combination of geopolitical, geo-economic, and geo-transit roles in Southwest Asia (Noorali & Ahmadi, 2023). Iran has 5894 km of land borders and a total of 2440 km of sea borders in the Persian Gulf and the Caspian Sea. These positions have made the country a pivotal player in the region with extraordinary potential for regional interactions with global

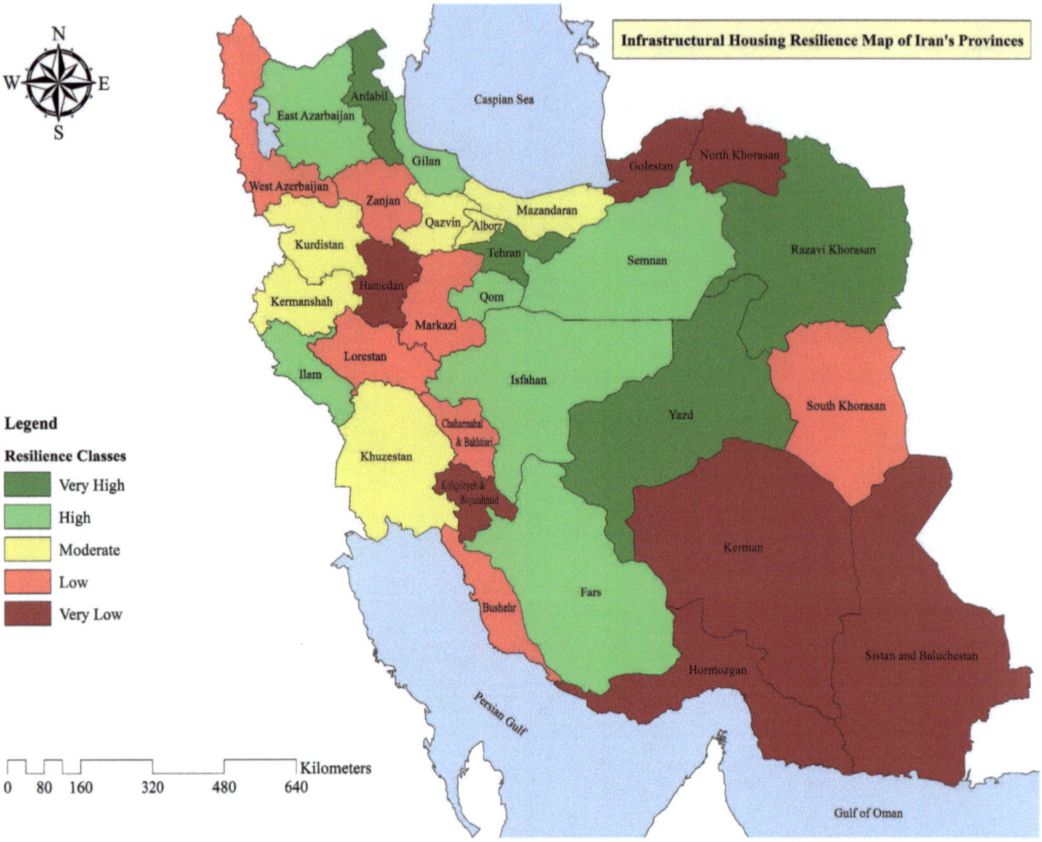

Fig. 32 Infrastructure and housing resilience situation in Iran's provinces. *Source* Javadpoor et al., 2021

impacts (Soltanieh, 2023). Iran is generally mountainous and semiarid, with an average elevation ranging from −40 to 5670 meters above sea level. The country's climate is influenced by its diverse topography, with annual temperatures ranging from −20 to +40 °C (Darand & Zandkarimi, 2019). Iran's precipitation regime varies across the country. The northern areas of the country exhibit a more uniform precipitation regime, while the other parts experience a seasonal regime characterized by a longer dry season (Asakereh et al., 2024). The establishment and evolution of human settlements in Iran have been intricately shaped by the interplay of natural and socio-economic factors. These areas have adapted and transformed within the context of the interconnected nature-population-economy ecosystem (Baburin et al., 2017).

Rural settlements in Iran have been influenced by macro developments, leading to greater access to modern amenities and the adoption of modern changes, which can contribute to improved living conditions and social development (Shakoori, 2019). Also, the literature and observations in rural Iran reflect the participation of rural communities in national trends of sociocultural and economic development, indicating the integration of rural areas into the broader cultural and economic landscape of Iran (Sheikhzadegan & Meier, 2017). Despite the economic and environmental difficulties, rural settlements in Iran are at the forefront of dealing with widespread environmental changes. While some residents have refused to migrate, indicating a strong connection to nature (Mohammadi & Khanian, 2021), a large group of rural residents migrate to cities due to vulnerability to environmental problems, climate change, hydro-climatic disasters, human maladaptive activities, and unfavorable governmental policy-making and planning (Mianabadi et al.,

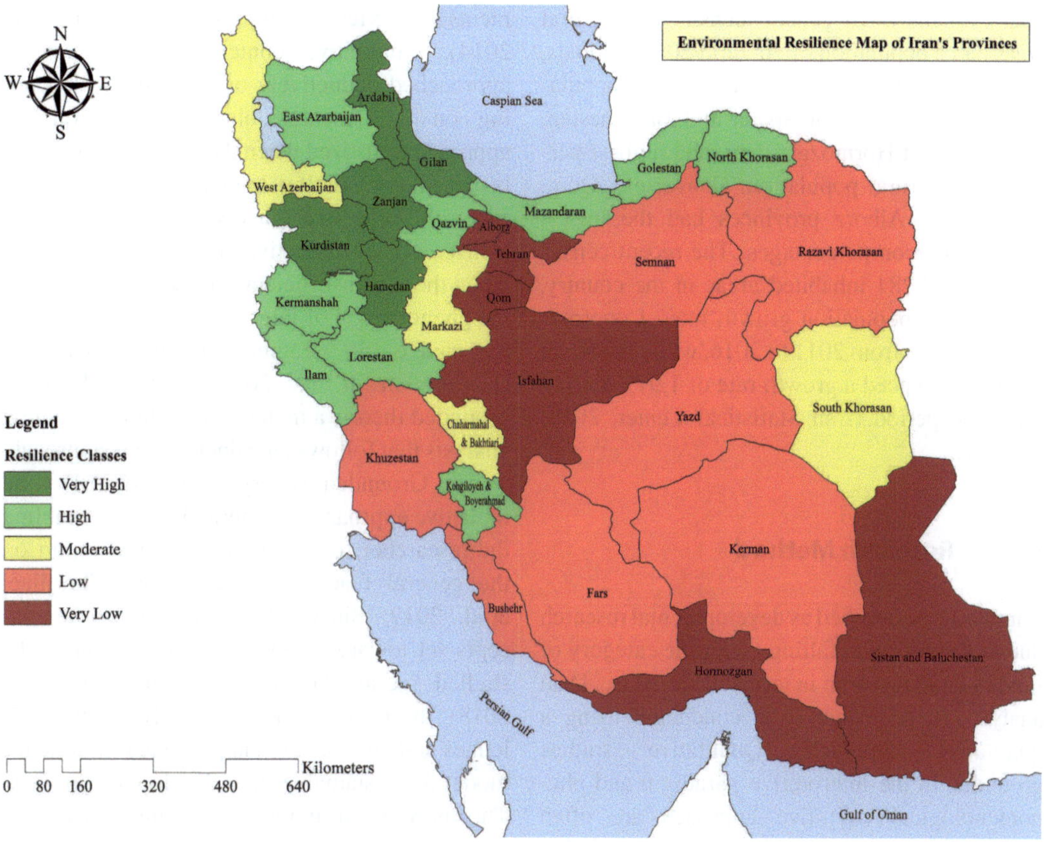

Fig. 33 Environmental resilience situation in Iran's provinces. *Source* Javadpoor et al., 2021

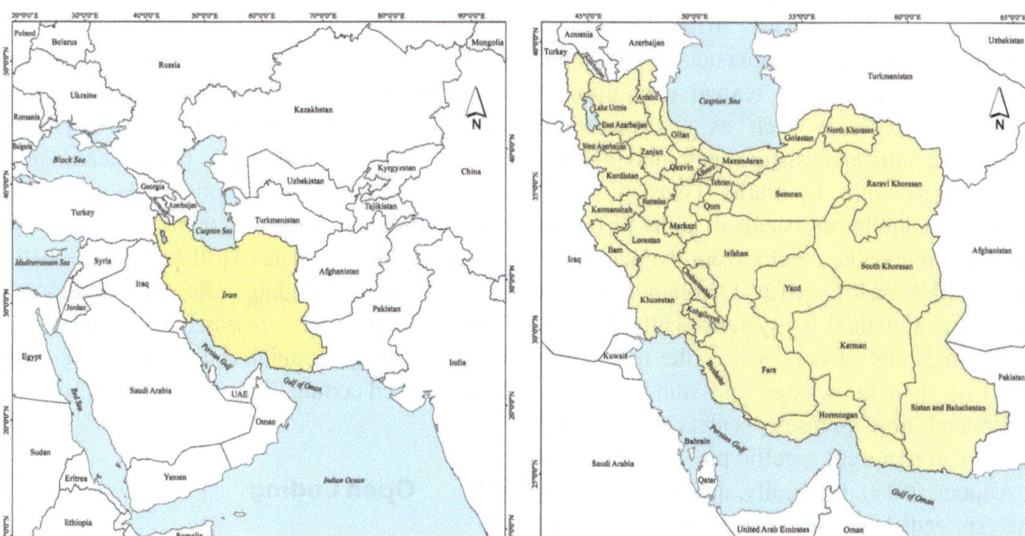

Fig. 34 Study area map

2023). In the 2016 general census, Iran's rural areas had a population of 20,730,920 individuals, constituting 29% of the country's total population. The provinces of Sistan and Baluchestan, Golestan, and Hormozgan have the highest percentages of rural population. Conversely, Qom, Tehran, and Alborz provinces had the lowest rural population percentages. The recent census recorded 62,284 inhabited areas in the country, with a rural population growth rate of approximately 0.68% from 2011 to 2016, while the urban areas experienced a growth rate of 1.97% during the same period (Iran Statistical Center, 2016) (Fig. 34).

3.2 Research Method

This study is classified as developmental research due to its objective, falling under the category of fundamental research in terms of its nature. Data analysis in this study was conducted using a qualitative approach. Qualitative studies grounded in the interpretive paradigm and phenomenological cognitive approach are often undertaken to comprehend the meaning of phenomena and events. In qualitative research, there is a holistic and comprehensive perspective on phenomena, aiming to gain insight into meanings through the discovery of their interpretations (Ghaedi & Golshani, 2016). Qualitative content analysis can be understood as a mental content interpretation of textual data through systematic processes such as categorization, coding, thematic analysis or theme-making, and designing recognized patterns (Ghasemi et al., 2020). Essentially, this analysis aims to obtain valid and replicable results from extracted text data (Shiri & Azimi, 2012). Qualitative content analysis is a method for systematically describing the meaning of data, where the researcher, based on their knowledge and study, seeks to understand how phrases, sentences, or paragraphs can represent specific messages (Taslimi & Alipour, 2019). Generally, qualitative content analysis endeavors to explore concepts, terms, and the relationships between these concepts, aiming to reveal latent patterns in written documents, interviews, and observations (Momeni, 2014). Qualitative content analysis can be approached deductively or inductively, depending on the research objective. A deductive approach is utilized when the analytical structure is based on existing knowledge or when the research objective involves theory testing. Conversely, an inductive approach is employed when there is insufficient prior knowledge about the phenomenon of interest or when the existing knowledge is fragmented and disjointed (Karimmian et al., 2019). The study used grounded theory, a method of qualitative content analysis that follows an inductive logic approach.

The Grounded Theory Method (GTM) is an inductive approach of pattern discovery, enabling the researcher to present a theoretical report on the general features of the subject (Tavallaei et al., 2019). Since GTM aims to integrate concepts set towards theoretical clarification of the studied social phenomena (Varmazyari et al., 2018), the present study has analyzed the challenges and approaches and presented a resilience model for sustainable rural development in Iran. The analysis unit in this study comprises a set of upstream documents, national reports, books, articles, and specialized meetings (see Table 28).

4 Result

Coding is a crucial step in qualitative data analysis, involving the process of narrowing data into themes or categories to identify patterns (Tsang et al., 2019). GTM is a phenomenological approach that involves various levels of coding: open, axial, and selective coding, which aim to sort and categorize data (Atrill & Lindley, 2019). In other word, Coding data and developing themes in qualitative research based on GTM begins with open/initial coding and ends with axial/focused coding (Terosky et al., 2023).

4.1 Open Coding

In the GMT, the initial step before open coding is the extraction of selected propositions from the

Table 28 Unit of analysis in the study

Type	Author	Title
Upstream documents	ICA (1989)	First Five-Year Development Plan (1989–1993)
	ICA (1995)	Second Five-Year Development Plan (1995–1999)
	ICA, (2000)	Third Five-Year Development Plan (2000–2004)
	ICA, (2005)	Fourth Five-Year Development Plan (2005–2009)
	ICA, (2011)	Fifth Five-Year Development Plan (2011–2015)
	ICA (2017)	Sixth Five-Year Development Plan (2017–2021)
	EDC (2005)	Twenty-Year Vision Document (2005–2025)
	CIIMP (2018)	Islamic-Iranian Model of Progress Document (2018–2065)
	SSCSP (2021)	National Spatial Planning Document (2021–2046)
	DRDDA (2017)	Rural Development Document (2017–2021)
	NCMO (2021)	National Crisis Management Strategy Document (2021–2032)
National reports	Rajabpour (2018)	Program Planning System Pathology: A Review of Planning System Developments in Iran
	Afshar et al. (2023)	Pathology of Policy-Making in Iran: With an Emphasis on the Social Issue
	Karimi and Dastvareh (2023)	Analysis of the Policy-Making Process in Optimal Crisis Management
	Khorasani & Babaei, (2022a, (2022b)	Exploration and Pathology of the Rural-to-Urban Transformation Process in the Territory Management System of Iran
	Ebrahimi (2016)	Critical Analysis and Pathology of Arrangement Plans (Spatial Planning) in Iran
	Inanloo (2021)	Pathology of the Country's Planning Documents
	Geraeinejad and Vafaei (2022)	Root Cause Analysis of Planning System Challenges for Use in formulating an Appropriate Planning System
	Varmazyari (2014)	Challenges Exploration in the Monitoring and Evaluation System of Agricultural and Rural Development in Iran
	Gholipour and Ahangar (2011)	Public Policy-Making Process in Iran
	Ghasemi and Zamani (2010)	Pathology of Development Plans in Iran
	Hoseini et al. (2023)	Strategic Analysis of the Economic and Social Situation of the Country on the Threshold of the Seventh National Development Plan
	Maham (2023)	Spatial Planning, Social Changes, and Sustainable Progress: Emphasizing the Seventh Five-Year Development Plan
	Rezaei and Heidari (2023)	Necessity of National Action to Enhance Climate Change Adaptation
	Nasr and Kohi (2022)	Analysis of Global Megatrends and Future Trends in Iran for the Seventh Development Plan
	Varmazyari et al. (2023a, 2023b)	Leadership in Knowledge-Based Entrepreneurship Development in Rural and Nomad Regions
	Varmazyari and Kalantari (2021)	Expert Opinion on the Comprehensive Plan for the Development and Rural Construction
	Varmazyari et al. (2022)	Key Axes of Reforming the Rural Local Management System in Iran
	Mirbagheri and Ismail (2022)	Conceptology of Participatory Governance
	Kaviani and Ghofrani (2021)	Administrative System and Development Management in Iran
	Sheikhi (2023)	An Introduction to Rural Development Planning in Iran
	Kaviani and Ghofrani (2023)	Designing a Framework for Managing Structural and Organizational Changes
	Esmailpour et al. (2023)	An Overview of Democratic Governance

(continued)

Table 28 (continued)

Type	Author	Title
Specialized books	Firoznia and Eftekhari (2004)	The Place of Rural Area in the National Development Process from Experts Point of View
	Rezvani (2015)	An Introduction to Rural Development Planning in Iran
	Moulai and Alizadeh (2019)	An Overview of the Pathology of Rural Management in Iran
	Eftekhari et al. (2015)	Strategies and Policies for the Physical Development of Rural Settlements: Emphasizing Global and Iranian Experiences
	Firoznia et al. (2014)	Prospects and Planning Management System for the Physical Development of Rural Regions in Iran
	Anabestani and Javanshiri (2014)	Physical Planning of Rural Settlements in Iran
	Eftekhari and Sadeghloo (2019)	Resilience of Local Communities Against Environmental Hazards
	Eftekhari and Sarban (2021)	Rural Empowerment: Fundamental Principles, Approaches, Theories, and Experiences
	Ebrahimzadeh et al. (2021)	Local Communities Planning and Empowerment
	Farajirad and Kazemian (2012)	Local and Regional Development from an Institutional Perspective
	Ziari and Mohammadi (2018)	Regional Development Management with Emphasis on Iran
	Faraji et al. (2018)	An Introduction to the Regional Management Model in Iran
	Ebrahimzadeh and Mousavi (2022)	Principles and Fundamentals of Territorial Spatial Arrangement
	Mosalanejhad (2016)	Policy-Making in the Power Structure of Iran
	Rezvanifar (2021)	Sustainable Development Policy-Making in Iran: Necessities and Strategies
Scientific articles	Varmazyari et al. (2018)	A Grounded Theory Approach to Developing a Model for Challenges of Policy-Making and Planning of Rural Development in Iran
	Badri et al. (2019)	Designing a Pattern for Challenges of Spatial Policy-Making Formulation in Iran's Villages
	Badri et al. (2020)	Analysis of Factors and Processes Affecting the Formulation of Spatial Policy-Making of Physical Development in Iran's Rural Areas
	Ahmadinejad et al. (2021)	Challenges of Development Policies Implementation in Iran: A Thematic Analysis
	Kamali et al. (2022)	Challenges of Formulation of Development Policies in Iran, Five-Year Development Plans
	Noori and Ahmadi (2013)	The Rural Development Strategy in Iran, An Analysis of Viewpoints
	Farajirad et al. (2013)	Pathology of Regional Development Policies in Iran: An Institutional Approach
	Mohammadi and Kamali (2013)	Analysis on the Institutionalization of Sustainable Development Policy in Iran: Examining Barriers and Proposing a Model
	Ahmadi and Mottaghi (2022)	Pathology and Monitoring of Rural Development Policies in Development Plans
	Kashef and Abootalebi (2020)	Improving and Agiliting of the Development Planning Process in Iran
	Sameni and Zebardast (2023)	Feasible Pathology of Spatial Planning in Iran, Case Study: The Fourth Round of Spatial Planning Studies
	Seifoori and Taghavi (2018)	Pathology of Iran's Development Programs Before and After the Revolution, and Presentation of Pivotal Strategies
	Rahnavard (2022)	Mega-Policies of The Development Plan: A Case Study of Iran
	Kamali (2023)	Identifying and Ranking the Challenges of Evaluating Development Policies in Iran (Five-Year Development Plans)
	Mohebati et al. (2020)	Designing a Phased Model of Crisis Management with the Empowerment Approach of Local Communities Using Meta-synthesis Approach
	Fazeli (2011)	An Introduction to the Role of Strategic Planning in National Vision Research, Case Study: Perspective 2026 of the Islamic Republic of Iran
	Ziari and Mohammadi (2015)	A Comparative Analysis of Regional Development Management Features in Iran Before and After the Islamic Revolution, From 1967 to 2015
	Motevaseli et al. (2017)	An Opportunity to Deal with the Concept of Planning in Iran
	Nazmfar et al. (2018)	Inequality of Development in the Provinces in Order to Achieve Sustainable Development
	Alizadeh and Amanpour (2023)	Explaining The Barriers to The Realization of Strategic Spatial Planning in Iran
	Ghavam-Maleki (2018)	International Approaches and Policy and Strategic Barriers to Rural Development in Iran
	Kazemian et al. (2019)	A Pattern of Role and Structure of Local Governance in Iran's Management System
	Moghimi et al. (2017)	Describing Elements of Good Governance in Policy-Making of Budgeting System in Iran
	Ghaffaryfard (2019)	Review the Process of Development Policy-Making and Regional Balance During Iran's Development Plans

(continued)

Table 28 (continued)

Type	Author	Title
Specialized meetings	Haqshanas et al. (2023)	Intervention Strategies in Vulnerable Rural Areas to Natural Disasters
	Anabastani et al. (2024)	Sustainable Strategies for Population Stability in Rural Areas of the Country
	Azizpour et al. (2022)	Rural Development Pathology in Iran with Emphasis on Development and Advancements Projects in Rural Systems
	Monfared et al. (2022)	A Experience Review of Rural Development and Advancements Projects in Rural Systems
	Ghorbani and Afarakhte (2023)	Local Governance in Rural Planning System Development
	Jabari et al. (2022)	Situation of Rural Development in the Seventh Five-Year Development Plan
	Jalalian et al. (2023)	A Review of Successful Global Experiences in Rural Planning Systems
	Moridsadat et al. (2023)	Identification of Challenges and Development Solutions for Rural Entrepreneurship in Iran
	El-Khoury et al. (2023)	Combatting Inequality for a Resilient Future
	Babaeian (2023)	International Day Against Climate Change: Iran's Climate Future Image and Presentation of Plans for Resilience

unit of analysis. Open coding is an analytical process through which concepts are identified, and their features and dimensions are discovered in the data (Lee, 2001). In other words, open coding is a process in which data come in as separate meaningful units and can be used at the beginning of the study. The main goal of open coding is to conceptualize and label the data. In the data analysis stage and the search for codes, open coding becomes possible. During open coding, researchers engage in the briefing, analysis, comparison, labeling, and categorizing of the data. Coding leads to direction determination before selection and transforms the data into analytical parts so that they take a conceptual form. Concepts are the basic foundations of theory construction. Open coding in GMT is a process in which concepts are identified and expanded based on their characteristics and dimensions. Open coding is part of the analysis that involves identifying, naming, categorizing, and describing phenomena found in the text. Every sentence and paragraph that is read is to answer two repetitive questions: What is this content about? And what does it refer to?

Open coding is a two-stage process. In the first stage, meaningful units are broken down into smaller parts, and the researcher identifies the concepts that these smaller parts represent. These are called conceptual labels, propositions, or initial codes, which should accurately show the data content. The researcher and reader should be able to understand the meaning of sentences by observing them. The second stage, focused coding, involves using the most meaningful or frequent initial codes to sort and reduce a significant amount of data. In this stage, the researcher consolidates codes or concepts, placing similar codes into a unified concept. The study extracted 866 initial codes after reviewing their data overlap, repetition, ambiguity, and irrelevance. Categories and concepts were named based on the researcher's understanding of the subject as well as the shared content of the open codes. A set of concepts that have comparable meanings were recognized to form sub-categories. In addition, main categories were created to represent sub-categories with similar concepts. This helped increase the abstract nature of the concepts. The study resulted in the identification of 44 sub-categories (dimensions or features), 13 categories, and six axial categories.

4.2 Axial Coding

Axial coding is a fundamental step in the GTM used to develop a deeper understanding of the relationships between categories and concepts identified during open coding. It involves two main stages: Identifying relationships between sub-categories and core categories and developing a theoretical framework. The first step of axial coding is to establish connections between the categories and sub-categories identified during open coding. This is done by examining the data and identifying common themes, patterns, and relationships. The researcher then begins to build a network of related categories using visual tools such as diagrams or matrices. This network helps visualize the relationships between the categories and how they contribute to the general studied phenomenon.

In the second stage of axial coding, the researcher selects one or more categories to serve as the core category of the theory. The core category represents the studied axial phenomenon or process. The researcher then examines how the remaining categories and sub-categories relate to the core category. This involves identifying the conditions that contribute to the core category, the context in which it occurs, responding strategies to it, intervening conditions, and its consequences. The researcher then uses these relationships to build a theoretical framework that explains the studied phenomenon. This framework should be grounded in the data and be able to account for the relationships between the different categories. Axial coding is an iterative process that involves continually refining and developing the theoretical framework as new data is collected and analyzed. Through this process, the researcher can develop a well-grounded theory that provides a deep understanding of the studied phenomenon (Bryant & Charmaz, 2016).

4.2.1 Causal Conditions

The term "causal categories," stemming from causal or antecedent conditions, refers to events, occurrences, and phenomena that contribute to the onset, growth, or expansion of a core phenomenon (Vollstedt & Rezat, 2019). The causal categories of this study include the weakness of the intellectual and theoretical foundations of rural policy-making, which consist of two categories: Contradictions among dominant approaches to rural development and ambiguity in the definition and position of the village and rural development (Table 29).

Table 29 GTM levels in the causal conditions

Category	Sub-category	Concept label
Contradictions among dominant approaches to rural development	Predominance of centralized approach in rural planning system	Failure to consider the rural population in the national planning process Failure to properly understand the needs of the rural community at the time of decision-making Failure to pay attention to the specific characteristics and issues of villages while approving laws, policies, and plans Neglecting local communities' cultural and historical heritage, beliefs, and attitudes in rural planning The increasing attention of policy-makers and planners to densely populated areas and cities due to lower costs and neglecting villages Lack of sufficient attention to the gap and the distance between the city and the village in the previous national development and construction plans Lack of local attitude toward rural development in centralized planning The centralized management system growth (top to bottom) despite the adoption and formulation of regional and provincial development plans The establishment of a centralized spatial structure in the planning system and finding the importance of urban centers compared to rural settlements The lack of transparency of laws and macro-national policies in the rural development sector Compilation of strategies, macro goals, policies, and executive measures from top to bottom according to the tasks of sectoral institutions and without regard to rural issues and problems Underestimating the role of the rural community in the process of national development and the growth of a passive attitude towards these areas among officials and policy-makers Lack of attention to rural development by decision-makers in the country's planning structure Lack of clear definition of the village as a planning unit in the country The dominant view of executive institutions towards rural development in theory and not in practice

(continued)

Table 29 (continued)

Category	Sub-category	Concept label
		Lack of sufficient attention to rural development in the country's Five-Year Development Plans Ambiguity in rural development credits in the country's annual budget laws Failure to pay attention to the diversity and differences of villages in regional development plans The government's lack of attention to villages and the allocation of small budgets for these areas' development Lack of comprehensive support for the rural society to expand profitable activities, such as tourism Lack of attention to mobility of resources at local levels and disorder in land management Gradual reduction of employment opportunities in the industry and services sector; Paying attention to big projects instead of micro plans and employment on a regional scale Planning deficiency on a local scale and incomplete planning levels in the country Lack of accurate division of responsibilities planning between national, regional, and local levels Coordination scarcity between executive institutions in designing, and implementing rural development plans Lack of coherence between the organizations and institutions in charge of rural development and the overlap of their duties together Failure to delegate the consultative, executive, and supervisory affairs to the people and the private sector due to utilitarianism Lack of coherent and effective communication between executive institutions to achieve the goals of rural development plans Absence of specific, long-term, and practical plans for the well-considered organization of country's rural centers and spaces
	One-dimensional insight into rural development plans	Paying too much attention to the physical dimension of villages in rural planning and development The role of the government as an executive and not a guide of plans The predominance of the modernization approach to rural development in the construction period The predominant project and short-term view of rural development The political issues intervention in rural development plans and forgetting spatial attitude Point planning and lack of constellations view Lack of integrated and all-round development approach to rural issues Failure to properly use the capabilities and social capital for rural deprivation Creating physical and service infrastructures in villages regardless of production capacities (needs-based approach) Prioritizing urban projects over rural plans in the executive institutions Budget and credit differences between cities and villages Lack of proper concentration of rural development investments and credits at local levels Lack of bank credits for agricultural production to meet the needs of villagers Emphasis on providing public services and improving the level of people's benefit from them The attitude of only depriving in rural development plans The trend towards industrialization regardless of the agricultural infrastructure
	Expediency and planning of rural affairs with urban glasses	The transformation of the physical structure of the villages into urban landscapes and the destruction of the cultural heritage and historical identity of these settlements Inadequacy of local management methods in villages Lack of detailed knowledge of officials and experts of executive institutions about the issues and management of villages The competence of consulting companies in the field of urban planning and urban development, not rural development Using urban planning experts and civil engineers to prepare rural development plans and not interacting with villagers to make these plans fruitful The general similarity in looking at the rural municipality and its organization with the look at the municipality Prioritizing service and civil affairs (road construction, rural furniture, and garbage collection) for rural municipalities like municipalities

(continued)

Table 29 (continued)

Category	Sub-category	Concept label
		Emphasis only on the traditional and costly approaches of providing energy for the villagers through gas supply and neglecting new energies Expanding social security coverage for rural people with the same criteria as urban workers working in the formal economy The existence of a relative surplus of labor in the agricultural sector and their lack of concentration in other activities of the rural economy The gradual growth of the social and welfare distance between the city and the village Lack of sufficient knowledge of officials and experts of executive bodies regarding the different nature and functions of villages Attempting to link the resources needed by rural municipalities to construction fees and sales density like municipalities Neglecting the villages as a productive part of the society and not trying to equalize the living standards in the villages with the cities Slowness and stagnation in the implementation of educational and construction projects in villages Development of rural roads through creating large and main roads
	Inconsistency of rural development plans	Separate examination of activity and physical elements in Five-Year Development Plans Non-implementation of the Ministry of Agricultural Jihad Law and the Ministry's sole focus on agricultural matters Nonrecognition of the agricultural sector in providing nonproductive functions for poverty alleviation or based on the supply of environmental goods Lack of provision or dispersion of financial resources needed for development and investment in villages The lack of proper role of rural municipalities in the process of improving the rural economy, especially in the agricultural sector and related industries Lack of connection between the social insurance of villagers and farmers with agricultural exploitation systems in order to provide job security Failure to use social capital and networking in the way of strengthening production Predominance of technocratic and productionist approaches to agricultural development policies Lack of a targeted strategy for rural development and its long-term fruition Lack of diversification of rural jobs to reduce the pressure on national and agricultural lands Changing the use of fertile agricultural lands as a result of Rural Guide Plans Lack of support for cooperatives and production organizations Lack of proper marketing system and pricing policy Instability in planning and implementing rural development plans
Ambiguity in the definition and position of the village, and rural development	Absence of rural development policies	Simultaneous attention to the agricultural sector and rural development in the system's policies Absence of executive and specific plans and policies for rural development Lack of policy-making for rural development in the real sense in the country Differences in rural development priorities among high-level managers Hasty planning without theoretical and intellectual support for rural development Lack of a successful and stable strategy for rural development in the country The isolation of the rural society of Iran from the political system as a whole The impossibility of the beneficiaries claiming their inalienable rights in the perspective and goals of the policies The inefficiency of endogenous rural management and lack of macro-rural management Lack of strategic plan for rural development at regional levels Adopting a social welfare approach based on the distribution of services to villages Lack of comprehensiveness and targeting of programs to remove deprivation in rural areas

(continued)

Table 29 (continued)

Category	Sub-category	Concept label
	Instrumental view to rural society	Lack of attention to villages in the planning system De-ruralization by government agencies Attention to villages only during elections Lack of national and media coverage of rural society challenges Project as the main tool of rural development organizationalism The governance of a sectoral view in the institutions responsible for rural development and the lack of synergy and communication between them The governance of the physical view with a focus on the construction and development of the physical fabric of the villages Focusing rural development plans and actions on the hardware and physical-service dimension The relative weakness of the physical-functional structure of rural settlements The only focus of local management based on maintaining environmental health, collecting waste, creating green space, creating concrete curbs for roads, reconstructing and creating irrigation canals, and asphalting village roads Failure to pay attention to the employment of villagers through income-generating businesses such as tourism The attitude of political economy governing the economic issues of villages The excessive emphasis of the Ministry of Agricultural Jihad on the issue of food security and the weakening of the issue of rural development Mastery of economic and quantitative perspective with a focus on increasing production The focus of placing the growth of agricultural crops in the formulation of agricultural development policies of the Ministry of Agricultural Jihad Continuation of agricultural production growth policies without regard to farmers' employment security by the Ministry of Agricultural Jihad Instrumental approach to farmers in the production plans of the Ministry of Jihad Agriculture Focusing on one dimension of development and neglecting the influence of other dimensions in the decision-making process and implementation of rural plans Establishing the technocracy approach of rural development on the model of modernization The tendency towards industrialization of villages without considering the infrastructure of the agricultural economy of these areas Failure to establish a logical connection between the industry and agriculture sector Ambiguity in giving priority to industries or agriculture Lack of attention to small industries for the development of the rural economy Continuation of giving priority to urban communities Neglecting the development of human resources and cultivating facilitators and multi-professionals in rural areas Continued misperception of villages as backward and unsafe communities for investment Failure to define the place of social-organizational innovation and the attitude of local people in the rural planning system, despite the existence of diversity in local views and knowledge Limiting the participation of the residents of the villages in local management issues to the selection of members of the Rural Islamic Council Failure to invite professors, elites, entrepreneurs, representatives of guilds and local groups in the decision-making meetings of Islamic Councils focused on rural development Failure to employ experts from different disciplines in the formulation of rural development plans such as geography, and sociology Failure to establish the possibility of polling and generating ideas for the design of various civil, economic and social projects and plans by rural municipalities Failure to use the capabilities and expertise of local people in the process of implementing rural construction and service projects The migration of rural elites to urban areas

(continued)

Table 29 (continued)

Category	Sub-category	Concept label
	Lack of indigenous definition of village, and rural development	The absence of a theoretical, conceptual and operational definition of the village in the country
		The existence of a mere official and administrative definition of a village
		Considering rural development and agricultural development as synonyms
		Lack of a codified definition of rural development
		Attention to the village as a residential unit instead of a production, cultural and heritage unit
		Lack of a scientific definition of the village in accordance with the country's native requirements
		Lack of a specific pattern for rural development
		Lack of a structure to define and design rural development patterns based on geographical areas
		Uncertainty of the position of the village in the ideology and economic thinking ruling the country
		Lack of suitable conditions to resolve the conflict between tradition and modernism in the villages
		The predominance of the distinction between the city and the village based on the demographic criteria and the size of each region
		Attitude towards the village as a center of production and a residential unit instead of a settlement in the system's policies
		Ignoring the demographic, sociocultural and economic structure and structure in the village definition
		Ignoring the distinctive characteristics and dynamics of villages and preventing targeted interventions for local development
		Lack of accurate and real knowledge of rural communities about their needs and rights and lack of attention to indigenous knowledge
		Lack of villagers' awareness of the methods and principles of achieving sustainable development
		Lack of coordination of rural development programs with regional and national policies
		Making decisions based on taste and without new scientific support on rural issues
	Lack of recognition of the fundamental problem of villages	Belief in universal understanding in rural issues and the return of management from executive decision-making to theorizing
		Failure to recognize villages based on the diversity of their unique characteristics
		Failure to identify current needs, capabilities and opportunities in rural areas
		The lack of a specific mechanism for villagers to interact with the executive institutions in charge of the country's rural development
		The lack of ability of the administrative system in problem definition of rural development due to lack of time and cumbersome procedures
		Lack of knowledge of decision-makers and implementers about the intellectual and theoretical foundations of rural development
		Lack of understanding of the changes in the external environment and its effects on the rural society
		The failure of the country's rural planning and policy-making system to use academic-scientific studies to identify and find the roots of rural problems
		Lack of coordination of rural development policies and programs with the sociocultural, economic and environmental needs and characteristics of the local community
		Lack of effectiveness of rural development policies and plans due to a misunderstanding of the basic needs of the local community
		Determining the place, time and cost of rural development programs by the government without considering the fundamental problem of villages and the ability to implement and complete the plans
		Paying attention only to service indicators of rural development in monitoring reports
		Lack of access to up-to-date information and reliance on statistical data from the last decade to determine the level of rural development
		Lack of database and geodatabase for accurate monitoring and evaluation of rural issues
		Failure to establish decision-making support systems for the intervention of local community stakeholders and elites in solving rural issues
		Single decision-making for the country's villages despite regions' sociocultural, economic, and environmental differences

Table 30 GTM levels in the main phenomenon

Category	Sub-category	Concept label
The dominance of government-oriented policies in the rural planning system	Sectoral and centralized policies	The institutionalization of sectoral thinking in the policy-making and planning system The importance and priority of sectoral approach in codified trans-sectoral policies The sectoral nature of planning, policy-making and decision-making structures and organizations Sectoral activity of ministries and organizations in the policy-making process of rural development Institutional development and structural and organizational expansion preceding rural development Absence of firm determination to realize rural development The idealistic nature of rural development plans Absence of a specific legal position for rural planning Lack of capable human resources in organizations to create creativity and innovation in rural areas The prescriptive, top-down and nonspecialist nature of policies and plans affecting the rural society Weakness and successive changes in policies and nonadherence to codified and valid regulations in order to advance decentralization policies to reduce the tenure of government institutions The predominance of centralized policies on development plans and their incompatibility with the sociocultural and environmental characteristics of villages The neglect of rural development planners and policy-makers to develop development plans based on the function and socio-economic relations governing rural society The predominance of modernization thinking and economic growth in the structure and parts of the country's policy and planning system The loss of the competitive atmosphere, and the tendency towards the polarization of management caused by the combination of planning, implementation, and supervision in an executive institution Lack of communication and separation between the rural development chapter and the agricultural chapter in the country's Five-Year Development Plans Continuation of ambitious plans based on eliminating deprivation and providing services regardless of scientific foundations, comprehensive development and sustainable development Noncompliance of rural development programs with specific and expert policies due to changing conditions and the personal taste of managers in the past decades The predominance of the attitude of only physical-physical development in the plans and executive actions related to villages and neglecting the all-round transformation in the dimensions of the development of these areas The attitude of passivity and indifference of the rural community towards their fate Lack of space for cooperation and nonacceptance of local community participation by government organizations Non-responsiveness of the institutions responsible for rural development to the village's issues Failure to take advantage of local knowledge in rural areas for the organization of local management Lack of organic and comprehensive communication between organizations responsible for rural development Failure to adopt participatory planning methods in confronting villages with issues, crises, and risks Lack of equal opportunities and fair distribution of income among different regions Stability of the country's economic structure based on urban, service-oriented, and productive plans The lack of attention of the planning system to strengthening the production base, increasing income, and improving the living standards and fueling unemployment and inequalities in villages Inequality between urban and rural living standards due to nonoptimal allocation of resources and spatial differences in investment The predominance of positivist methodology on rural development plans and policies Lack of attention to the principle of meritocracy in the selection of officials to hold and manage institutions affecting rural development Weakness in group cohesion and lack of synergy between organizations and local communities

(continued)

Table 30 (continued)

Category	Sub-category	Concept label
	Multiplicity and variety of institutions responsible for rural development	Lack of a specific and responsible institution for rural development Dispersion of departments responsible for rural development in different organizations Excessive involvement of government organizations in rural affairs Lack of formation of single and coherent rural management after land reforms Lack of coordination of executive institutions active in the field of rural development with local management Interference, conflict, and non-specialization of the duties of organizations responsible for rural development Failure to correctly define and recognize the manager and management responsible for rural development for the local community Disproportion between the responsibilities assigned to rural managers The noninherent nature of local management responsibilities and excessive concentration of responsibilities at regional and national levels Reducing the quality of decisions made at the upstream level for rural areas and dealing with unnecessary issues in executive institutions The illogicality of the tenure of management of executive institutions related to rural areas Lack of legality and meritocracy in the promotion, dismissal, and appointment of managers and employees of institutions responsible for rural development The administrative system's inability to attract, maintain, and encourage capable, efficient, and expert employees of institutions responsible for rural affairs Failure to establish a competitive system in the recruitment and selection of human resources for institutions responsible for rural development Lack of participation and formation of two-way communication between the institutions responsible for rural development and the local community Lack of different channels to connect villagers with institutions responsible for rural development
	Rent distribution of capital and natural resources	The government's reliance on oil revenue and the allocation of subsidies to basic goods Planning and policy-making based on the distribution of oil revenues Absence of representatives with rural influence in rent distribution centers The injection of oil revenues into the country's governing body and the national economy and its impact on the lower levels of society Lack of coordination of executive institutions with different levels in the decision-making process Dependence on oil and lack of attention to production sectors, including rural production The impact of the interventions of political officials on the budget allocation mechanism for different sectors of the country's development Lack of capital and banks' profit-oriented approach to villages Disruption in the market caused by the government's excessive interference in economic activities and the formation of the rent-seeking phenomenon Payment of extensive subsidies for fuel products and the emergence of incorrect consumption patterns in the domestic, commercial, manufacturing, and transportation sectors Nonoptimal allocation of resources and multiple pricing of goods due to the payment of various types of subsidies in the economic sectors of the country Providing goods at a false price and cheaper than the real production price to the society and harming the rural producing communities Formation of cultural conflict between subcultures and local society Lack of synergistic interaction between the private sector and the local community The spread of the culture of consumerism and economic activists being deprived of equal conditions for initiative and innovation in this sector An increase in the access of the rich to luxury goods, along with a significant decrease in the access of the poor to essential goods Inadequacy of executive actions with villagers' expectations due to the short-sightedness and haste of officials in decision-making and resource allocation The injection of capital in the society due to oil reserves and not work and production Increase in inflation in the land sector and limited supply of housing

(continued)

Table 30 (continued)

Category	Sub-category	Concept label
		The profitability of intermediaries in the market of buying and selling agricultural products and intensifying price fluctuations for consumption by keeping the price of these products low The increase in the price of consumer goods, the decrease in the purchasing power of the middle and deprived sections of the society, and the increase in demand for second-hand goods Destruction of fertile agricultural lands and wetlands with the establishment of industrial facilities Changing the use of rural land in favor of oil industry land uses Attracting young and active workforce of the agriculture sector to industrial jobs Dissemination of air-polluting particles and reduction of air quality in rural settlements adjacent to industrial areas and activities Creating oil pollution in soil and water and affecting terrestrial and aquatic organisms Destruction of the natural ecosystem and impact on the distribution of plant and animal communities Destruction of the natural landscape of villages located in industrial areas
Continuous changes in rural development policies	Absence of a planning and policy-making institution for rural areas	Lack of a specific trustee for rural development policy-making The superiority of sectoral and vertical management over regional, horizontal, and trans-sectoral management in the country Weakness of governance institutions in separation of powers and decentralization of power and inability to respond to people's needs Failure to delegate responsibility and authority from national to regional and local levels Conflict of interests and interference of activities between institutions in rural development affairs The existence of dictated and independent programs for each organization related to the rural areas The lack of a valid executive plan among governmental institutions for the development of rural areas according to the remoteness of the parliament from the country's executive sphere and the distance of the government from the legislative sphere Confusion of villagers in pursuing their needs, issues and problems between executive institutions Lack of coherence in the follow-up of rural issues due to the incorrect separation of duties between different institutions Interception of top-down and authoritarian management and planning approach instead of bottom-up and democratic management and planning Lack of a clear and transparent legal framework for monitoring local management actions Ignoring the residents' right to access information on contracts, budgets and reports related to villages Lack of design and implementation of villagers' investigation and response mechanisms and petitions and complaints Failure to design and use mechanisms to quickly handle petitions and complaints of rural residents Failure to consult with the stakeholders of local communities in the field of decision-making and implementation of plans Lack of effort to promote the ethics of serving the rural residents Failure to pay attention to removing unnecessary obstacles to promote job security and rural self-employment Lack of proper understanding of the requirements of infrastructural development and activation of socio-economic capacities Severe reduction of natural resources needed for rural production The continuation of agricultural land use change and the weakness of the monitoring systems for its proper control Failure to take the necessary measures to prevent the shrinking of agricultural lands

(continued)

Table 30 (continued)

Category	Sub-category	Concept label
	Adopting reactive measures and policies	Weakness of rural policy-making position at the macro level
		Absence of rural strategic policy-making based on long-term perspective
		Failure to educate the public on how to deal with the damages of natural disasters
		Not knowing the spatial distribution of settlements and rural population centers to deal with risks
		Failure to set boundaries for rivers and wadi (dry channels or valleys, except during rains) to prevent massive construction
		Failure to establish ecological reservoirs to absorb, transfer, and store runoff in low-water areas
		Failure to create secondary routes to access villages during disasters and dysfunction of freeways
		Failure to take the necessary measures to reduce the concentration of road traffic in places prone to natural hazards
		The lack of balance and observance of hierarchy in the distribution of services in rural areas and the creation of areas deprived of development
		Failure to disperse additional facilities assigned to high-security areas to other vulnerable areas during natural hazards
		Lack of identification of migrant-sending villages and the emptying of the villages from the active population of the agricultural sector
		Establishment of polluting industries in the vicinity and upstream of rivers, lakes and wetlands
		The destruction of plant and animal ecosystems and damage to the agricultural sector due to the lack of attention to environmental principles in locating dam site
		Forced displacement of villages due to sudden floods or drying up of rivers due to the construction of huge dams in recent decades
		Creating industrial projects and reducing the rural farmer population despite the possibility of income generation and social welfare for the villagers themselves
		Failure to use durable materials in different climatic conditions and retrofitting rural housing
		Failure to provide the necessary solutions to modify the agricultural pattern due to climate changes in the coming years
		Lack of income diversity to deal with the decrease in the price of agricultural products
	Changing structures and policies at the same time as changing governments	Failure to provide a comprehensive executive framework in accordance with the realities and requirements of the executive institutions for the development of the country's rural areas
		Conservatism under the pretext of preventing the wastage of financial resources and a tasteful desire to satisfy immediate interests in facing rural issues
		Lack of benefit from the model and knowledge of modern regional management and misuse of resources due to the inefficiency of managers and their short management periods
		The slowness of the decision-making process, the passing of responsibilities to each other and the abandonment of some policies and programs affecting rural development
		Nonacceptance of responsibility by institutions involved in rural development
		Lack of accurate forecasting of the future needs of villagers and country's rural studies
		Failure to draw the desired vision and coordinates for rural development plans
		Failure to determine the tools, responsibilities and executive measures to make plans related to rural development fruitful
		Failure to foresee arrangements and requirements for realizing plans related to rural development
		Absence of long-term rural development policy-making due to continuous changes in the country's structures, institutions and planning approaches
		Inconsistency of programs with organizational goals and the needs of local community stakeholders
		The noncompliance of strategies with the capacity of executive institutions, the environmental conditions of rural areas and the requirements and capabilities of the country
		Decreasing the quality of rural products and services as a result of price instability and increasing consumer goods costs
		Failure to use the opinions of stakeholders and experts in the selection of rural development strategies
		Multiplicity of institutions responsible for rural development with parallel and similar tasks
		Dissolution of institutions in one state and their revival in another state
		The different tastes of the ministers in the creation, deletion and merger of the sub-departments of the ministries

(continued)

Table 30 (continued)

Category	Sub-category	Concept label
		Lack of participation of the local community, private sector and cooperatives together in solving rural issues and increasing the effectiveness of plans
		Lack of attention to the theoretical support and based on the appropriate theory or idea for the preparation of rural development plans
		Delayed implementation or non-completion of the development plans of deprived areas and lack of commitment and responsibility of the executive institutions towards them
		Failure to create technical, structural and cultural infrastructures to continue the planning process after the governments change
		Lack of integrated and coherent pathological methods to reorganize and determine the maturity of reforms of organizations with tasks related to rural and agricultural development
		The incompatibility of the goals of the officials with the perspective and coordinates of the rural development policies and programs defined by the executive institutions
		Lack of managerial stability in different sectors to coordinate affairs and accelerate the process of implementing plans related to rural development
		The unplannedness of institutions in determining socio-economic policies due to the different tastes of the ministers in the creation, elimination, and merger of the sub-departments of the ministries
		Non-implementation of policies by some ministers and the cancellation of plans due to their personal opinions not being aligned with the planning groups of previous periods
		Merging or separating ministries in different governments and delaying or stopping the progress of construction, service and production fields
		The existence of political will to solve the specific problems of the society and not the public ones and try to convert commitments into action only at that level
		Changing or repeating the policies and guidelines of the country's development by the government ministries or the commissions of the legislative assembly by changing their members

4.2.2 Main Phenomenon

The main phenomenon in grounded theory refers to the central idea or event around which a set of actions or reactions are observed (Bryant, 2020). The main phenomenon of this study includes centralized, sectoral, and top-down policy-making, which consists of two categories: The dominance of government-oriented policies in the rural planning system and the continuous changes in rural development policies (Table 30).

4.2.3 Intervening Conditions

Intervening conditions in GTM can be seen as structural conditions that facilitate or limit the intervention of other factors (Andersen, 2023). These conditions have a causal and general nature, similar to the concept of causation in causal conditions (Damonte, 2021). The intervening condition in the study is the incoherent in the process of compiling and implementing rural development plans, which consists of two categories: Compilation of incoherent and disunite plans and ignoring the implementation and results of plans (Table 31).

4.2.4 Contextual Conditions

Contextual conditions in GTM refer to the specific situational and environmental factors that influence the phenomena under study (Fleet & Reeves, 2023). They are the structural factors that shape how individuals respond to situations, circumstances, or issues using actions, interactions, and emotions. These factors can range from the macro to the micro level (Ashtari et al., 2022). Contextual conditions of the study, include the territorial and institutional incoherence of rural development policies, which consist of two categories: Weakness in institutionalization and local networking and the territorial elements imbalance (Table 32).

4.2.5 Action or Interaction Strategies

Strategies for action or interaction in GTM aim to handle, overcome, perform, or react to a phe-

Table 31 GTM levels in the intervening conditions

Category	Sub-category	Concept label
Compilation of incoherent and disunited plans	Disregarding the spatial attitude in the preparation of plans	Paying attention to rural development spatially and not pointwise The non-complementarity of sectoral plans due to the lack of a spatial attitude in planning Lack of connection and coordination between rural development plans of different organizations Role-playing the general knowledge of national-level decision-makers in rural development and not relying on the experiences and knowledge of the local community Nonalignment of rural physical development measures with the system's goals and policies Islamic Revolution Housing Foundation's emphasis on physical measures, in the form of guiding plans, organization, valuable structure, separation, improvement and strengthening of housing, and not all-round rural development measures Low attention to local and regional capacities in policies, plans, and decisions Failure to reflect the needs of the local community in the rural development approach considered by upstream institutions Weakness and fragmentation of participation and role-playing of local community groups and actors Lack of attention to local management policies and plans and the tendency to follow national goals and measures to advance rural development affairs The development of large-scale industrial projects, the creation of agricultural hubs, and the decline of villages with weak agricultural and tourism economies The need for financial support and tax exemptions to encourage economic investment and participation in rural and underprivileged areas

(continued)

Table 31 (continued)

Category	Sub-category	Concept label
Compilation of incoherent and disunited plans	Preparing plans without regard to the facts	The predominance of irregular essay writing approach and without real statistics of the current situation instead of plan writing of pattern-oriented in the matter of rural development Failure to pay attention to the complex and dynamic nature of rural areas Mere imitation of internal and external patterns of rural planning without recognizing the specific characteristics of each region The idealism of the plans and insufficient attention to the cultural, socio-economic and environmental aspects of rural development Weakness of regional planning thinking in formulating rural development plans Lack of connection between the policy of plans and its intellectual and philosophical framework Formulating the policy of plans related to rural development without paying attention to objective and field evidence Formulating laws in previous programs without allocating sufficient financial resources Lack of planning approach in the rural budgeting system Absence of endogenous patterns for rural development in the planning and policy-making system Lack of credibility and necessary means of plans implementation related to rural development Lack of attention to inter-sectoral relations and cooperation for developing rural development plans Lack of a specific mechanism to resolve sectors conflicts in inter-sectoral relations Lack of prioritization in allocating resources to institutions affecting rural development Absence of a planning monitoring system outside the organizations to monitor and evaluate the content of rural development plans The noncompliance of rural development programs with each other and the unclear them agenda Weakness in the legal writing of the country's development plans and the damage of the stages of preparation to approval and extensive changes in the bill to the stage of becoming a law Willingness to extensive interventions in the market mechanism through import substitution policies Attempting unprincipled targeting in all aspects of development without respecting the priority in meeting current needs Lack of political will to implement rural development plans Lack of concentration and unity in decision-making and absence of necessary authority to formulate, implement and monitor rural development plans The complexity of the political structure and the multiplicity of planning institutions to create inter-sectoral coordination
	Inconsistency of institutions in preparing plans	Unpreparedness of the institutions to accept the trans-sectoral approach Failure to pay attention to the principles of integrated development in formulating and implementing rural development plans Predominance of bureaucratic and technocratic approach in designing, implementing and monitoring rural development plans Weakness of theoretical coherence of country's development plans and neglect of various goals and dimensions of development and sustainable development Incompatibility between the goals of the country's development plans with the tools and structural, procedural and legal requirements for their realization Lack of interactive participation of local and regional level institutions and stakeholders in the preparation and regulation of rural development plans The difference in the perception of formulators of rural development plans compared to the executive officials and the willingness of the executive officials to show a brilliant record devoid of effective activities for the development of these settlements Providing plans in a short period and their noncompliance with the local realities and requirements Lack of a comprehensive and up-to-date statistical system in formulating rural development policies Lack of coordination between the objectives of the plans with the duties and executive capacities of institutions associated with rural development The disbelief of provincial officials in holding coordination meetings between executive institutions and eliminating interference in their duties and functions Failure to determine the relationship and hierarchy of the content of rural development plans with the overall policies of the country's development plans

(continued)

Table 31 (continued)

Category	Sub-category	Concept label
		The lack of a suitable mechanism to receive the opinions of development policy implementers to increase the feasibility of implementing plans Neglecting to resolve the conflict of interests of the actors in formulating development policies Absence of inter-ministerial and inter-power council formations during the development of rural development policies Imbalance between goals and costs and subsequently increasing the allocation of funds for plans Lack of a specific, legal and executive mechanism to link the solutions of the programs with the provisions of the country's annual budget Failure to complete plans due to sectoralism, parallelism and dispersion of resources and credits Lobbying the executive institutions with the Plan and Budget Organization, the Parliament and the Government to obtain more funds Inserting the tasks and provisions of the development plans law with the aim of obtaining more funds
Ignoring the implementation and results of plans	Paradox between processes and results of planning	Noncompliance with the plans implementation period with the responsibility period of different governments The lack of a problem-oriented approach while being comprehensive and not focusing on the key sectors of rural development Failure to pay attention to the political, economic, and sociocultural structures of the community in the formulation process of development plans The lack of a local development pattern in accordance with the structures, norms and values of the society and compatible with the conditions and goals of the country The large volume of plan provisions and the reduction of the possibility of full implementation of the plans due to the lack of energy and power of the expert body Relying on management reports in advancing the plans and neglecting the bottlenecks and the main axes that guide planning for the mobility of other sectors Lack of proper knowledge of planning institutions about strategies, policies and implementation of rural development measures Non-implementation of plans by the officials of the new governments due to excess of the duties of promises and commitments expressed in group and factional competitions Conducting monitoring and inspection after implementation and not during the formulation and implementation of rural development plans Failure to adopt a process approach for the preparation, implementation, monitoring and continuous evaluation of development plans and the occurrence of deviations from the goals set in them Increasing public sector ownership and implementing plans regardless of private sector and local community participation The absence of executive institutions downstream of the Vice President of Rural Development and Underprivileged Areas in provinces and counties and its passive role in rural decision-making, planning and policy-making The wideness of the study areas in the development plans and the lack of sufficient monitoring mechanisms in the preparation of plans for these regions Lack of attention to executive mechanisms and sufficient accuracy in the examination and approval stage of rural development plans Instability in the management of organizations responsible for preparing and implementing rural development plans Lack of planning perspectives and implementation measures in rural development plans Lack of hierarchical structure in the preparation and implementation of rural development plans The sameness of the service description of the development plans and neglecting the structures and functions of the regions Lack of necessary integration between regional and local management institutions in making effective decisions for rural development Failure to stop management changes while approving and implementing development plans Exerting influence and individual and group decisions in the approval of development plans

(continued)

Table 31 (continued)

Category	Sub-category	Concept label
		Lack of integration and horizontal and vertical connection between regional and local development plans Lack of flexibility in decision-making and limited measurement of development indicators in plans only for prioritizing villages The increasing acceleration of polluting and destructive factors of natural resources and the inadequacy of laws and regulations regarding the agricultural land use Lack of comprehensive participation of local community activists, including the government, private sector and people, during the preparation and implementation of development plans
	Failure to evaluate of plans results	Lack of sufficient recognition of the results of previous development plans Lack of systematic evaluation of rural development policies and plans Ambiguity in the amount of return of costs resulting from the implementation of development plans Failure to evaluate the effects of resource and budget allocation in the development of villages Failure to impact assessment of rural development plans and modify them based on the feedback received Lack of attention to the convergence of plans and weak decisions in rural areas of the country Deficiency and inefficiency of statistics and data required for rural development plans Lack of reliance on rural development plans on maps and up-to-date spatial data The existence of the same and similar views of different regions despite the existence of regional differences Ignoring the progress of implementation of the ongoing development plans The long process of preparing, reviewing, approving, and communicating rural development plans Failure to employ specialized and committed forces in institutions responsible for preparing and implementing rural development plans Lack of knowledge of some managers and experts with the content and nature of rural development plans Lack of a mechanism to review and reform rural development plans The emergence of financial and credit problems during the revision of plans and the lengthening of the interval between formulation and implementation Imbalance of financial and credit facilities allocated between urban and rural areas of the country to improve facilities and infrastructure Decreasing the quality of rural infrastructure and services due to the neglect of these settlements in the country's development plans The lack of a suitable sewage and surface water disposal system in most of the country's villages and the reduction of the quality and environmental health of these areas The low quality of rural houses and their extreme vulnerability to accidents and natural hazards Direct damage to agricultural activities and rural economy as a result of natural hazards Lack of legal and financial support and plan for the implementation of rural development plans Lack of monitoring systems to make decisions appropriate to the changing conditions in rural development plans

Table 32 GTM levels in the contextual conditions

Category	Sub-category	Concept label
Weakness in institutionalization and local networking	Absence of participatory evaluation system	Insufficient attention to human and economic factors in rural development Insufficient attention to the position of the people as the elements of the country's transformation and development Failure to employ skilled people in rural affairs and lack of cultural and economic excellence Weakness in the use of social and participatory capacities of Rural Islamic Councils Failure to empower villagers to monitor rural development plans Refusal to recognize villagers as elements of the transformation of the national economy and promoters of indigenous cultures in the country's monitoring and evaluation system Lack of an institution at the local level to collect the opinions and criticisms of the villagers Lack of credible field and verified information on the factual impacts of rural development plans Failure to edit and update rural development plans to improve people's access, use, and satisfaction with facilities and the society's state after implementation Lack of delegating appropriate authority to farmers in the field of production and markets The predominance of the bureaucratic administrative structure in the society and the managers' little acceptance of collaborative approaches due to their individualism, tyranny in their personal opinion, and self-interest Lack of attention to the spread of interference and administrative corruption in the planning and management structures of the social, economic, and political sectors of the community
	Marginal position of rural grassroots structures	Increasing instrumental participation and activity challenges for the people, not with the people A marginal view of the rural community and nongovernment and grassroots institutions in the country's planning and development system Incomplete participation of Islamic councils and rural municipalities in managing rural affairs Lack of inclusive participation of people in the design phase and use of their cooperation in the implementation phase of rural development plans The sense of isolation, futility, and emptiness of the villagers and their lack of contribution and role in the production of national added value Ambiguity in the position of the grassroots and nongovernmental organizations in the country's development planning The weakness of grassroots and nongovernmental organizations in establishing a relationship with executive institutions responsible for preparing and implementing rural development plans The low bargaining power and the voice of the villagers in the country's policy-making assemblies and centers The low importance of socio-economic organizations in villages Failure to play the real role of rural socio-economic systems in the development of the country Governmentalization of local organizations such as rural cooperatives The insignificant contribution of villagers in the administration of production and management units of the country Failure to activate the private sector to invest in rural development of the country Negligence of the media towards socio-economic and environmental problems of rural society
	Inadequacies of local management	Overcoming patriarchal structures in rural areas Weakening of local rural management in a centralized system Lack of unified and powerful local management aligned with rural characteristics Absence of cohesive, clear, and coordinating rural affairs structure Neglecting the management traditions of rural areas Absence of comprehensive rural management system post-land reform Lack of efficient social and productive management in rural areas Lack of coordination among local authorities Parallel and scattered work of rural development organizations; Increasing overlap of tasks and inefficiencies, delays, and gaps in providing services due to the multiplicity and variety of organizations responsible for rural development Reducing the hope, and trust of the villagers in the capabilities of government institutions due to the lack of response to their needs

(continued)

Table 32 (continued)

Category	Sub-category	Concept label
		Lack of coordination and information sharing of executive institutions with the provincial, and local organizations Absence of organizational relationships and division of labor at the national, regional, and local levels to solving rural issues Neglecting the role of rural municipalities to help the rural development policy-making system The disparity between the activity of the rural development policy system and the management of rural municipalities Lack of sufficient power and capacity of village councils in policy-making Paying attention only to the population threshold in the formation of rural municipalities Entrusting numerous tasks to village assistants without necessary capacity building Failure to employ expert staff in rural management and deviation from correct diagnosis of villagers' needs Lack of attention to the management pattern based on indigenous knowledge Continuous changes in the administrative structure of rural management Decision-making about villages by urban institutions Predominance of planning based on action and activity, not result and performance in rural areas Failure to allocate development tools to rural areas by the country's planning system The excessive superiority of municipalities compared to rural municipalities in terms of receiving public financial credits, capital facilities for projects, and assistance in special circumstances from the government The excessive superiority of municipalities over rural municipalities in terms of power, decision-making, planning and implementation for physical development Failure to designate the institution responsible for the physical development of the villages from among the rural municipalities and the Islamic Revolution Housing Foundation and increasing the wandering of the villagers to solve their problems
	Non-return of the added value of rural production	Disintegration of economic circles in villages Dependency of villagers on products manufactured in cities Low profitability of rural productions compared to other economic sectors Lack of adequate marketing system for rural products Dependence of rural livelihoods and farmers on government support Inadequate investment in villages and agricultural sector Low productivity of production factors in rural areas Non-economic nature of agricultural production due to high production costs Inappropriate pricing system in the guaranteed purchase of agricultural products Impact of macroeconomic indicators such as unemployment, recession, and inflation on villages Prominent role of brokers and intermediaries in stealing economic value in villages Transfer of value added from the trade of raw agricultural products of villages to others' pockets Lack of economic stability and decreased gross domestic product (GDP) and per capita income Existence of profit-oriented perceptions of investors by the public and officials Lack of social cohesion among small-scale producers in rural areas Inadequate provision of services to rural residents by banks High waste and food insecurity in rural areas The unrealistic low price of rural products

(continued)

Table 32 (continued)

Category	Sub-category	Concept label
Non-integration of territorial elements	Lack of spatial attitude in rural development	Lack of appropriate design of executive institutions for spatial planning
		Lack of coordination and cooperation of ministries and organizations responsible for spatial planning regarding rural development issues
		Absence of a designated authority for managing and implementing regional development plans
		Focusing more on rural policy at the national level and overlooking it at provincial and local levels
		Weak implementation of spatial planning policies
		The challenge in effectively executing spatial planning policies
		Absence of policies based on spatial data at both national and provincial levels
		Insufficient national-level spatial policies for rural development
		Inconsistency in regional development and neglecting stakeholders' rights and interests caused by a shortage of experts in the fields of spatial planning
		Underutilization of modern technologies in the fields of spatial planning related to rural areas
		National focus on production processes, employment, and investment instead of comprehensive spatial considerations
		Delays in implementing approved regional development plans
		Absence of suitable models for spatial planning at the provincial level
		Insufficiently established structures and entities for carrying out spatial planning activities
		Lack of robust enforcement mechanisms for implementing spatial planning decisions
		Ambiguity regarding the position of plans and spatial considerations in legal documents guiding decisions and projects
		The weakness of the prepared spatial plans in providing an efficient model for sustainable or integrated regional development
		Limited spatial planning to recommended plans without strong institutional and legal foundations
		Failure to adhere to spatial planning directions by sectoral implementing institutions
		Loss of economic and functional justification of regional development plans due to the prolonged process of responding to inquiries
		Inadequacy of the rules and regulations of spatial planning interventions to local conditions
		Imbalance, exhaustion, and insecurity of physical and infrastructure conditions in rural areas
		Rise in low-population cities influenced by government policies and merging densely populated rural areas into urban centers or converting them into cities
	Spatial centralism at national and regional levels	Confusion in the paradigm and doctrine of national development
		Formation of unbalanced relations between urban and rural areas
		Reduction in planning, management, and accountability abilities at regional and local levels
		Ambiguity in the role and function of rural spaces in national development
		Centralized planning and implementation for the farthest regions of the country
		Imbalance in large-scale investment projects across different regions of the country
		Incoherence and logical connection between infrastructure projects of regional development
		Limited growth of cooperative and people's economic organizations
		Neglect of social and human capital in provinces
		The impact of the economic activities of provinces on national political and economic decisions
		Continuation of imposed and inappropriate policies for aggregating dispersed villages
		Allocation of increasing credits to regional centers and regression of rural and surrounding areas
		Political pressure from provincial and national authorities to implement construction projects without economic, technical and practical justification
		Lack of clarity in the mechanism for coordinating projects and adherence of local and downstream plans to upstream and national plans
		Weakness in the legal status and continuous hierarchy of plans
		Lack of attitude or mechanism to adjust plans based on executive power, capacities and facilities of the country

(continued)

Table 32 (continued)

Category	Sub-category	Concept label
	Ignoring the territorial structure and function	Incompatibility and lack of alignment of investments with conditions of different regions from a spatial planning perspective Dominant governmental attitude to spatial planning instead of a governmental-grassroots attitude The predominance of political attitude on the locating of economic and industrial activities at the border provinces Lack of an economic approach to the development of tourism activities in rural areas Lack of necessary policies regarding the balanced distribution of population Concentration of population and activities in the central regions of provinces A provincial view of the organization of the hierarchy of settlements instead of a macro-regional view of territorial areas Increase of unemployed specialized and academic human resources in the provinces Weakness of participatory foundations of local institutions in the field of sociocultural activities Weak knowledge and geographic ignorance of political, economic, social, and technical actors of the local community Imbalance in the syntax and arrangement of capital, population, activities, and structures in the country's geographical locations due to inefficient political management of space on a territorial scale Lack of sufficient knowledge of regional managers in attracting and allocating internal or external credits, incomes, and investments for various economic, social, and environmental sectors of the provinces Weakness in the production capacity or productivity of the agricultural sector and leaving most of the final value of agricultural products out of rural areas The incompatibility of the national development process with resilient development and the formation of an unbalanced pattern of development and geographical and spatial injustice Absence of formulating a spatial distribution pattern of territorial functions and roles according to the capabilities of different regions of the country The discrepancy between spatial planning and development plans, and the formation of duality in spatial decisions Imbalance in territorial exploitation and unfair allocation of activities and investments in different regions The movement of capital to several provinces and the formation of industrial hubs Centralization of development around the capital and consolidation of the existing situation as a normal picture of the spatial development process The reliance of the allocation system of resources and services on the bargaining power of local institutions with the capital or provincial centers instead of following the spatial planning attitude
	Spreading environmental and human crises	Neglecting the localization of spatial planning according to the different climatic and geographical conditions of the regions Disrupting the balance of the ecosystem of geographical spaces caused by loading economic and social development activities beyond its capacity The increase in open and hidden unemployment and the dramatic stagnation of the business environment following the economic, social, and environmental crises The lack of communication between universities and society and the industry, and consequently, the qualitative inefficiency of the employment of the existing human staff Limitation of surface water and groundwater and their improper exploitation Aggravation of physical problems and functional limitations and neglect of native architecture compatible with the climate Weak vegetation, increasing soil salinity, the occurrence of sudden floods, and lack of attention to watershed operations The increasing and unwanted connection of elements of politics, economics, and engineering and their ruthless interference with the natural infrastructure of the country's ecosystems The increase of the land-use change of agriculture and gardens and the construction in the river boundaries Weakness of educational and promotional plans related to environmental protection Lack of integrated management of water basins and the growing crisis of limited water resources

(continued)

Table 32 (continued)

Category	Sub-category	Concept label
		Ignoring the principles of watershed management and the occurrence of destructive floods Weak treatment of illegal well drilling and waste of water resources Failure to use proper methods to dispose of surface water and waterlogging of roads during rains Pollution of rivers due to the discharge of domestic and industrial sewage Destruction of pastures, and forests due to the creation of power transmission lines and factories Destruction of a large part of pastures and forests by handing them over to farmers and gradually annexing them to agricultural lands The heavy dependence of the soil of agricultural lands on chemical fertilizers and poisons and its degradation over time Failure to deal decisively with poachers of wild animals and migratory birds Leaving waste by tourists in the surroundings of villages and ancient and cultural sites Increasing the release of construction debris and burning of household waste around villages Air, water, and soil pollution due to the dumping of waste around the villages and the lack of use of sanitary waste disposal methods

Table 33 GTM levels in the action or interaction strategies

Category	Sub-category	Concept label
Spatial and Administrative Decentralization	Formulation of the rural development doctrine	Formulating a rural development and progress pattern for homogeneous geographical areas Choosing specific strategies by the sociocultural, economic, and environmental characteristics of the villages Changing the attitude of officials and recognizing the rights of villagers and rural areas in decision-making processes Extracting, documenting, and strengthening technical-indigenous knowledge to design suitable approaches for the development of the country's rural areas An accurate understanding of policy-makers thoughts and perceptions of the village and rural development by experts and creating a mechanism for grounding or adjustment and unity in understanding concepts, issues, and trends Changing the attitude of managers, policy-makers, and planners towards the position of the village in national, regional, and local development Expanding village culture honoring rurality and villagers through mass media and making life in the village a value Coordinating the institutions responsible for rural development with the cultural and media system to introduce villages as a meeting place and platform for the growth and upliftment of the society Respecting the hierarchy of the national development doctrine through formulating theories and perspectives, designing plans, and setting executive policies and legal measures in a package Determining indicators and creating a single network to evaluate and monitor changes in rural development variables Achieving concrete, objective, and better knowledge of rural areas due to their geographical, climatic, natural, ethnic, and cultural diversity Helping villages get out of isolation by facilitating access and creating functions and activities in different areas Increasing transparency in the service description rules of rural development plans to reduce the inconsistency of data, information, and regulations

(continued)

Table 33 (continued)

Category	Sub-category	Concept label
		Increasing transparency in the service description rules of rural development plans to reduce the inconsistency of data, information, and regulations
		Establishing the possibility of directly dismissing village headmen and council members by the residents of the villages if they are not satisfied with the official's performance
		Redesigning the local management system of villages based on the economics of scale principles, regional management, and flexibility
		respecting public interests and relying on the scientific, technical, social, and economic values of the local community
		Decentralization and use of local and regional resources and capabilities in the process of formulation and implementation of rural development policies and plans
		Formulation of rural development policies based on group decision-making and field evidence
		Upgrading the regional management system and equipping it with new information technologies
		Trying to remove the sectoral attitude of the organizational layers of the management and planning institutions of the provinces
	Organizing the management of rural affairs	Creating peace and structural stability in organizations responsible for rural development
		Amending laws, policies and duties of organizations responsible for rural development instead of expanding new administrative structures
		Reducing management levels and organizational positions to save the country's public expenses
		Liquidation or merger of organizations and transfer of some tasks of executive institutions to local management
		Transparency, accountability and legality in the process of reforming institutions and downsizing the government in order to properly address regional and local problems
		Involving regional managers in national decision-making related to rural development planning
		Revision and amendment of internal and extra-sectoral laws affecting rural development
		Creating the necessary platforms to increase the efficiency of individuals, groups and local organizations
		Effective use of the relative and competitive advantages of rural areas
		Coordination and implementation of development activities in private sector, cooperative and government organizations
		Strengthening stability in decision-making institutions and avoiding politicizing their expert structure with management changes
		Duties division of regional institutions to follow up and implement rural development plans
		Providing opportunities for users to participate in the processes of preparation, examination, approval, implementation, and monitoring of the implementation of rural development plans
		Organizing the intra-provincial executive institutions in the framework of their missions, duties, and the extent of their participation in rural development affairs
		The requirement to separate policy-making objectives for each of the institutions responsible for rural development affairs
		Organizing and limiting government activities and tasks to policy-making, planning, monitoring, and evaluation and handing over management to the people and the private sector
	Strengthening the local management system	Elevating the personality of society members through public education, strengthening the sense of belonging to the place and the spirit of responsibility
		The evolution of indigenous knowledge by enhancing the flow of ideas, thoughts, experiences, and information in rural areas and adapting to rapid technological changes
		Strengthening a critical attitude towards development plans and rational decision-making in the field of managing rural areas
		Identifying key roles in the field of local management and using delegation of authority to skilled and competent people
		Creating flexibility in the local management structure and providing public services based on the size of the population, the geographical features, deprivation index, and economic capabilities
		Providing a legal framework for the presence of representatives of various stakeholders in rural council meetings to express opinions in decision-making and present public demands

(continued)

Table 33 (continued)

Category	Sub-category	Concept label
		Analyzing and reflecting the economic, sociocultural, and infrastructural problems of the villages through the establishment of the local planning system
		Approving, monitoring, and evaluating rural development programs through the mobilization of villagers in the process of designing, implementing, and presenting creative and society-oriented solutions
		Establishment of selection, appointment, training, evaluation, punishment, and encouragement systems for regional and local management
		Creating a system for compiling local management performance reports to record the details of activities, complaints, objections, suggestions, and rating villages to the institution's performance
		Obligation to clarify financial aid and income sources of local management institutions in villages
		Establishing the possibility of directly dismissing villager head and council members by the residents of the villages due to not being people satisfied with their performance through the local management performance reporting system
		Revision of government organizations by transferring necessary and sufficient responsibilities and duties to local officials and creating noncentral organizational units to compensate for the backwardness of deprived areas
		Not selecting people with a history of committing crimes or abusing people's power and resources or lacking the necessary knowledge to manage local affairs
		Increasing public motivation to cooperate with the rural management system for transformation and development
		Reducing the government's ownership by focusing on the supervisory role and using the private sector's participation to carry out activities related to regional and local development
		Enhancing the insight, awareness, and ability of people to accept heavy responsibilities and effectively perform regional management duties through the expansion of inclusive education
		providing the training ground for regional and local managers through holding short-term and long-term courses to develop their technical, managerial, and scientific abilities
		Organizing and properly selecting regional and local managers with experience and people-oriented through the implementation of the meritocracy selection system
		Increasing the participation of regional and local managers in the national decision-making process for the proper use of territorial capacities and relative advantages of provinces
		Developing the capacities and capabilities of the regional and local management system for greater interaction and increasing public participation
		Attracting the cooperation and companionship of local society stakeholders in the development, improvement, and management of rural affairs
		Supporting the local management system while performing its guiding and active monitoring role
		The fair and effective division of the role of routing, targeting, and granting the necessary authority to determine the responsibilities of employees between regional and local managers
		Decentralization in the rural management system through the wise distribution of responsibilities and institutionalization of decentralized processes in decision-making
		The institutionalization of voluntary, grassroots, and nongovernmental organizations through the adoption of policies and laws at the regional and local levels
		Legalizing the local management system to strengthen the spirit of hope in the society, legitimacy to the structure, and facilitating the decision-making process in the institutions responsible for rural development
		Legalism, spreading the culture of socio-administrative responsibility, respecting and responding to people, and avoiding personal and tasteful treatment in all activities
		Institutionalization of conscientiousness, social discipline, culture of self-control, trustworthiness, frugality, simplicity, and preservation of treasury in the regional management system
		Creating a rural development organization with a cross-sectoral perspective with the membership of experienced professionals in this field to continuously follow up and quickly organize local affairs and respond to the people
		Improving the criteria and updating the methods of selecting human resources to attract capable, and committed experts and avoid narrow-mindedness and unprofessional and tasteful attitudes

(continued)

Table 33 (continued)

Category	Sub-category	Concept label
		Determining and delegating duties, responsibilities, and approaches governing the division of duties of provincial and local headquarters and executive units with an emphasis on improving the regional management level Creating an agile, appropriate, and logical administrative system structure to achieve the goals of the vision at the regional level Flexibility and administrative-organizational decentralization at the regional level with the approach of increasing the effectiveness, speed, and quality of services creating alignment between the service hierarchy system and the rural management system Analyzing the existing spatial structure, determining the description of services for development plans, and preparing an optimal spatial organization plan to improve people's lives Increasing transparency and awareness of the mutual rights and duties of the people and the administrative system, emphasizing the easy and regular access of the people to correct organizational and service information Regulation of administrative relations based on psychological, cultural, social, economic, health security, and relative well-being of society members Protecting the rights of the people and compensating the damages caused to natural and legal persons as a result of negligence or fault in decisions and actions against the law and regulations in the administrative system Making changes in the regional management structure and emphasizing result-oriented and process governance until the control and improvement of the situation and its continuation Promoting the health and growth of the moral values of the administrative system at the regional level through the reform of legal-administrative processes, the use of cultural facilities, and the effective system of prevention and dealing with violations Saving public expenses with emphasis on decomposing the duties and responsibilities of managers and experts, rationalizing the government size and reducing parallel and unnecessary institutions at the national and regional levels and eliminating unnecessary expenses Organizing, managing, and continuously monitoring the process of regional and local progress and development through creating a network of residential areas, expanding services and activities, and defining their special role and functions Formulating a management method for the interaction and participation of public institutions and local people and synergies as a result of joint action and intervention
The necessity of territorially integrated policy-making	Integration in policy-making, planning and governance	Identifying the roots of problems and strategic issues in formulating rural development plans Developing a strategic and operational plan and understanding the current and desired situation in preparing rural development plans Gradual handover of the responsibilities of preparing, reviewing, approving, implementing, evaluating, and monitoring rural development programs to local institutions Creating a connection between rural development plans and national development plans Aligning regional and local planning and management actions with upstream and downstream documents Creation of coordination and sharing of development literature with operational plans at the regional and local levels Monitoring, implementation, and evaluation of regional strategic documents to identify the socio-economic and environmental consequences of plans on rural areas Adopting measures and implementing appropriate programs to cultivate and distribute income and wealth fairly and reduce the existing income gap between different parts of the local society Developing a system for preparing and implementing construction projects to exploit and organize them optimally and to encourage people to participate in these projects Clarifying the role of regional and local policy-making as the gateway for people to enter into decisions and realize planning and efficient management Creating a statistical and informatics system at the regional and local level to provide timely and reliable information needed for spatial decision-making Benefiting from the capacities of the people and local institutions in obtaining information to formulate effective rural development plans

(continued)

Table 33 (continued)

Category	Sub-category	Concept label
		The need to revise the responsibility for developing rural development plans, especially consulting companies, in terms of service description Requiring organizations responsible for rural development to coordinate with local administration before implementing plans The need to pay attention to problem-oriented and comprehensive planning in formulating rural development plans and linking measures with leading challenges Paying attention to the polygon-based or regional approach and avoiding a point view in the planning system and implementation of rural development plans The necessity of relying on legal bases to carry out local development plans and the alignment of executive institutions at the operational level of villages The necessity of benefiting from the capabilities and capacities of the private sector, and grassroots and nongovernmental organizations in rural development plans The need to formulate plans and operational measures in rural development plans based on priorities and budget line codes Creating a communication network between the stakeholders of the local community and the country's rural development officials for inclusive participation in the stages of planning and implementation until the effectiveness of the plans
	Convergence of rural development policies and sustainable development dimensions	Protecting the natural and cultural resources of villages and using them intelligently Pursuing the goals of integrated development of rural areas within the framework of sustainable development dimensions The necessity of formulating rural development programs in accordance with the current situation and paying attention to the future perspective Revising the pattern and content of rural development plans to increase their efficiency according to the new needs of the villagers Creating an eco-friendly, coherent and aligned rural areas with the protection of natural, historical, and cultural heritage Protection and effective use of cultural, historical, natural and tourism attractions in the direction of sustainable rural development goals Implementation of the regional and local planning and policy system based on the new approach of people's right to settlements, citizenship rights and network interaction method Maintaining public property rights over natural resources and national wealth, along with observing conservation, environmental and intergenerational considerations in their exploitation Promoting and implementing vegetation restoration plans, and sustainability in the protection, exploitation, and management of natural ecosystems and forests Access to social justice, gender equality and the creation of effective, responsive and inclusive institutions at all levels to create peaceful societies and sustainable development at the regional and local levels
	Coherence and integration of agricultural and rural development policies	Creating coherence between agricultural development policies and rural development in novel approaches Establishing a close relationship between the development of nonagricultural economy and agricultural development in rural areas Determining the heading of government credits for rural development programs, separating and monitoring the allocation of credits, and appointing a specific responsible person Establishing the optimal pattern of regional cultivation based on the ecological capabilities of watersheds Targeted support for the production of inputs and agricultural production factors by the capacities and capabilities of the land Establishing a legal ban on changing the use of land suitable for various agricultural activities Protecting and strengthening soil fertility and preventing its destruction in agricultural fields Ensuring food security in rural areas by emphasizing self-sufficiency in the production of basic agricultural products

Table 33 (continued)

Category	Sub-category	Concept label
		Organizing rural spaces and settlements to coordinate and balance the development of agricultural and tourism activities with industry and services Taking advantage of the capacity of villages with the ability to develop agricultural activities and protect agricultural lands and gardens located inside and on the outskirts of villages The need to prevent the transfer of duties and management of village affairs to institutions in charge of civil affairs, housing construction, and road construction instead of rural development Creation and development of agricultural exploitation organizations and systems to strengthen the dimensions of water resources protection based on demand management Joint investment with non-governmental sectors in the field of agricultural development, water resources management, and development of rural areas Designing a system for recording, updating, and sharing comprehensive general and specialized information and a geodatabase for monitoring development actions
	Communicating between spatial elements and rural development policies	Paying attention to the rural space as a platform for the interaction of the environment, activity and people Formulating long-term rural development strategies based on integrated regional development studies Adopting integrated policies of balanced urban and rural growth and development Considering social, economic, physical, environmental, institutional and technical dimensions in rural development policy-making Increasing coherence in rural development policy-making through determining common goals and priorities, coordinating policies and plans, using collaborative methods, and continuous evaluation and monitoring of projects Organizing and guiding rural-urban migration flows based on the strategy of targeted population distribution and activity in the territory Paying attention to the presence and role-playing of people in the decision-making process and providing the context for all-round participation to improve the condition of villages The necessity of connecting the identity and architecture of villages with economic and sociocultural issues Moving from a point view to a polygonal or regional view in rural physical plans Expanding communication corridors to increase the flow of transfers of people, goods, energy and information between places Establishing a monitoring system and regular evaluation of the development and progress indices of rural areas and appointing local people in the pillars of this system The active participation of villagers to fill the gap between spatial elements and rural development policies by explaining the existing realities, the needs of the local community and insisting on the presence of elected people in formulating, implementing and monitoring policies and plans
	Adaptability and resilience against environmental hazards	Formulating and implementing rural development plans compatible with environmental and climate changes Strengthening and developing binding, encouraging, and supportive governance mechanisms to implement the principles of safety, prevention, and dealing with natural disasters Planning to attract the participation of public and private sectors to implement risk reduction plans in rural areas Diversification and development of insurance coverage of movable and immovable assets in the private and public sectors against all types of risks Defining plans to support the livelihood of rural and low-income households and vulnerable groups, compensation for damages, and development of insurance and support subsidies The need to change the local administration's view of securing the physical fabric of villages scientifically Common recognition of risk, local capabilities, and scientific measures for prevention or resilience against natural hazards Identifying, prioritizing, and determining the method of intervention in villages exposed to natural hazards Compliance with the principles of resilience in the design, implementation, and maintenance of rural infrastructure along with continuous monitoring

(continued)

Table 33 (continued)

Category	Sub-category	Concept label
		Retrofitting houses and securing the environment of villages to deal with natural hazards The necessity of zoning rural areas at risk with a multi-hazard approach through the organization of a geographic information system for these areas Establishing a disaster risk management system for rural areas of the country and updating their vulnerability maps against natural hazards Determining and revising regulations and criteria for changing the use of rural lands according to ecological characteristics Increasing the safety factor of infrastructure facilities and rural buildings by considering different criteria of geology, hydrology, and topography in evaluations Identifying the key factors and strategies for managing water resources affecting rural settlements, considering the increasing tensions of limited water resources in the country Prohibition of indiscriminate and unsustainable harvesting of surface water and groundwater resources to reduce the risk of subsidence in villages Providing accurate and timely information to villagers in the field of harvesting water resources, changing the cultivation pattern and livestock grazing to manage the risk of drought Compliance with the principles of comprehensive management of watersheds according to the conditions of climate change in the country Compliance with the principles of comprehensive management of watersheds according to the conditions of climate change in the country through educational plans, community participation, adaptation strategies, and the use of traditional knowledge and indigenous practices Establishment and promotion of intelligent flood management in watersheds and villages at risk Combating desertification, stopping the process of land degradation, and preventing salinity and erosion of soil resources Continuous control and monitoring of polluting sources and destructive factors of natural and environmental resources Protection, restoration, and restoration of unique and valuable natural territories, such as wetlands The necessity of managing the critical foci of creating fine dust inside the country

nomenon, regardless of whether the study pertains to individuals, groups, or collectives. The phenomenon emerges within a specific context or under particular circumstances. The action or interaction categories relate to the acting person and other interactions (Vollstedt & Rezat, 2019). The strategic category of this study includes a focus on integrated sectoral-spatial policy-making, which consists of two categories: Spatial and administrative decentralization and the necessity of territorially integrated policy-making (Table 33).

4.2.6 Consequences

Actions or interactions can have consequences that are not always predictable or intended. These consequences can be fact or hypothetical and may occur in the present or future. The consequences of an action or interaction can also change over time, as it may become part of the causal conditions for another event (Vollstedt & Rezat, 2019). The consequences category of this study includes relative reduction of sectoralism and attention to spatial policies, which consist of three categories: Rural economic development, social welfare expansion, and physical-spatial development of rural areas (Table 34).

4.3 Selective Coding

Selective coding involves the integration of categories and their properties, linking them to the core category to form and develop a comprehensive theory (Atrill & Lindley, 2019). The results obtained from the stages of open coding and axial coding suggest that the weakness of the intellectual and theoretical foundations of rural policy-making as a core category of the study has led to the formation of centralized, sectoral, non-regional, and top-down attitudes in rural policy-making and planning approaches in Iran. Such deficiencies led to the policies continuation that are often fragmented and incoherent and fail to provide an integrated cover from all aspects of rural development. Consequently, several significant socio-economic, environmental-physical, and institutional challenges ensued. This weakness emphasizes the vital importance of the country's planning system's focus on integrated regional and place-based policies through spatial-administrative decentralization, the result of which is the relative reduction of sectoralism, the implementation of spatial policies driving rural development, increasing resilience, and sustainable rural development.

4.4 Overview of Results

The study used a qualitative content analysis method to interpret implicit and explicit meanings. The current research has investigated challenges and resilience approaches for sustainable rural development in Iran. In pursuit of this objective, 866 concept labels were utilized, subsequently organized into six axial categories, 13 categories, and 44 sub-categories. The results of the qualitative data analysis showed that:

- Among the axial category concepts of the weakness of the intellectual and theoretical foundations of rural policy-making, the sub-category predominance of centralized approach in rural planning system with 36 concepts or 4.16%, and the sub-category of instrumental view to rural society within the category ambiguity in the definition and position of the village and rural development with 34 concepts or 3.93%, allocated the highest number of concepts in causal conditions to themselves, depicting the challenges and approach of resilience in rural sustainable development of Iran.
- Among the concepts within the axial category of centralized, sectoral, and top-down policy-making, the sub-category of sectoral and centralized policies, with 32 concepts or 3.70%, as well as the sub-category of changing structures and policies at the same time as changing governments in the category of the continuous changes in rural development policies, with 28 concepts or 3.23%, obtained the highest number of concepts in the main phenomenon

Table 34 GTM levels in the consequences

Category	Sub-category	Concept label
Rural economic development	Rural entrepreneurship development	Changing the status of the village from a settlement to a center of activity for the development of new rural businesses
		The movement of rural development policy from the predominance of the physical-infrastructure approach towards the socio-economic development approach
		Creating constellation, ecosystem-driven, and partnership-based governance for entrepreneurship development in rural areas
		The formation of an entrepreneurial system of marketing and selling rural products
		Creating institutional mechanisms at the local level based on entrepreneurial ideas and field realities to encourage the private and cooperative sector to invest
		Development of leading and entrepreneurial companies on a provincial and national scale for poverty alleviation in rural areas
		Strengthening infrastructure for the establishment and development of rural businesses
		Designing a system of rules and regulations for the formation of rural business
		Prioritizing the interests of rural producers to target the payment of subsidies
		Strengthening value chains and developing and making rural production competitive
		Providing a benchmarking approach to use experiences, compare activities according to local conditions and improve their status
		Using benchmarking techniques to improve productivity and increase the efficiency of economic activities in villages
		Economic and managerial bargaining to attract resources and financial and instrumental support to create and develop business in local areas and increase the sale of rural products
		Applying appropriate tariff policies to support rural productions and businesses
		Supporting small and family businesses through the establishment of a rural financing system
		Preventing the migration of rural youth and teenagers to cities by providing low-interest facilities and employment development
		Creating funds and local institutions to partner with the private and cooperative sector, develop export-oriented enterprises and create employment in rural areas
		Facilitating the process of obtaining a license to invest in rural employment-generating activities
		Supporting knowledge-based companies and rural economic or social entrepreneurs
		Creating a technological knowledge system and empowering the society to produce industrial, and technologically scalable products in the villages
	Diversification of the rural economy	Defining the specific role and function for villages with relative and competitive advantage
		Development of technical-skill training in border and rural areas
		Creating the groundwork for the development of businesses that suit local requirements
		Branding, marketing and market development of rural products
		Clustering and prioritizing economic activities in rural areas
		Development of businesses based on agricultural activities and rural tourism
		Income generation of villagers by improving technology and increasing productivity of rural activities
		Developing employment opportunities through the development of small industries and rural workshops and completing the production process of agricultural products
		Paying serious attention to the tourism sector by introducing natural attractions and strengthening the cultural-historical heritage of villages
		Creating a context for private sector investment in nonagricultural sectors with priority on tourism, handicrafts, conversion and complementary
		Branding of counties and provinces in terms of various natural, social, economic and tourism capitals, potentials and opportunities
		Establishing sales centers for agricultural products and handicrafts of villages on the way leading to cities and border markets
		Handing over the administration and maintenance of border trade facilities to the villagers
		Creating stalls and shops in markets and assigning a quota of shops and stalls at the international border to the villagers of these areas
		Holding festivals and art festivals based on native rituals and ceremonies
		Increasing services and amenities, accommodation, and hospitality at the level of tourism target villages
		Creating brochures and replicas to advertise the tourism potential of villages in urban centers to guide tourists to those areas
		Educating villagers through local institutions to cooperate in management and acquaint them with the potential of rural areas

	Development of rural production value chains	Paying attention to the relative advantages of the rural agricultural sector through increasing exports, foreign exchange income, and increasing the food security factor Generating wealth, developing production value chains and large-scale business through the implementation of efficient entrepreneurship development plans in rural areas Integrating agricultural lands to increase production capacity, adjusting the cultivation pattern to meet the needs of the villagers Creating a base for investment of added value from rural activities in the village itself Currency exchange of agricultural products and the possibility of entering the global markets for their export The transition from punitive and mandated pricing to fair, guaranteed, and supportive pricing of agricultural products and creating an obligation in the government to purchase products above the market price to increase consumer and producer protection and reduce product imports Management of chains of economic activities and formation of business clusters Strengthening the performance and value addition of rural border markets and spilling its benefits to border villages Improving the performance and value of rural border markets and spilling its benefits to border villages Creation, restoration, and investment of added value of rural productions Improving the processes towards completing the chains and networks of rural activities Preventing excessive import of products and its effect on rural production
Social welfare expansion	Empowering local societies	The growth of ideas and the continuation of efforts, pursuits, and actions of villagers to improve their society Environmental, institutional, and individual empowerment through the development of capacities and capabilities within rural areas Strengthening the social security system by allocating revenues from handing over government activities and enterprises to the private and cooperative sectors Creation of micro-financing funds and participation of cooperatives as strategies to support rural businesses Financing development projects for empowerment, institution building, education, employment, implementation of basic projects, and protection of natural resources and environment Bringing the capacity of the traditional credit system closer to the new economic rules at the rural household level to legalize rural business development Development of production cooperatives in villages by encouraging the private sector to invest in activities Designing a specific plan for skill training of rural youth to create an efficient productive social management system Investing in comparative advantages to create economic transformation and employment and permanence of the active stratum of the agricultural sector in villages Supporting small industries in villages and continuously monitoring the way financial credits are spent to track the effectiveness of facilities Granting damages through agricultural insurance to increase the risk-taking power of farmers to grow crops in adverse climatic conditions and lack of water Increasing public trust in local management decisions by inviting interested and influential rural households to meetings to encourage villagers to participate in inclusive Establishing technical and professional training centers in central villages to acquaint residents especially young people with specialized skills and accept new agricultural technologies Issuing official documents of land ownership to villagers and helping them get loans from banks

(continued)

Table 34 (continued)

Category	Sub-category	Concept label
	Strengthening social capital	Creating rural and people-oriented development policies through strengthening the policy discourse based on the networked local community
		Development of participation and growth of constructive ideas of self-help and cooperation of rural communities and strengthening of self-confidence in them
		The necessity of defining social organization and organized participation with the approach of local planning and implementation and participatory evaluation
		Crystallization of social capital elements in the form of councils, grassroots and nongovernmental organizations, voluntary groups, cooperative companies, development funds, and family funds
		Increasing the motivation of villagers to pursue their individual and social rights by strengthening the spirit of trust and mutual convergence of officials and local people
		Promoting a culture of positive interaction and proper resolution of disputes to facilitate the decision-making process among local groups
		Better organization of social institutions through participation in consensus and decision-making at the local level
		Changing the passive attitude of local institutions towards decision-making and their access to development benefits by encouraging villagers to participate
		Protection and sustainable management of rural natural resources through strengthening the sense of collective responsibility, active participation in projects and measures to control crises and environmental pollution, using social networks to inform and emphasize cultural commonalities
		Increasing the benefits of local people from implementation projects and investments by creating transparency in practice, accountability, and responsibility
		Return of material and spiritual capital to villages through creating sustainable economic and welfare incentives for living in villages
		The prosperity of rural tourism according to relative and competitive advantages to strengthen social, cultural, and political convergences
		Strengthening national, regional, local, religious, and ethnic unifying symbols and events in the geographical regions of the country
		Preserving and revitalizing the identity, rituals, traditions, and special ceremonies of different ethnic groups for the development of community-oriented tourism
		Promoting mental well-being and social vitality in rural areas through participation and synergy among rural community members
		Creating a platform for the creation of cultural, social, and economic organizations to accelerate the implementation of rural development plans
	Improving the quality of life	Development of the attitude of modernization and reconstruction of rural areas based on new intellectual, educational, technological, and production approaches
		Organizing rural development programs based on scientific ideas, individual-collective ideas, and future Insights
		Strengthening the spirit of teamwork by encouraging the local society to consensus, cooperate, and solidarity to solve rural issues
		Emphasis on the anti-poverty spirit and promotion of citizenship rights to reduce rural deprivation
		Holding training courses for technical-skill empowerment of villagers
		Adapting the content of public education in rural schools to the needs of the rural community
		Development of sociocultural infrastructure and increasing the penetration rate of the Internet, mobile phones and social media
		Basing and developing virtual education methods to improve the quantity and quality of general and skill training in rural areas
		Investing in infrastructure to improve access to basic facilities such as drinking water and electricity supply
		Allocation of credit required for the construction, improvement, maintenance, and asphalting of rural roads
		Improving the physical conditions of villages through the use of resistant materials in the construction of residential units
		Establishment of bank branches to facilitate access of villagers to financial and credit services
		Equitable access to health services through the design of the rural health and treatment network
		Identifying, preserving, revitalizing, and introducing cultural heritage, handicrafts, and traditional arts of rural areas
		Identifying and promoting tourism capacities in villages and benefiting the local community from the benefits of its development
		Creating incentives to improve the employment and livelihood situation and increase the life expectancy of the villagers
		Providing retirement income for villagers and ensuring employment security for young people and rural farmers
		Determining a strategic road map to improve the food security of villagers

Physical-spatial development of rural areas	The balance between spaces and activities	Passing from cross-sectional approach to regional approach in rural development plans Passing from sectoral planning to integrated regional planning and sustainable rural development Decentralization of the planning and decision-making system by establishing the hierarchical distribution of services in territorial spaces Integration of economic, social, ecological, and physical dimensions in territorial spaces Creating the necessary foundation for reducing the development gap between regions and creating a relative balance in regional development Creating geographical balances in territorial spaces and leakage of economic and social benefits in the border and small villages to remove deprivation Eliminating spatial discrimination in the allocation of public resources to rural areas Balanced settlement of the population in territorial spaces with the priority of ruralization through reverse urban-rural migration Organizing the settlement network to improve the productivity of territorial spaces based on environmental considerations and reducing existing problems and challenges Appropriate use of the capacity of the special economic zones of the provinces with the aim of balanced regional growth, export of early-yielding agricultural products in susceptible counties, and development of border and rural areas Organizing complexes, towns, and industrial areas by the requirements and capacities of territorial spaces with the priority of employment of local people in those places Spatial organization and determining the functions of rural tourism destinations in the country Equitable distribution of services based on demographic considerations and regional capacities Strengthening the infrastructure and superstructures of regional development for deprived and less privileged provinces and counties Adopting the necessary measures to preserve and improve the environment and natural resources of villages according to their value and ability to access them Organization of processes and procedures for the transformation and annexation of rural centers to the city with the aim of permanence of the productive population in the villages and preservation and productivity of cultivable lands Coordinated development of border villages through the strengthening of infrastructures and the use of comparative advantages to develop tourism, educational and commercial capacities, and cross-border interaction Targeted strengthening of factors of progress and development in border rural areas concerning defense, security, cultural, social, economic, and environmental considerations Development of infrastructural, economic, and social activities, energy, and other vital resources according to the capacity, rights, and biological ethics in deprived and rural areas of the country Organization of service levels and population settlement according to the principles of balanced regional development, natural disaster management, and water and soil resource management Observing the ecological capability and environmental tolerance capacity in the establishment of development activities in rural areas, especially tourism and agricultural activities Coordinated development of the villages near the coasts based on the ability of the regions, geographical location, infrastructure, marine and energy resources Diversifying the tourist attractions of the host societies according to the expectations of tourists Presenting and implementing the model of optimal cultivation based on ecological capability Making the regional distribution systems of goods and services efficient by using the mechanism of clarification and information and reducing unnecessary and inefficient intermediaries

(continued)

Table 34 (continued)

Category	Sub-category	Concept label
	Organization of relations among settlements	Paying attention to the correlation of places and operationalizing spatial planning on a local scale
		Coherence in the organization of territorial spaces by creating a multi-level, multi-centered, and multi-functional network
		Paying attention to the network approach for synergism and developing the rural capacities
		Strengthening convergence, integration, and regional links based on the level of development of different regions
		Determining local and regional poles and development clusters
		Paying more attention to the rural community during decision-making and policy-making
		Creating and strengthening spatial attitude through rural constellations
		Preparing and implementing the Rural Constellations Sustainable Development Plan (RCSD) as an agile and strategic plan for rural and regional development
		Development of cooperative, Synergistic, and interconnected relations between villages and cities based on rural constellations
		Rational support for small and medium-sized economic enterprises with a focus on rural tourism
		Establishing multilateral relations between rural constellations and centers of entrepreneurship to formulate rural development plans
		Creating suitable conditions for the natural connection of villages with other settlements
		Modifying and balancing the spatial structure and inter-regional flows to facilitate, complete, improve, and move the transport fleet, goods, and resources
		Establishing a dynamic and people-oriented institutional structure to monitor and solve rural development issues in sync with the existing developments in urban areas
		Maintaining a dynamic and constructive relationship between the urban and rural areas base on the institutional, physical-environmental, socio-economic, and considerations of the settlements
		Strengthening rural-urban linkages within the framework of local networks
		Division of work among territorial spaces instead of the attitude of each village as a product
		Setting separate development priorities for each province and county based on their territorial capabilities and opportunities
		Establishing a service hierarchy system between settlements based on the RCSD plans
		Distribution of services to the villages through the socio-economic capacities of the small towns

Table 35 Frequency and percent of the concepts in GTM levels of the study

Paradigm	Axial categories	Categories	Sub-categories	Concepts	
				Frequency	Percentage
Causal conditions	The weakness of the intellectual and theoretical foundations of rural policy-making	Contradictions among dominant approaches to rural development	Predominance of centralized approach in rural planning system	36	4.16
			One-dimensional insight into rural development plans	18	2.08
			Expediency and planning of rural affairs with urban glasses	16	1.85
			Inconsistency of rural development plans	14	1.62
		Ambiguity in the definition and position of the village and rural development	Absence of rural development policies	12	1.39
			Instrumental view to rural society	34	3.93
			Lack of indigenous definition of village, and rural development	18	2.08
			Lack of recognition of the fundamental problem of villages	16	1.85
Main phenomenon	Centralized, sectoral, and top-down policy-making	The dominance of government-oriented policies in the rural planning system	Sectoral and centralized policies	32	3.70
			Multiplicity and variety of institutions responsible for rural development	16	1.85
			Rent distribution of capital and natural resources	28	3.23
		The continuous changes in rural development policies	Absence of a planning and policy-making institution for rural areas	20	2.31
			Adopting reactive measures and policies	18	2.08
			Changing structures and policies at the same time as changing governments	28	3.23
Intervening conditions	The incoherent in the process of compiling and implementing rural development plans	Compilation of incoherent and disunited plans	Disregarding the spatial attitude in the preparation of plans	12	1.39
			Preparing plans without regard to the facts	22	2.54
			Inconsistency of institutions in preparing plans	20	2.31
		Ignoring the implementation and results of plans	Paradox between processes and results of planning	26	3.00
			Failure to evaluate of plans results	22	2.54
Contextual conditions	The territorial and institutional incoherence of rural development policies	Weakness in institutionalization and local networking	Absence of participatory evaluation system	12	1.39
			Marginal position of rural grassroots structures	14	1.62
			Inadequacies of local management	26	3.00
			Non-return of the added value of rural production	18	2.08
		The territorial elements imbalance	Lack of spatial attitude in rural development	22	2.54
			Spatial centralism at national and regional levels	16	1.85
			Ignoring the territorial structure and function	20	2.31
			Spreading environmental and human crises	22	2.54

Action or interaction strategies	A focus on integrated sectoral-spatial policy-making	Spatial and administrative decentralization	Formulation of the rural development doctrine	20	2.31
			Organizing the management of rural affairs	16	1.85
			Strengthening the local management system	18	2.08
			Strengthening the regional and local management system	30	3.46
		The necessity of territorially integrated policy-making	Integration in policy-making, planning, and governance	20	2.31
			Convergence of rural development policies and sustainable development dimensions	10	1.15
			Coherence and integration of agricultural and rural development policies	14	1.62
			Communicating between spatial elements and rural development policies	12	1.39
			Adaptability and resilience against environmental hazards	24	2.77
Consequences	Relative reduction of sectoralism and attention to spatial policies	Rural economic development	Rural entrepreneurship development	20	2.31
			Diversification of the rural economy	18	2.08
			Development of rural production value chains	12	1.39
		Social welfare expansion	Empowering local societies	14	1.62
			Strengthening social capital	16	1.85
			Improving the quality of life	18	2.08
		Physical-spatial development of rural areas	The balance between spaces and activities	26	3.00
			Organization of relations among settlements	20	2.31

describing the resilience challenges for sustainable rural development in the country.

- Among the concepts of the axial category the incoherent in the process of compiling and implementing rural development plans, the sub-category of preparing plans without regard to the facts from the category of the compilation of incoherent and disunited plans with the number of 22 concepts or 2.54% and also the sub-category of the paradox between processes and results of planning from the category of ignoring the implementation and results of plans with 26 concepts or 3.00%, obtained the highest number of concepts in intervening conditions describing resilience challenges for sustainable rural development in Iran.
- Among the concepts in the axial category the territorial and institutional incoherence of rural development policies, the sub-category of inadequacies of local management from the category of weakness in institutionalization and local networking with the number of 26 concepts or 3.00%, as well as two sub-categories of lack of spatial attitude in rural development and spreading environmental and human crises from the territorial elements imbalance category with an equal share of 22 concepts or 2.54%, the largest number of concepts in contextual conditions describing resilience challenges for sustainable rural development of the country.
- Among the concepts in the axial category a focus on integrated sectoral-spatial policy-making, the sub-category of strengthening the regional and local management system from the spatial and administrative decentralization category with the number of 30 concepts or 3.46%, also the sub-category of adaptability and resilience against environmental hazards from the category the necessity of territorially integrated policy-making, with 24 concepts or 2.77%, obtained the highest number of concepts in action or interaction strategies related to resilience for sustainable rural development of Iran.
- Among the concepts in the axial category relative reduction of sectoralism and attention to spatial policies, the sub-category of rural entrepreneurship development from the category of rural economic development with the number of 20 concepts or 2.31%, the sub-category of improving the quality of life from the category of social welfare expansion with the number of 18 concepts or 2.08%, as well as the sub-category of the balance between spaces and activities from the physical-spatial development of rural areas category with 26 concepts or 3.00%, obtained the highest number of concepts related to the consequences of resilience approaches for sustainable rural development in the country (Table 35).

5 Conclusion

Iran's rural areas face various challenges, including water scarcity, climate change impacts, economic fluctuations, limited access to services, migrations, and abandonment. The ability of rural communities to maintain their socio-economic, and environmental systems while withstanding and recovering from challenges is known as rural resilience. The results of this study demonstrate that the centralized, sectoral, non-regional, and top-down policy-making system functions as the main phenomenon in the current pattern of rural development, which is a consequence of the weakness of intellectual and theoretical foundations in rural policy-making in Iran. Moreover, contextual conditions such as "the territorial and institutional incoherence of rural development policies" and intervening conditions of "the incoherent process of compiling and implementing rural development plans" exacerbate this type of policy-making. To achieve rural development and resilience, it is suggested that integrated sectoral-spatial policy-making should be the focus, and spatial policies should receive more attention with a relative reduction of sectoralism. Also, to achieve rural sustainable development, it is recommended to implement a spatial planning and policy system based on a network and hybrid approach to rural development. This system should cater to the unique needs of rural societies and be in harmony with their conditions. A network and hybrid approach will improve the current rural development model. This can be accomplished by following the successful examples set by member countries of organizations such as the Economic Cooperation Organization (ECO), the Shanghai Cooperation Organization (SCO), and the Cooperation Organization Islamic Organization (OIC). In addition, it is important to continue formulating and implementing new rural development models that have been successful in recent years, such as the Rural Constellations Sustainable Development Plan (RCSD), the Economic Development and Rural Employment Plan (EDRE), the Underprivileged Foundation Development and Progress Plan (UFDP), the Comprehensive Model for the Progress of Deprived and Rural Areas with the Kowsari development approach of the Barakat Foundation (Sahab, Asman, Aftab, and Mahtab), and the Roostatish projects (economic, social, and socio-economic). These new patterns emphasize investment in places rather than sectors and improving socio-economic and environmental-physical status. To move beyond the current situation, a strategy of territorial good governance and integrated spatial policy-making should be prioritized and formulated as a research strategy. The outcome of this strategy should focus on the principles of spatial planning and sustainable rural development, encompassing socio-economic, institutional, physical-environmental, and spatial dimensions. The study's findings provide valuable insights for policy-makers, planners, and researchers on how to enhance the resilience and sustainability of rural communities in Iran.

References

Afrakhteh, H. (2019). *Rural geography of Iran* (4th ed.). Samt Publishing.

Afshar, M. T. Z., Sheikhi, S., & Gheidarlou, K. (2023). *Pathology of policy making in Iran: with an emphasis on the social issue*. Islamic Parliament Research Center.

Afsharipour, A., Barghi, H., & Ghanbari, Y. (2021). Appropriate policy-making for rural regions management in Iran. *Land Use Policy, 109*, 105669.

Ahmadi, M. A., & Mottaghi, S. L. (2022). Pathology and monitoring of rural development policies in development plans: Lessons for developing rural development executive policies in the seventh development plan. *Journal of Management and Development Process, 35*(3), 129–164.

Ahmadian, R., & Mohammadi, H. (2012). *Rural histology of the country: General criteria for the formation of rural physical elements* (2nd ed.). Islamic Revolution Housing Foundation.

Ahmadinejad, F., Kamali, Y., & Sheikhzadeh, S. (2021). Challenges of development policies implementation in Iran: A thematic analysis. *Journal of Management and Development Process, 34*(2), 29–62.

Alizadeh, H., & Amanpour, S. (2023). Explaining the barriers to the realization of strategic spatial planning in Iran. *Journal of Town and Country Planning, 15*(2), 365–379.

Amiraslani, F., & Caiserman, A. (2018). Multi-stakeholder and multi-level interventions to tackle climate change and land degradation: The case of Iran. *Journal of Sustainability, 10*(6), 2000.

Anabastani, A. A., Ziarati, E., & Fathi, E. (2024, October). *Sustainable strategies for population stability in rural areas of the country*. Natural Disasters Research Institute (NDRI).

Anabestani, A. A., & Javanshiri, M. (2014). *Physical planning of rural settlements in Iran* (1st ed.). Tehran.

Andersen, H. K. (2023). Running causation aground. *The Monist, 106*(3), 255–269.

Asakereh, H., Masoodian, S. A., & Tarkarani, F. (2024). Identification of Iran's precipitation regimes. *Theoretical and Applied Climatology, 155*(1), 273–288.

Ashtari, R., Darvishi, M., Bakhshandeh, G., Hemati, M., Darvishib, M., & Grounded, A. (2022). A grounded theory study: Representing a new model to exploring transnational capabilities for the steel pipe manufacturing companies attaining competitive advantages. *Petroleum Business Review, 6*(2), 48–62.

Atrill, P., & Lindley, L. (2019). *Issues in accounting and finance* (1st ed.). Taylor and Francis.

Azizpour, F., & Shamsi, R. (2014). The role of environmental factors in the spatial organization of rural settlements, Case study: Small village Lavasan. *Quarterly of Geographical Data, 23*(89), 106–112.

Azizpour, F., Varmazyari, H., Garakani, A. H., Neshat, S. N., Tahmasebi, A., & Heydarian, A. (2022, November). *Rural development pathology in Iran with emphasis on development and advancements projects in rural systems*. Natural Disasters Research Institute (NDRI).

Babaeian, I. (2023, November). *International day against climate change: Iran's climate future image and presentation of plans for resilience against it*. University of Sistan and Baluchestan.

Baburin, V. I., Danshin, A. I., & Katrovskij, A. P. (2017). Transformation of natural-economic systems of northern Iran. *Lomonosov Geography Journal, 1*(2), 49–58.

Badri, A., Rezvani, M. R., & Khodadadi, P. (2019). Designing a pattern for challenges of spatial policy-making formulation in Iran's rural regions. *Journal of Majlis and Rahbord, 26*(99), 5–34.

Badri, A., Rezvani, M. R., & Khodadadi, P. (2020). Analysis of factors and processes affecting the formulation of spatial policy-making of physical development in Iran's rural areas. *Journal of Physical Social Planning, 7*(1), 25–43.

Badri, S. A., Kazemi, N., Khodadadi, P., & Mohammadnejad, A. (2021). Why rural development policies have not contributed to rural development in Iran. *Journal of Rural Society, 30*(2–3), 84–100.

Barakat Foundation. (2021). *A comprehensive model for the progress of deprived and rural areas: Kowsari approach* (1st ed.). The Novel Idea Publications.

Bongaarts, J. (2009). Human population growth and the demographic transition. *Philosophical Transactions of the Royal Society B: Biological Sciences, 364*(1532), 2985–2990.

Bryant, A. (2020). The grounded theory method. In P. Leavy (Ed.), *The oxford handbook of qualitative research* (2nd ed., pp. 167–199). Oxford University Press.

Bryant, A., & Charmaz, K. (2016). *The SAGE handbook of grounded theory* (1st ed.). SAGE Publications.

Center for the Islamic-Iranian Model of Progress: CIIMP. (2018). *Islamic-Iranian model of progress document (2018–2065)*. Secretariat of Expediency Discernment Council.

Cheng, C., Jinlong, G., & Hui, C. (2023). A literature review of spatial distribution and function of rural settlements and its research prospects: From urbanization to urban-rural integration in China. *Journal of Geographical Research, 42*(6), 1480–1491.

Cupples, J., Palomino-Schalscha, M., & Prieto, M. (2018). *The routledge handbook of Latin American development* (1st ed.). Taylor and Francis.

Cutter, S. L., & Derakhshan, S. (2020). Temporal and spatial change in disaster resilience in US counties, 2010–2015. *Environmental Hazards, 19*(1), 10–29.

Czulda, R. (2022). Iran's water security: An emerging challenge. *Journal of Middle East Policy, 29*(2), 113–123.

Dadashpour, H., & Alidadi, M. (2020). *Integrated local development planning: A theoretical and operational framework for Iran*. Islamic Revolution Housing Foundation.

Damonte, A. (2021). Modeling configurational explanations. *Italian Political Science Review/Rivista Italiana Di Scienza Politica, 51*(2), 182–197.

Darand, M., & Zandkarimi, F. (2019). Identification of atmospheric boundary layer height and trends over Iran using high-resolution ECMWF reanalysis dataset. *Theoretical and Applied Climatology, 137*(1–2), 1457–1465.

Darvishi, H., Akbari, M., & Ehterami, T. (2019). The identification of effective factors on deficiencies in regional development policymaking system in Iran: An interpretative-structural modeling. *Journal of Majlis and Rahbord, 25*(96), 237–264.

Deputy for Rural Development and Deprived Areas: DRDDA. (2017). *Rural development document (2017–2021)* (1st ed.). Legal Deputy of the Presidency.

Ebrahimi, R. (2016). *Critical analysis and pathology of arrangement plans (Spatial Planning) in Iran*. Center for the Islamic-Iranian Model of Progress.

Ebrahimzadeh, E., Eskandari, M., & Kashefidust, D. (2021). *Local communities planning and empowerment* (1st ed.). University of Sistan and Baluchestan Pulisihing.

Ebrahimzadeh, I., & Mousavi, M. (2022). *Principles and fundamentals of territorial spatial arrangement* (2nd ed.). Samt Publishing.

Eftekhari, A. R. R., Purtaheri, M., & Badri, A. (2015). *Strategies and policies for the physical development of rural settlements: Emphasizing global and Iranian experiences* (2nd ed.). Housing Foundation of Islamic Revolution Pulisihing.

Eftekhari, A. R. R., & Sadeghloo, T. (2019). *Resilience of local communities against environmental hazards* (2nd ed.). Tarbiat Modares University Pulisihing.

Eftekhari, A. R. R., & Sarban, V. H. (2021). *Rural empowerment: Fundamental principles, approaches, theories, and experiences* (2nd ed.). Tarbiat Modares University Pulisihing.

El-Khoury, G., Veerassamy, L., Wanco, E. M., Zebardast, E., & Kipour, H. (2023, October). *Combatting Inequality for a resilient future*. Natural Disasters Research Institute (NDRI).

Esmailpour, T., Nazarian-Mofid, M. T., Sheikhi, S., Nodeh-Farahani, I., Abdolhosseinzadeh, M., Karimi-Fard, H., Naseri-Jazeh, M., & Abdulmalik, M. (2023). *An overview of democratic governance*. Islamic Parliament Research Center.

Expediency Discernment Council: EDC. (2005). *Twenty-Year Vision Document (2005–2025)*. Secretariat of Expediency Discernment Council.

Faraji, A., Nazari, A. R., & Babaei, H. R. (2018). *An introduction to the regional management model in Iran* (1st ed.). Jahade Daneshgahi Organization Pulisihing.

Farajirad, K., Kazemian, G., & Eftekhari, A. R. (2013). Pathology of regional development policies in Iran: An institutional approach. *Journal of Management and Development Process, 26*(2), 27–58.

Farajirad, K., & Kazemian, G. R. (2012). *Local and regional development from an institutional perspective* (1st ed.). Jahade Daneshgahi Organization Pulisihing.

Fathi, E., & Javaid, N. M. (2018). *Economic-social and demographic characteristics by gender based on the information of the general population and housing census 2016*. Research Institute of Statistics, Iran Statistics Center.

Fazeli, A. (2011). An introduction to the role of strategic planning in national vision research. case study: Perspective 1404 of the Islamic Republic of Iran. *Journal of Majlis and Rahbord, 18*(67), 253–287.

Fedai, M. S., & Khalili, M. (2011). *Agricultural policy* (2nd ed.). Payam Noor University Pulisihing.

Fekete, A., Asadzadeh, A., Ghafory-Ashtiany, M., Amini-Hosseini, K., Hetkämper, C., Moghadas, M., Ostadtaghizadeh, A., Rohr, A., & Kötter, T. (2020). Pathways for advancing integrative disaster risk and resilience management in Iran: Needs, challenges and opportunities. *International Journal of Disaster Risk Reduction, 49*, 101635.

Firoznia, Q., & Eftekhari, A. R. R. (2004). *The place of rural area in the national development process from experts point of view* (1st ed.). Rural Development Institute of Iran Pulisihing.

Firoznia, Q., Eftekhari, A. R. R., & Badri, A. (2014). *Prospects and planning management system for the physical development of rural regions in Iran* (2nd ed.). Housing Foundation of Islamic Revolution Pulisihing.

Fleet, D., & Reeves, A. (2023). Adapting Strauss and Corbin's grounded theory in clinical case study research in the field of counselling. *Counselling and Psychotherapy Research, 23*(4), 874–881.

Geraeinejad, G. R., & Vafaei, E. (2022). *Root cause analysis of planning system challenges for use in formulating an appropriate planning system*. Center for Development Research and Foresight.

Ghadermarzi, H. (2015). Employment strategic planning in rural area of javanrood. *Space Economy and Rural Development, 4*(11), 109–130.

Ghadiri-Masoum, M., Faraji-Sabokbar, H. A., & Cheraghi, M. (2019). *Population geography of Iran*. University of Tehran Publishing.

Ghaedi, M. R., & Golshani, A. R. (2016). Content analysis method: From quantity-orientation to quality-orientation. *Psychological Models and Methods, 7*. University of Tehran Publishing.

Ghaffaryfard, M. (2019). Review the process of development policy and regional balance during development plans in Iran. *Strategic Studies of Public Policy, 9*(23), 57–82.

Ghasemi, A., Ghezelsofla, M. T., & Mehni, Y. M. (2020). Good urban governance as the referential of urban policy: analysis of the content of the fifth and sixth development plans the Islamic Republic of Iran. *Journal of Majlis and Rahbord, 27*(101), 195–221.

Ghasemi, M., & Zamani, R. (2010). *Pathology of development plans in Iran*. Islamic Parliament Research Center.

Ghavam-Maleki, H. R. (2018). International approaches and policy and strategic barriers to rural development in Iran. *Journal of Political Strategy, 1*(3), 119–148.

Gholipour, R., & Ahangar, E. G. (2011). *Public policy making process in Iran*. Islamic Parliament Research Center.

Ghorbani, M., & Afarakhte, H. (2023, June). *Local governance in rural planning system development*. Natural Disasters Research Institute (NDRI).

Haberman, D., & Bennett, E. M. (2019). Ecosystem service bundles in global hinterlands. *Environmental Research Letters, 14*(8), 084005.

Haqshanas, E., Khoini, A., Neshat, S. N., & Falahati, F. (2023, July). *Intervention strategies in vulnerable rural areas to natural disasters*. Natural Disasters Research Institute (NDRI).

Hejazi, S. J., Sharifi, A., & Arvin, M. (2022). Assessment of social vulnerability in areas exposed to multiple hazards: A case study of the Khuzestan Province, Iran. *International Journal of Disaster Risk Reduction, 78*, 103127.

Helbling, M., & Meierrieks, D. (2021). How climate change leads to emigration: Conditional and long-run effects. *Review of Development Economics, 25*(4), 2323–2349.

Heshmati, M., Gheitouri, M., Sheikhvaisi, M., Arabkhedri, M., & Hosini, M. (2017). Combating the forest mortality crises in Zagros regions, Iran through adaptive approaches solutions. *Journal of Geography and Environmental Hazards, 6*(3), 125–141.

Hoseini, J., Shahbazi Ghiyashi, M., Amiri, M., Soleimani Pourlak, F., & Banitaba, M. (2023). *Strategic analysis of the economic and social situation of the country on the threshold of the seventh national development plan*. Islamic Parliament Research Center.

Hosseini-Abri, H. (2004). *An introduction to the rural geography of Iran* (2nd ed.). Isfahan University Publishing.

Hosseini, S. B., Faizi, M., Norouzian-Maleki, S., & Karimi Azari, A. R. (2015). Impact evaluation of rural development plans for renovating and retrofitting of rural settlements: Case Study: Rural Districts of Tafresh in Iran. *Environmental Earth Sciences, 73*(7), 3033–3042.

Inanloo, H. (2021). *Pathology of the country's planning documents*. Center for the Islamic-Iranian Model of Progress.

Iran Statistical Center. (2016). *General census of population and housing*. Iran Statistics Center Publication.

Islamic Consultative Assembly: ICA. (1989). *First five-year development plan (1989–1993)*. Islamic Parliament Research Center.

Islamic Consultative Assembly: ICA. (1995). *Second five-year development plan (1995–1999)*. Islamic Parliament Research Center.

Islamic Consultative Assembly: ICA. (2017). *Sixth five-year development plan (2017–2021)*. Islamic Parliament Research Center.

Islamic Consultative Assembly: ICA. (2000). *Third five-year development plan (2000–2004)*. Islamic Parliament Research Center.

Islamic Consultative Assembly: ICA. (2005). *Fourth five-year development plan (2005–2009)*. Islamic Parliament Research Center.

Islamic Consultative Assembly: ICA. (2011). *Fifth five-year development plan (2011–2015)*. Islamic Parliament Research Center.

Islamic Parliament Research Center. (2018). *Law of the 6th five-year economic, social and cultural development plan of the Islamic Republic of Iran*. Islamic Parliament Research Center.

Issawi, C. P. (2010). In Y. Ajand (Ed.), *The economic history of Iran from 1800 to 1914* (3rd ed.). Gostareh Publishing.

Jabari, H., Rezvani, M. R., Varmazyari, H., Azizpour, F., & Neshat, S. N. (2022, November). *Situation of rural development in the seventh five-year development plan*. Natural Disasters Research Institute (NDRI).

Jalalian, H., Khalili, A., & Pourjohari, A. H. (2023, October). *A review of successful global experiences in rural planning systems*. Natural Disasters Research Institute (NDRI).

Javadpoor, M., Sharifi, A., & Roosta, M. (2021). An adaptation of the Baseline Resilience Indicators for Communities (BRIC) for assessing resilience of Iranian provinces. *International Journal of Disaster Risk Reduction, 66*, 102609.

Kalantari, K., Varmazyari, H., & Askari Nodoushan, A. (2018). The future perspective of rural population and challenges of sustainable development management in Iran. *Journal of Village and Development, 16*(2), 1–23.

Kamali, Y. (2023). Identifying and ranking the challenges of evaluating development policies in Iran (Five-Year development plans). *Journal of Majlis and Rahbord, 30*(115), 91–125.

Kamali, Y., Sheikhzadeh-Jooshani, S., & Hosein Askari, F. (2022). Challenges of formulation of development policies in Iran (Five-year development programs). *Journal of Program and Development Research, 2*(4), 95–126.

Karimi, H., & Dastvareh, F. (2023). *Analysis of the policy-making process in optimal crisis management*. Islamic Parliament Research Center.

Karimmian, Z., Mohammadi, M., Ghazinoori, S. S., & Kermani, M. M. Z. (2019). Classification of governance features through policy networks using meta-synthesis method. *Journal of Public Administration, 11*(3), 377–402.

Kashef, M., & Abootalebi, M. (2020). Improving and agiliting of the development planning process in Iran. *Journal of Program and Development Research, 3*(2), 151–178.

Kaviani, H., & Ghofrani, F. (2021). *Administrative system and development management in Iran: Evaluation of preparedness and bureaucratic quality on the threshold of the seventh five-year development plan*. Islamic Parliament Research Center.

Kaviani, H., & Ghofrani, F. (2023). *Designing a framework for managing structural and organizational changes in executive agencies*. Islamic Parliament Research Center.

Kazemian, G. R., Ghorbanizadeh, V., Vaezi, R., & Shahmohammadi, M. (2019). A pattern of role and structure of local governance in Iran's management system. *Journal of Public Administration, 11*(2), 179–202.

Khorasani, M. A., & Babaei, M. (2022a). *Exploration and pathology of the rural-to-urban transformation process in the territory management system in Iran*. Islamic Parliament Research Center.

Khorasani, M. A., & Babaei, M. (2022b). *Scrutiny and pathology of the process of transformation from village to city in the land management system in Iran*. Islamic Parliament Research Center.

Kristensen, I., Dubois, A., & Teräs, J. (2018). In I. Kristensen, A. Dubois, & J. Teräs (Eds.), *Strategic approaches to regional development* (1st ed.). Routledge.

Lambton, A. K. (2013). In Y. Ajand (Ed.), *Continuity and change in medieval Persia : aspects of administrative, economic, and social history, 11th–14th century* (4th ed.). Ney Publishing.

Lee, J. (2001). *A grounded theory: Integration and internalization in ERP adoption and use* (1st ed.). University of Nebraska Publishing.

Leimgruber, W., & Chang, C. D. (2019). *Rural areas between regional needs and global challenges: Transformation in rural space* (1st ed.). Springer Nature.

Li, Y. (2023). A systematic review of rural resilience. *China Agricultural Economic Review, 15*(1), 66–77.

Madani, K., AghaKouchak, A., & Mirchi, A. (2016). Iran's socio-economic drought: Challenges of a water-bankrupt nation. *Iranian Studies, 49*(6), 997–1016.

Mafi-Gholami, D., Zenner, E. K., & Jaafari, A. (2022). Mapping recent (1997–2017) and future (2030) county-level social vulnerability to socio-economic conditions and natural hazards throughout Iran. *Journal of Cleaner Production, 355*, 131841.

Maham, M. (2023). *Spatial planning, social changes, and sustainable progress: Emphasizing the seventh five-year development plan*. Islamic Parliament Research Center.

Mahdavi, M. (2022). *An introduction to rural geography in Iran: The study of rural geographical circumstances* (Vol. 1, 16th ed.). Samt Publishing.

Maurel, M., & Tuccio, M. (2016). Climate instability, urbanisation and international migration. *The Journal of Development Studies, 52*(5), 735–752. https://doi.org/10.1080/00220388.2015.1121240

Mcneely, J. A. (2020). Today's protected areas: supporting a more sustainable future for humanity. *Integrative Zoology, 15*(6), 603–616.

Mianabadi, A., Davary, K., Mianabadi, H., Kolahi, M., & Mostert, E. (2023). Toward the development of a conceptual framework for the complex interaction between environmental changes and rural-urban migration. *Frontiers in Water, 5*, 1142307.

Mir-Bagheri, V., Beraderan, M., Emami, J., Hosseini, M., & Asadi, H. (2016). *Investigating the situation of villages in Iran* (1st ed.). Islamic Parliament Research Center.

Mirbagheri, M., & Ismail, A. (2022). *Conceptology of participatory governance*. Islamic Parliament Research Center.

Moghimi, M., Pourezzat, A. A., Danaeefard, H., & Ahmadi, H. (2017). Describing elements of good governance in policy making of budgeting system in Iran. *Strategic Studies of Public Policy, 6*(21), 31–52.

Mohammadi, H. R., & Kamali, Y. (2013). Analysis on the institutionalization of sustainable development policy in Iran: Examining barriers and proposing a model. *Journal of Management and Development Process, 26*(2), 161–186.

Mohammadi, S., & Khanian, M. (2021). Staying in crisis: Choice or coercion a review of the reasons of rural-to-urban migrations due to environmental changes in Iranian villages. *Space and Culture, 26*(4), 598–617.

Mohebati, H., Yaghoubi, N.-M., Mohammadi, M., & Washan, M.-M. (2020). Designing a phased model of crisis management with the empowerment approach of local communities using meta-synthesis approach. *Journal of Public Administration, 12*(2), 175–203.

Momeni, A. (2014). Qualitative content analysis in research tradition: nature, stages and validity of the results. *Quarterly of Educational Measurement, 4*(14), 187–222.

Monfared, S. S., Varmazyari, H., & Garakani, A. H. (2022, April). *A review of the experience of rural development and advancements projects in rural systems.* Natural Disasters Research Institute (NDRI).

Moridsadat, P., Mirghasemi, A., & Khedri, J. (2023, December). *Identification of challenges and development solutions for rural entrepreneurship in Iran.* Natural Disasters Research Institute (NDRI).

Morris, J. (2019). In R. Rezazadeh (Ed.), *History of urban form: Before the industrial revolutions* (15th ed.). Iran University of Science and Technology Publishing.

Mosalanejhad, A. (2016). *Policy-Making in the Power Structure of Iran* (1st ed.). University of Tehran Publishing.

Motevaseli, M., Momeni, F., Ranjbar, M., & Lajevardi, R. (2017). An Opportunity to Deal with the Concept of Planning in Iran. *Journal of Majlis and Rahbord, 24*(92), 255–278.

Moulai, N., & Alizadeh, A. H. (2019). *An overview of the pathology of rural management in Iran* (1st ed.). Islamic Azad University, Central Tehran Branch Publishing.

Namdar, R., & Sadighi, H. (2013). Investigation of major challenges of rural development in Iran utilizing delphi technique. *Journal of Agricultural Science and Technology, 15*(3), 445–455.

Nasr, A. R., & Kohi, A. (2022). *Analysis of global megatrends and future trends in Iran for the seventh development plan.* Islamic Parliament Research Center.

National Crisis Management Organization: NCMO. (2021). *National crisis management strategy document (2021–2032).* Natural Disasters Research Institute.

Nazmfar, H., Hamidi, S.-M., & Akbari, M. (2018). The inequality of development in the provinces in order to achieve sustainable development. *Journal of Majlis and Rahbord, 25*(94), 297–340.

Nilsen, T., Grillitsch, M., & Hauge, A. (2023). Varieties of periphery and local agency in regional development. *Regional Studies, 57*(4), 749–762.

Noghani, H., & Rahnama, M. (2020). Is income inequality reflected in consumption inequality in Iran? *Middle East Development Journal, 12*(2), 284–303.

Noorali, H., & Ahmadi, S. A. (2023). Iran's new geopolitics: Heartland of the world's corridors. *GeoJournal, 88*(2), 1889–1904.

Noori, H., & Ahmadi, M. A. (2013). The rural development strategy in Iran, An analysis of viewpoints. *Journal of Research and Rural Planning, 2*(2), 93–123.

Palmer, S. (2014). *Suicide: Strategies and interventions for reduction and prevention* (1st ed.). Taylor and Francis.

Pascariu, G. C., Ţigănaşu, R., Kourtit, K., & Nijkamp, P. (2023). *Resilience and regional development: New roadmaps.* Edward Elgar Publishing.

Pourtaheri, M., Eftekhari, A. R., & Badri, A. (2014). *Development strategies and policies for the physical development of rural settlements (with emphasis on global and Iranian experiences)* (2nd ed.). Housing Foundation of Islamic Revolution.

Quintá, F. X. A., González, R. C. L., & Arce, X. C. M. (2018). Advanced internet services: New opportunities for rural territorial development. *Anales de Geografía de La Universidad Complutense, 38*(2), 271–287.

Rahimi, E., & Dong, P. (2022). What are the main human pressures affecting Iran's protected areas? *Journal of Environmental Studies and Sciences, 12*(4), 682–691.

Rahnavard, F. (2022). Mega-policies of the development plan: A case study of Iran. *Journal of Management and Development Process, 35*(3), 3–56.

Rajabizadeh, Y., Ayyoubzadeh, S. A., & Ghomeshi, M. (2019). Flood survey of Khuzestan province in 97-98 and providing solutions for its control and management in the future. *Iranian Journal of Ecohydrology, 6*(4), 1069–1084.

Rajabpour, H. (2018). *Program planning system pathology: A review of planning system developments in Iran* (2nd ed.). Islamic Parliament Research Center.

Ren, K., & Wu, T. (2024). Analyzing the evolution of a rural construction community in China from the perspective of cultural landscape. *Journal of Buildings, 14*(1), 97.

Rezaei, M., & Heidari, M. (2023). *Necessity of national action to enhance climate change adaptation.* Islamic Parliament Research Center.

Rezapour, M., Moshfeghi, V., & Bahrami, Z. (2019). *Residential system and spatial distribution of population: Volume II, Investigating and analyzing the existing situation and extracting the spatial structure and organization.* Center for Development Research and Foresight.

Rezvani, M. R. (2015). *An introduction to rural development planning in Iran* (6th ed.). Ghoomes Publishing.

Rezvanifar, M. M. (2021). *Sustainable development policy in Iran: Necessities and strategies* (1st ed.). Majd Publishing.

Riahi, V., & Ghadermarzi, H. (2018). *Crisis management in rural areas of Iran* (1st ed.). Iranian Geographical Association Publisher.

Riahi, V., & Masoumi, M. (2020). Challenges of rural management and its role in the underdevelopment of suburban villages, Case study: Shiraz. *Journal of Urban Pripheral Development, 2*(3), 65–84.

Rizvani, A. A. (2018). *Interrelationships between the city and the countryside with emphasis on Iran* (13th ed.). Payame Noor University Publishing.

Roostatish. (2022). *Three-year report on the performance of roostatish*. Roostatish Publisher.

Sadeghi, S. H., & Hazbavi, Z. (2022). Land degradation in Iran. In R. Li, T. L. Napier, S. A. El-Swaify, M. Sabir, & E. Rienzi (Eds.), *Global degradation of soil and water resources: Regional assessment and strategies* (1st ed., pp. 287–314). Springer Nature.

Saidi, A. (2009). *Zoning of the country's villages* (2nd ed.). Shahidi Publications.

Saidi, A., & Hosseini, S. (2009). *The foundation of locating and establishing new villages*. Islamic Revolution Housing Foundation.

Saleh, I., Salehnia, N., Mirbagheri, S. S., Akbarpoor, H., & Bastani, M. (2022). Investigating the role of livelihoods diversity in resilience and welfare level of the rural community in the face of climate change, with emphasis on drought. *Journal of Water and Sustainable Development, 9*(1), 75–84.

Sameni, A., & Zebardast, E. (2023). Feasible pathology of spatial planning in Iran (Case study: The fourth round of spatial planning studies). *Journal of Majlis and Rahbord, 30*(115), 127–162.

Saraie, H. (2022). *Demography: Basics and principles* (9th ed.). Samt Publishing.

Schultz, T. W. (2007). In M. S. Nouri-Naini (Ed.), *Transforming traditional agriculture* (2nd ed.). Ney Publishing.

Secretariat of the Supreme Council of Spatial Planning: SSCSP. (2021). *National Spatial Planning Document (2021–2046)*. Center for Development Research and Foresight.

Seifoori, B., & Taghavi, R. (2018). Pathology of Iran's development programs before and after the revolution, and presentation of pivotal strategies. *Journal of Humanities Horizons, 11*(22), 1–24.

Shahi, A. (2019). Drought: The achilles heel of the Islamic Republic of Iran. *Asian Affairs, 50*(1), 18–39.

Shahraki, S. Z. (2013). Explanation on conversion of rural areas into town as a challenge in national scale and formation of Khamshahr. *Journal of Rural Research, 4*(3), 535–557.

Shakoori, A. (2019). Rural development in Iran: A survey of policies and outcomes. *Journal of Developing Societies, 35*(3), 346–366.

Shakuri, A. (2014). *Agricultural Development Policies in Iran* (7th ed.). Samt Publishing.

Sheikhi, S. (2023). *Examining the challenges of democratic governance in Iran and proposing solutions to enhance the public's role in governance*. Islamic Parliament Research Center.

Sheikhzadegan, A., & Meier, A. (2017). *Beyond the Islamic revolution: Perceptions of modernity and tradition in Iran before and after 1979*. De Gruyter Pulisher.

Shiri, T., & Azimi, N. (2012). A comparative study of qualitative content analysis and hermeneutics. *Journal of Sociological Studies, 5*(15), 79–99.

Soltanieh, A. (2023). Threats for peace and security: Asia vs. West: Developments oblige the new iranian government to give a higher priority to its neighbouring countries. In J. C. Peña & J. M. Ramírez (Eds.), *Threats to peace and international security: Asia versus West* (1st ed., pp. 168–204). Springer.

Steiner, A., & Markantoni, M. (2014). Unpacking community resilience through capacity for change. *Community Development Journal, 49*(3), 407–425.

Tahbaz, M. (2016). Environmental challenges in today's Iran. *Iranian Studies, 49*(6), 943–961.

Taslimi, M. S., & Alipour, M. (2019). Talent research approaches in Islamic Republic Iran's talent policies. *Journal of Public Administration, 11*(1), 123–150.

Tavallaei, R., Nezafati, N., & Ahmadi, M. M. (2019). The role of gamification in knowledge sharing using grounded theory and MAXQDA qualitative modeling. *Journal of Human and Information Interaction, 6*(1), 1–16.

Terosky, A. L., Baker, V. L., & Sun, J. C. (2023). *A practical guide to teaching research methods in education: Lesson plans and advice from faculty* (1st ed.). Routledge Publisher.

The Islamic Republic News Agency. (2022). *The development and progress plan of underprivileged foundation is being implemented in 40 deprived areas of the country*. Irna.

Torkian, M., Daneshfard, K., Valikhani, M., & pilevari, N. (2022). Presenting a model of effective factors on the implementation of Iran's crisis management policies with an interpretive structural approach. *Journal of Housing and Rural Environment, 41*(178), 35–48.

Tsang, K. K., Liu, D., & Hong, Y. (2019). *Challenges and opportunities in qualitative research: Sharing young scholars' experiences*. Springer.

Van Der Cruysse, D. (2017). In H. Akhwan-Tagvi (Ed.), *Journey to Persia: Jean Chardin's portrait of a seventeenth-century empire* (3rd ed.). Farzan Roz Publishing.

Varmazyari, H. (2014). *Exploration of challenges in the monitoring and evaluation system of agricultural and rural development in Iran*. Islamic Parliament Research Center.

Varmazyari, H. (2021). *Expert opinion on the comprehensive plan for the development and construction of rural areas in the country*. Islamic Parliament Research Center.

Varmazyari, H., Babaei, M., & Doosti-Sabzi, B. (2022). *Key axes of reforming the rural local management system in Iran*. Islamic Parliament Research Center.

Varmazyari, H., Doosti-Sabzi, B., & Babaei, M. (2023a). *Analysis and evaluation of sustainable development plans of rural systems*. Islamic Parliament Research Center.

Varmazyari, H., Doosti-Sabzi, B., & Babaei, M. (2023b). *Leadership in knowledge-based entrepreneurship*

development in rural and nomad regions. Islamic Parliament Research Center.
Varmazyari, H., & Kalantari, K. (2021). *Expert opinion on the comprehensive plan for the development and rural construction of the country*. Islamic Parliament Research Center.
Varmazyari, H., Kalantari, K., Lavaei Adaryani, R., & Samadi, M. (2018). A grounded theory approach to developing a model for challenges of policy-making and planning of rural development in Iran. *Journal of Village and Development, 21*(2), 97–123.
Vollstedt, M., & Rezat, S. (2019). An introduction to grounded theory with a special focus on axial coding and the coding paradigm. In G. Kaiser & N. Presmeg (Eds.), *Compendium for early career researchers in mathematics education* (1st ed., pp. 81–100). Springer.
Welivita, I., Willcock, S., Lewis, A., Bundhoo, D., Brewer, T., Cooper, S., Lynch, K., Mekala, S., Mishra, P. P., Venkatesh, K., Vicario, D. R., & Hutchings, P. (2021). Evidence of similarities in ecosystem service flow across the rural-urban spectrum. *Journal of Land, 10*(4), 430.
Yang, M., Jiao, M., & Zhang, J. (2022). Spatio-temporal analysis and influencing factors of rural resilience from the perspective of sustainable rural development. *International Journal of Environmental Research and Public Health, 19*(19), 12294.
Ziari, K., & Mohammadi, A. (2015). A comparative analysis of regional development management features in Iran before and after the islamic revolution (From 1967 to 2015). *Journal of Majlis and Rahbord, 84*(22), 73–98.
Ziari, K., & Mohammadi, A. R. (2018). *Regional Development Management with Emphasis on Iran* (2nd ed.). University of Tehran Publishing.

Embeddedness of Environmental Politics in Tropical Rural Resilience: Analytical Innovation and Application in Brazil

Tian Yan Baxter

Abstract

This chapter theoretically advances the analytical framework of how environmental politics affects tropical rural resilience, which is further applied in the scenario of Brazil. The empirical analysis of Brazil unequivocally underpins that the development of environmental politics effectively strengthens the compatible interrelationship between ecology and the social economy. Through the discourse of ecological modernisation as an advanced environmental political theory and an approach to understand environmental politics, the chapter critically analyses the embeddedness of the elements of the theory in Brazil's tropical rural resilience, which include modern science and technology, ecological economisation or economy ecologisation, ecological politicisation and societal power. The tropical rural resilience is explained through the elements defined as tropical capacity building, tropical embeddedness, agroecosystems, tropical-social integrated systems and tropical-economic integrated systems. This research found out that Brazil is competent in employing science and technology, ecological politicisation and societal power in improving tropical rural resilience. There is also an established integrated system between tropical rural resilience and ecological economisation, which is, however, incipient and competing with policies developed in traditional economic sectors such as agriculture. Overall, the process of mainstreaming environmental politics in Brazil helps the rural areas to gain resilience. The simultaneous development of green policy and development policy is together contributing to Brazil's rural resilience.

Keywords

Environmental politics · Tropical rural resilience · Innovation · Brazil

1 Introduction

Tropical developing countries with agricultural competitiveness and abundant valuable ecological resources face increasing pressure to feed the world's growing population and to simultaneously halt agriculture-embodied habitat conversion. This can be best exemplified by Brazil occupying six biologically diverse biomes in its territory, including the well-known lung of the planet—the Amazon rainforests. This fertile land provides Brazil a salient position in leading global environmental governance and climate

T. Y. Baxter (✉)
China Representative of Aprosoja Mato Grosso, Cuiabá, Brazil

Y. Li (ed.), *Global Perspectives on Building Resilience for Sustainable Rural Development*, Sustainable Development Goals Series, https://doi.org/10.1007/978-981-96-6149-7_6

change negotiation. It further enables the country to become a great agricultural power. Brazil has been the world's third largest exporter, distributing 34 agricultural and food commodities to 222 countries and territories that are produced by over 15 million people in its rural establishments, accounting for the 15% of national labour force (Valdes, 2022). However, the competitive relationship over forested lands and farmlands, especially the implications of logging and land grabbing on biodiversity loss and climate change in tropical zones, has attracted the attention of environmental activists and politicians in those countries and in countries around the world. The disappearance and degradation of tropical biomes is responsible for 12–20% of global emissions (Watson & Schalatek, 2020).

With the concern of agriculture-embodied deforestation and habitat conversion becoming part of the worldwide mainstream discourse, they have been enshrined in the political rhetoric and behaviour of tropical countries such as Brazil, which, to a large extent, rationalises their administrative-regulatory architecture and normalises agri-business operation and structure. What remains loosely delineated is the causal relationship between contemporary Brazil's environmental politics, as a prime constructivist result of global-local spatial interacting dynamics, and on-the-ground rural agrarian resilience. This chapter, correspondingly, explores whether Brazil's farming communities are gaining resilience or vulnerability while being embedded in the centre of environmental politics internally and universally. Precisely, it tests the hypothesis that environmental politicisation is an effective approach to help improve balanced development in rural Brazil as a protagonist of global capitalism and neoliberalism, while also reversing global warming and protecting vibrant biodiversity.

Following this logic, this chapter firstly streamlines the theoretical approach to resilience, creating and conceptualising the notion of tropical rural resilience. It highlights the crucial role undertaken by the biodiverse biomes in the tropics, tying in well with both local and national socio-economic prosperity and those newly extended worldwide security threats, including climate change and food security. To grasp the politicisation of the biomes in Brazil's rural development, this article adopts the discourse of ecological modernisation as an advanced theory and pragmatic exercise in environmental politics. It channels the analytical framework by innovatively attempting to integrate the theoretical framework of ecological modernisation into traditionally well-established dimensions of rural resilience comprising the ecological, social and economic aspects. Dissecting their interrelationship establishes the political account of bottom-level practical agrarian resilience.

The chapter proceeds sequentially with empirically manifesting the application and influence of environmental politics on Brazil's countryside. It adopts the research method of policy document analysis and analyses the linkage between politics and resilience through the established analytical framework. This helps interpret what agricultural practitioners in rural Brazil are supposed to do to better adapt to the prime environmental regimes and institutions, thereby transitioning to a new lifestyle that is politically sensitive and capable of engaging with orthodox environmental discourse. In addition to fostering politically resilient rural communities, mobilising environmental politics to steer social, productive and environmental rationales destined for the comprehensive, systematic resilience of tropical communities is also investigated in the empirical study.

2 Rural Farming Resilience in Global Tropics

Resilience is a malleable concept and penetrates into social science from a naturalistic biosphere (Barnes, 1997; Jessop, 2000). Resilience in rural localities can be portrayed as the self-organising or decision-making capacity of collective agencies for ecologically pertinent systems, mainly ecologically economic and social-ecological interdependent systems, that are simultaneously continued and transformed under permanent exogenous disturbances while reaching livelihood-climate-food sustainability (Walker & Salt, 2006; Jones & Diepart, 2015; Folke et al., 2016; Baird et al., 2014; Berkes & Ross, 2013). For farming-centred resilience, the capacity reveals the adaptability of farmers to surrounding economic, political, technological and social evolution over generations and even centuries of farming activity (Darnhofer et al., 2016). Learning is the identified key mechanism enhancing the resilience capacity or adaptability (Cutter et al., 2008), which is described as reconstituting and reconnecting the relationship of rural communities with their surrounding ecosystems (Walter, 2002).

Streamlining resilience approaches in the tropical scenario innovatively covers the loophole of placelessness and significantly narrows down the geographical boundaries (Shrivastava & Kennelly, 2013); as quite often the attention is given to globalisation, i.e. global countryside, macro-scale corridors and global exchange, in the binary reconstitutive process of local and global, i.e. global organisation (Johannisson & Olaison, 2007; Woods, 2007; Wilson, 2012). The tropics represent one group of localities that cannot be taken for granted as belonging to the ordinary global countryside. They possess distinct geographical characteristics, abundant resources, human-ecosystem connection and sense of affiliation, e.g. emotional empathy for biome conservation (Pilgrim et al., 2008; Vlasov et al., 2018). They embrace similar social-institutional establishments, agriculture-embodied multifunctionality, need to green economic transition and political sensitivity (Wilson, 2007).

Discussing tropical farming resilience under globalisation reflects a growing awareness of their interplay. The resilience approach acknowledges that exposure to climate change, natural hazards and ecological destruction forms part of stringent, urgent globalised destabilising events (Wilson, 2012). This is especially true for agro-ecosystems in sensitive tropical agriculture, which need to be resilient to warm and humid conditions of tropical latitudes, in comparison to temperate climate agriculture (Vlasov et al., 2018; Zhou et al., 2022). The disappearance of tropical biomes can cause erosion, temperature change and precipitation. It is indispensable for industries and farmers to grasp the uncertainties and constraints of their involved agroecosystems and determine business operations responsively (Darnhofer et al., 2016). In addition, as emissions released from logging in tropical countries are more significant than those originating from energy consumption, tropical countries are major emitters (OC, 2021). International communities expect the tropics to shoulder complex responsibilities, including eliminating global hunger, protecting Indigenous inhabitants and preserving native vegetation (Walker & Salt, 2006; Munoz & Cohen, 2017; Korhonen & Seager, 2008). It is tricky but crucial to reverse the fundamental antagonism between meeting the soaring demand for food security and reducing domestic ecological footprints (Darnhofer et al., 2016).

Therefore, in addition to treating ecology as one sort of perpetual, unanticipated crises, ecologically rooted resilience recognises the instrumental value of natural resources (Shepherd & Patzelt, 2011). The biological richness can be capitalised and provides the tropics yield, nutritional quality, industrial competitiveness and economic growth in a number of sectors such as agriculture, forestry, minerals and biotechnology. Accordingly, various decentralised modular innovations in economic, institutional and biological spheres are structured upon (Walker & Salt, 2006; Brown, 2016). Appreciating the value of ecosystems helps establish an intimate and eco-susceptible affection, care and wisdom of surrounding native habitats (Whiteman & Cooper, 2011; Morris & Kirwan, 2011; Biggs et al., 2010). This is essential so the externalities can be tailored in a manner that projects benign ecological consequences, which reflects the level of rootedness of ecological resilience (Vlasov et al., 2018; Walker & Salt, 2006).

Building ecologically oriented harmony is heavily intertwined with social and economic spheres. The embeddedness of tropical communities in global-local interconnection is getting ever closer due to the local harnessing of external provisions that help socialise and economise resource-based resilience (Folke, 2006; Walker & Salt, 2006; Folke et al., 2016; Vlasov et al., 2018).

Socially, internalising ecological resilience is to integrate climate, biodiversity and environmental considerations in social forces, contracts, human generations and capabilities (Brown & Westaway, 2011; O'Brien et al., 2009). The social-ecological foundation mediates, amplifies and rationalises the institutionalisation of ecology ranging from campaigning the green political agenda (Emirbayer & Sheller, 1999) to mobilising economic agents to be social carriers (Bell, 1998). Correspondingly, local perceptions, values, power relations, organisations (Folke et al., 2010) and lifestyles (Machado, 2019) are adjusted and empowered on the path to system persistence and evolution (Folke, 2006). Moreover, all-sector and all-scale participation is supposed to be envisioned when socialising ecological resilience (Brown, 2016), ranging from deliberative democracy (Goldstein et al., 2015) to techno-economy transition (Machado, 2019) right through to farmer cooperatives as a cohesive social relation (Darnhofer et al., 2016).

In purely economic means, the tropical resources prominently contribute to material-reliant industries and soft commodities, which form one shape of economy underneath the prevalent capitalist architecture. This, in turn, vertically radiates and constitutes development activities at national and lower grades in tropical countries (Gray, 2002). They experienced the so-called contradictory transitions (Pieterse, 1995), as sketched by the neoliberal re-division of responsibility, that is, setting up material structures in the South whereas deindustrialisation in the North (Baylis et al., 2008; Woods, 2007), although some Southern states relocalise resilience to deagararianisation and urbanisation (Bryceson, 2002). The salience of natural resources in economy and livelihood methods means that tropical countries can be sensitive to macroeconomic environment. Tropical capitalist rural economy is characterised by agro-industrial organisation, production intensification, agricultural internationalisation and marketisation (Pilgrim et al., 2008). Globalisation also catalyses trans-local, horizontal engagement between economically vibrant entities (Rofe, 2009) via infrastructure, capital mobility and information technology (Vlasov et al., 2018). This is exemplified by transnational land acquisition (Arezecki et al., 2012), cross-border logistics, business expansion, immigration (Woods, 2005, 2007), information exchange and technology transfer (Machado, 2019).

To converge horizontal and vertical resourcefulness, embeddedness is the process. It refers to pooling, attuning and customising exogenous transferable resources, skills, knowledge, measurements and activities to fit in the local agenda; while surpassing traditional knowledge, geography and biophysics to invigorate knowledge collision and localise innovations as well as entrepreneurship for resilience (Korhonen & Seager, 2008; Sonnino & Griggs-Trevarthen, 2013; Marsden et al., 1993). This is especially important for cross-fertilised infusion between technical breakthroughs and tropical heritage. The former can be either ecology-friendly or -detrimental, such as effective chainsaws for land clear and agrochemicals. The latter is filled with specialised, historically rooted features encompassing tropical planting culture, tropical climate and soil, bio-diversified resources, aboriginal inhabitants, colonial legacy and immigration (Machado, 2019). The appearance of the embedded process is pragmatic, which may involve exogenous forces persuading, negotiating, manipulating and applying power politics (Woods, 2007). As a result, external interests are blended in and a multifunctionality paradigm of rural resilience is fostered (Wilson, 2007; Pretty, 1995).

Globalisation also gives rise to disturbances undermining multi-dimensional and -functional resilience (Wilson, 2012). Pursuing ecological rationale can be counted as a source of conflict of interest when communities are reluctant to embed divergent norms and practices, such as frugal sourcing, pollution-free and regenerative agriculture (Seyfang & Smith, 2007; Mair & Marti, 2006; Light & Dana, 2013), into existing industry structures and inherent culture, e.g. diet habits, defensive localism and consumerism (Wilson, 2012; Vlasov et al., 2018; Seyfang & Smith, 2007; McNaughton & Gray, 2017). The fusion of space-based tropical knowledge and external diffusion also receives critiques (Jones & Diepart, 2015), as the former deviates from the externalities for intensifying economic embeddedness (Barthel et al., 2015; Woods, 2005). Indeed, the resilience narrative is often in tandem with anti-capitalism (Smith, 1990) and possessing behaviour-transformative property (Staggenborg & Ogrodnik, 2015). Besides, rural communities tend to be more sensitive to economic gains and adversities, e.g. price fluctuation, rather than social forces (Aggarwal, 2006). Over-reliance on the outside aggravates vulnerability to economic and political influence (Sonnino & Griggs-Trevarthen, 2013) and may accelerate the exploitation of internal social and natural capital (Korsgaard et al., 2016).

Through streamlining the theoretical framework of the resilience approach with a focus on system building and persistence in tropical environments, where the balanced development of agriculture and ecosystems is a priority concern affecting national- and global-level food security, climate change and biodiversity conservation, this section identifies five characteristics that describe rural resilience under the tropical scenario (Table 1). It demonstrates that the very pivotal elements of resilience, including capacity and embeddedness, need to be integrated in the tropical environment, underscoring the value and impacts of tropical climate, soil and resources upon rural agriculture and development. The characteristic of tropical agroecosystems further recognises the importance of ecosystems upon tropical agriculture. The final two features follow the traditional well-established social and economic spheres in the resilience approach.

Table 1 Characteristics of tropical rural resilience

Features	Explanations
Tropical resilience capacity	Self-organising capacity, decision-making and adaptability of the tropical farming communities in the changing ecosystems and ecologically embodied economic and social systems
Tropical embeddedness	The fusion of exogenous resources and endogenous tropical agriculture and agroecosystems to reach multifunctionality, e.g. food security and biodiversity conservation
Tropical agroecosystems	Understanding, capitalising and appreciating the value of ecosystems to community development, industrial competitiveness and global food security and environmental regimes
Tropical-social integrated systems	Internalising ecosystems in social relationships, values and capacity across sectors and scales
Tropical-economic integrated systems	Industrial organisation and innovation, materials and bio-technology innovation, production intensification, international marketisation and global distribution

(Source: Authors' own)

3 Loophole in Tropical Resilience and Brazil in Environmental Politics

Mainstream approaches interpreting ecologically economic and social-ecological systematic resilience leavened by global externalities touch upon environmental and technocratic fields, e.g. disaster management (Storr et al., 2015; Berkes & Ross, 2013; Brown, 2016), natural resource governance and rural development policy (Vlasov et al., 2018; Korsgaard et al., 2016; McNaughton & Gray, 2017; Morrison et al., 2017). In comparative terms, the effects on the daily lives of farming communities receive limited attention from political scientists in the denotations of politics, international relations and international political economy, especially on the environmental derivative of politics. That is, the political angle, i.e. the politicisation of ecology, may also threaten the self-organised ecosystem-based resilience in rural space, in addition to conventional economic, social and ecological aspects of provisions and disruptions (Marsden, 1996).

This level of politics does not merely refer to the fundamental tones set by international regimes such as *the Paris Agreement*; it can also be fragmented, less cohesive than global, outside the UN box. It can include the contemporary inter-state bi-/multi-lateralism politicising the environment through flexibly switching conflictual or collaborative standpoints and bringing environment-induced political amendments into nation states (Young, 1994). It is important to conceive a political space in resilience study within which regulatory innovation, institutional learning (Giddens, 2009), administrative agency restructuring, political cultures, political programmes (Jänicke, 1992), political networks, policy discourses and identity construction (Buttel, 2000; Cohen, 1997), in relation to the environment discipline at international and (sub-) national layers, stand a chance of maintaining, strengthening or hindering the status quo of resilience.

As to tropical countries, politics can be an extremely paramount external factor affecting rural continuity and transformation. By owning the biomes and the need to develop, tropical countries are at the frontier, sensitive to the politics of both international trade and the environment. The latter refers to global climate change regimes, biodiversity negotiations and deforestation-pertinent regulation from consumer markets. Tropical countries are all playing a pivotal role in maintaining carbon sink, atmospheric circulation, Indigenous human culture and flora and fauna species. Logging and habitat conversion release more carbon than they absorb, making the net-zero target unattainable (Dummett & Calder, 2023). Environmental politics can certainly exert great impacts on on-the-ground human-ecosystem interrelationships, bearing in mind that a considerable proportion of human resources is established in tropical rural areas with resource-dependent livelihoods. It is arguable that politics fosters balanced ecologically social-economic viable resilience systems. Therefore, as the resilience approach is largely constituted by the fluid ecological, social and economic interconnectedness of the global and local, this paper argues that global-local environmental political engagement should be at the centre of resilience study, in particular to the tropical rural resilience.

Although the position of government is transitioning its form of governance from intervention to neoliberalism and decentralisation (Bulkeley & Newell, 2010; Gouldson & Bebbington, 2007), environmental-political force is a fairly new intricate topic with intrinsically a political character (O'Brien & Williams, 2016). Hajer (1995) argues that environmental-political affinity relieves the governmental impulse for political greening and reform. Environmentally friendly political discourse tends to shore up the political and economic behaviour of a government and ultimately legitimate economic sanctions, market enlargement and/or investment in capital-intensive technologies (Hajer, 1995). Laferriere (1994) regards environmental problems as a discourse framed and internationalised by institutions, social movements and the non-territorial public-good nature of the environment, which generates either intra-state conflicts or coalitions.

With the emergence of both global neoliberal marketisation and international environmental institutions, Brazil's global-local interconnection has made its identity building inconsistent and wavered between a global environmental leader and an agricultural power. It has, nonetheless, always acknowledged the instrumental value of the biomes from both social and economic perspectives. In the meanwhile, it has enormously enriched its understanding of and fundamentally transformed its approaches to biodiversity conservation and Indigenous inhabitants, with the salient transition happened in 1988. Nonetheless, there is little doubt that Brazil has always been taking an ideological guiding position on the global arena, creating and articulating a discourse that represents national, regional as well as tropical-country interests.

Brazil's appearance can be traced back to the 1972 *United Nations Conference on the Human Environment (UNCHE)*, which bears witness to the division over discursive politics between the North and South, as well as Brazil's leadership in advocating the discourses of sovereign integrity, nationalism and the principle of non-interference in territory on behalf of the Global South. It legitimised the use of the biomes as an indispensable cost behind social-economic development and elevated the utilisation to the level of freedom and autonomy in pursuit of national sovereignty and security (Mahar, 1989). This identity was, therefore, reflected in the commencement of several large-scale development programmes in Brazil's Amazon rainforests disbursed either nationally or by multilateral banks between the 1970s and 1990s that comprised the *National Integration Plan*, Grande Carajas Project (PGC), *Northwest Development Pole (Polonoroeste)*, *Calha Norte Project (PCN)*, *Northern Corridor Highways (ECN)*, *Agricultural and Forestry Plan for Rondonia (Planafloro)*, *Brasil Avanca* and *Bidrovias*.

These programmes and projects, therefore, colonised and industrialised the Amazon and realised the social-economic values associated with the biomes through constructing on-site residential communities, heavy industry, hydroelectricity and road infrastructure (Mahar, 1989; Goodman & Hall, 1990; Hageman, 1996). There is limited doubt that these programmes and projects, together with financial incentives, contributed to immigration, land clearance and accessibility to further conversion in the Amazon (Binswanger, 1991). Most importantly, those migrant farm households held very restricted expertise to sustain production and thus protect the native vegetation (Hall, 1997), especially the cattle ranching sector, which substantially deforested 65% of the Amazon between 1966 and 1975 (Barbosa, 2000). This was also largely caused by the federal policy of legalising private property rights (Perz, 2002).

Brazil's discourse over sovereignty and socio-economic expansion, together with its development behaviour, attracted environmental critiques from international organisations and political leaders in developed countries. Both France in 1988 and the USA between 1989 and 1991 engaged bilaterally with Brazil and implicitly and explicitly persuaded Brazil to give away sovereignty integrity over the Amazon and start curbing land grabbing activities (McCleary, 1991; Barbosa, 2000). Germany spontaneously developed an aid programme that coordinated the G7 to help Brazil with conservation in 1990 (Kolk, 1996). The World Bank Group also modified its lending policy that had been accused of contributing to rural immigration, forest removal and soil erosion in Brazil (Hall, 1989; Hageman, 1996). The diplomacy, bilateral cooperation and pragmatic investment deviated conservation as a barrier from foreign investment. That is, conservation attracts international budgets and cooperation. These events together triggered the beginning of a compatible relationship between economic growth and ecological conservation in Brazil.

Brazil possessed strengthened learning capacity that enabled it to re-examine its approaches in its foreign policy to capitalising on and realising the instrumental value of the Indigenous and biomes. It became a recipient of the international norms of ecological conservation and Indigenous rights when it hosted *the 1992 Earth Summit of the United Nations Conference on Environment and Development (Rio-92)* (McDonald, 2010). We witnessed that Brazil abandoned the norms that it had promoted before to represent its identity and the interests of the Global South at *the 1972 UNCHE*, such as sacrificing the environment for development purposes and the principle of non-interreference. Instead, it strategically adopted a conciliatory stand toward the contemporary interdisciplinary discourse surrounding sustainable development, combating climate change and biodiversity conservation as framed by developed countries (Kolk, 1996). *The Convention on Biological Diversity (CBD)* and *United Nations Framework Convention on Climate Change (UNFCCC)*, *Rio Declaration*, and *Agenda 21*, as the results of *the Rio-92*, therefore, all showed the great legacy of its mediating and agenda resetting efforts at *the Rio-92*.

Domestically, Brazil re-meditated its relationship with the Indigenous and bio-diversified resources; specifically, it pursued the compatible development between the environment and economic growth. Brazil's image depiction, ecological institutionalisation and institutional repository have been closely related to its proactive involvement in global environmental negotiation, regime formulation, and bi-/multi-lateral engagement for decades since the *1992 Earth Summit*. Since the early 1990s, Brazil, even under the conservative Sarney administration (McCleary, 1991; Barbosa, 2000), re-conceptualised the biomes and the Indians from the conventional self-claimed ‘other’ to become part of the self in national identity and sovereignty (McDonald, 2010). The bio-diversified native vegetations and human culture were not depicted as an uncivilised, insecure or empty space that was excluded from Brazil’s national appearance and even required the military to manage and transform the remote population for security reasons (Filho, 1990).

Brazil’s hosting of *the UN Conference on Sustainable Development* in 2012 *(Rio + 20),* after two decades of *the 1992 Earth Summit,* consolidated its appearance of steering the sustainable development paradigm (GOV, 2023). It strategically and systematically designed a plan for executing the paradigm, from goal setting (i.e. the Sustainable Development Goals) to recapitalising the methods of green economy and financial institutions (UN, 2023; GOV, 2023). Tropical countries are strategically positioned to mobilise collaboration on technology transfer and to stimulate capital flows for common public goods. Brazil is certainly competent in taking advantage of and activating a common interest in constructing an equitably distributed global-scale fundraising architecture (E3G, 2023). It has received result-based compensation since 2009 from the international community, including the governments of Norway, Germany and the United Kingdom (UK) and donors such as *the Green Climate Fund* and *the Amazon Fund* through *the Reductions in Emissions from Deforestation and Forest Degradation (REDD+)* scheme (InfoHub, 2022). It also exerted its initiative by getting the UK government to fund its self-designed *Sustainable Rural Program (PRS)* since 2012 for over a decade till today and by establishing a global financial instrument—a 250 billion USD *Tropical Forests Forever Fund*—for almost 80 countries at *the UNFCCC COP28* in Dubai to further demonstrate the social-economic value of conserving the tropical biomes (Byrne, 2023).

Brazil also acts as a forerunner among developing and/or tropical countries in climate change negotiations at the UNFCCC COPs by making ambitious commitments exceeding the country’s historical emissions responsibility and the expected level of commitments from developing countries at large (UNFCCC, 2023). It also included the action of governing deforestation as its unique method of combating global warming, which is mostly exclusively suitable to tropical countries. Brazil published its *National Plan for Climate Change* at the COP 15 in Copenhagen in 2009. It also, for the first time, announced the country’s national emissions abatement targets and incentives to diminish agriculture-embodied land grabbing. It ratified the country’s *Nationally Determined Contribution (NDC)* to *the Paris Agreement* under the UNFCCC in 2016, which aimed to reduce emissions through sectoral reform such as by the instruments of low-carbon agriculture, renewable energy, afforestation and biofuels. Despite the second revised *NDC* in 2020 violating *the Paris Agreement* (OC, 2021), the third adjustment in 2023 was stringent and competitive, doubtlessly defended its determination to lead by example.

Brazil’s leadership was exercised in a democratic, fair and transparent manner through covering multiple thematic areas, involving civil society representatives and small island developing countries and encouraging hundreds of voluntary pledges at the 2012 *Rio + 20* (UN, 2023; GOV, 2023). It successfully established a multilateral alliance with seven other Amazon nations through convening *the Amazon Summit* in 2023 and formed a common interest as signed in *the Declaration of Belem* to promote sustainable development at the regional level and address irregularities (ACTO, 2023). Also, right after the enactment of the *European Union Deforestation-Free Regulation (EUDR)* in 2023, Brazil bolstered an alliance with 16 other like-minded Global-South producer countries and, together, wrote a letter to express their joint concern regarding the *EUDR* (MFA, 2023). It is expected that Brazil will continue exercising its leadership on the international agenda as an emerging giant, especially when chairing the high-profile *Group of 20* (G20) in 2024, the global-level *UNFCCC COP 30* in 2025 and multilateral *BRICS+* engagement in 2025.

Nonetheless, as shown by the second revised *NDC* in 2020 (OC, 2021) and the very recent contradictory announcements given by the Brazilian president, Luiz Inacio Lula da Silva, and the Minister of Energy in Brazil, Alexandre Silveira, at the Dubai *UNFCCC COP* in 2023 regarding conserving the tropical forests and cooperating with the business sector on oil extraction respectively, Brazil's international identity as a climate leader is still slightly inconsistent and divided over championing economic or environmental prowess (Watts, 2023). This, to a certain extent, would not help the coherence and consistency of long-term resilience building and maintenance, especially in maintaining and improving the ecological resilience dimension.

In this section, we start with identifying the research gap that loosened attention has been given to the impact of political environment on rural resilience. We, hence, provide explanations over the rationale for incorporating environmental politics into resilience research. By describing and analysing Brazil's involvement, leadership and identity building in green regimes, including climate change and environmental negotiations at regional and global scales, this section moves on and illustrates how a typical tropical country can be both environmentally and politically sensitive to ecological resilience. Countries like Brazil have a close association with environmental politics at different levels and scopes. To better grasp the linkage between environmental politics and the daily life and operation of farmers and agricultural industries in the tropics, the following sections are going to anatomise environmental politics through the discourse of ecological modernisation (Sect. 6.4) and establish a cross-subject analytical framework (Sect. 6.5) before proceeding with empirically applying the analytical framework to examine the political dimension of resilience in rural Brazil, in particular, how agriculture, rurality, economy, ecology and social force are shaped and manipulated by the governing entities in Brazil to achieve tropical rural resilience (Sect. 6.6).

4 Environmental Politics Through Ecological Modernisation Discourse

The institutionalisation of the care for the environment and championing the innovative compatibility of economic prosperity and a pleasant natural environment are well framed in the discourse atmosphere of *Ecological Modernisation* (EM) since the 1970s (Hajer, 1995; Jänicke, 1995; Mol, 1995; Simonis, 1989). Environmental politics analyses EM as a *political programme* or a *policy strategy* that is embedded into the political agenda and discourses to repair environmental institution (Anderson, 2002; Hajer, 1995; Weale, 1992). It brings stagnating reforms to the structural design of modernity (Dryzek, 1987), moving towards the interaction of ecology with technology, policy, economy and society (Mol, 1995; Simonis, 1989). Environmental improvements are technologically, economically, politically and socially feasible (Table 2). EM is also viewed as a *prescription* or a *method* by Christoff (1996), Fisher and Freudenburg (2001) and O'Neill (1998), through which the revolution of environmental policies of a nation state can be streamlined. It is, accordingly, explicitly revealed through pragmatic policy instruments in national reforms.

Reading off the disparateness facilitates our understanding of EM. Neo-Marxism, for example, deconstructs capitalist production in terms of its generation of ecological crises and engendering the impossibility of overcoming environmental devastation (Schnaiberg, 1980; Pepper, 1984). In contrast, the theory of EM demonstrates that capitalist political economy can be reconciled with ecological sustainability. The theory of deindustrialisation, or counter-productivity, is committed to analysing the devastating effects of technological-industrial breakthroughs on the environment (Bahro, 1984; Gorz, 1989). However, EM assumes that science and technology are essential institutions of modernisation, enabling environmental problems to be technologically attackable.

The three waves of incubations of the EM discourse describe three processes of social demands

Table 2 Characteristics of EM theory

Features	Explanations
Modern science and technology: from criticism in simple modernisation to central, prime institutions	Technological optimism and innovation legitimation, technologically tractable environmental problems, redirecting production and greening national economy, occupying overseas markets and lifting the bar for cross-market trade (Mol et al., 2009; Huber, 1982; Spaargaren, 1997) Examples: clean technology, carbon captures and storage and low-carbon materials (IEA, 2017)
Economisation of ecology: from antagonism to the marriage of economy and ecology	Reconciling eco-friendly capitalist political economy, internalising external effects on the environment into economic interests (Carter, 2007; Gibbs, 2000), greening the markets in neoliberal capitalism (Glasson, 2012) and immediate climate action being affordable and economically beneficial (Stern et al., 2006) Examples: emissions trading scheme, eco-tax, pricing natural resources, environmental standards, certifying organic and green merchandise, environmental audits, stimulating competition on environmental performance and offering financial stimuli for ecological production and consumption (Mol & Spaargaren, 1993; Mol, 2002)
Politicisation of ecology: from controlling to decentralising power, repairing institutions	Shifting the role of central governments from a controller to a facilitator (Black, 2002); decentralised, deliberative and participative decision-making triggering a competitive environmental sphere; private economic operators coordinating with governments and taking over control of market mechanisms (Mol, 2002; Paehlke & Torgerson, 1990); and governments steering social progress and including ecological institutions into production, consumption and trade (Mol, 2002) Examples: environmental law and policy, environmental bureau and agency, public-private partnership, legitimating economic sanctions and market enlargement (Hajer, 1995)
Societal power: social movements and changes by entrepreneurs, producers and consumers	Social movements introducing new concepts and mobilising consumers, new actors joining ecological forces and economic agents becoming social carriers and changing behaviour Examples: environmental campaigns, environment-related standards drafting by insurance companies, credit institutions, industrial customers and certification organisations, companies purchasing ecologically rational merchandise and creating niche markets (Buttel, 2000; Mol, 2000, 2002)

(Source: Authors' own)

and environmental policies throughout different time periods. The initial social concern in the 1970s articulated the consciousness of natural-environmental reservation, which was emerging against a background of industrialisation and urbanisation. Environmentalists initiated this concern as an imperative response to environmental deterioration and the devastation of the eco-balance. Their initiatives autonomously established the rudiment of an ecological mindset in societies. Nevertheless, the first wave neither questioned the ever-growing industrial development nor criticised existing institutional arrangements. The biosphere was a distinctive school of thought. It possessed its own rationality, in parallel with, rather than being internalised into, political-economic spheres (Mol, 1999, 2006; Spaargaren & Mol, 1992). The consequences of the first wave of EM were the creation of natural reserves and semi-protected areas in society, both of which accompanied the industrialisation process (Mol, 2000).

The second upsurge of environment-informed institutional reform in the following decade represented the beginning of integrating ecologically inspired issues into institutional reorganisation. Environmental policy was anchored institutionally to bring ecological objectives into effect (Jänicke, 1992). Several phenomena are teased out that internalise ecology into political agendas and affect the normal societal-order structure: i) founding an environmental ministry and departments and/or appointing a national environmental authority; ii) establishing and expanding environmental policy institutions and consolidating with legal frameworks and iii) operating environmental non-governmental organisations (ENGOs) (Mol, 2006; Jänicke, 1990, 2005). Accordingly, ecology in the second wave of EM was not a marginal factor because it had penetrated into the mainstream sphere of politics. Moreover, the business and economic architecture is affected; the distinctive division between ecological and economic spheres had been blurred in this wave. However, the relevance to institutional modernisation remained limited. Mol (2006, p. 50) asserts that most modern progressive institutions, such as those that take initiatives in industrial production, economic relations and scientific-technological innovation, have not been 'deterred from their devotion to a narrowly defined economic "progress"'. The second wave, therefore, could not provoke an in-depth revolution in challenged institutions.

The third wave in the 1990s was to be distinguished significantly from its predecessors. It concentrated on 'actual environmental-induced transformations of the institutional order of modernity' (Mol, 2000, p.52). It could not be comprehended as simply a discursive update or window-dressing as the first and the second waves. Instead, it has led to large-scale pragmatic and vigorous adjustments in behaviour that have enriched institutions and reorganised society (Cohen, 1997; Hajer, 1995; Weale, 1992). It witnessed industrial modernisation through: i) less pollution leading to cost efficiency; ii) a strong and urgent sense of mission to handle environmental problems; iii) anti-pollution bringing productive, healthier and happier workers (Langhelle, 2000); iv) dematerilisation that replaces resource-intensive manufacturing and consumption with the service sector (Inglehart, 1990); v) making profitable the production of green goods; and finally vi) making the business of pollution prevention and abatement profitable. It subverted and upgraded the normalicies of production and consumption, scientific-technologic progress and the capitalisation of market economy (Cohen, 1997; Mol, 1995; Mol & Spaargaren, 2000). The trajectory of this super-industrialisation interplayed with environmental issues because it could correct and internalise environmental flaws, catalyse eco-efficiency in the private sectors and eventually solve the ecological crisis (Mol, 1995; York & Rosa, 2003). Industry and ecology coexisted in harmony rather than playing a zero-sum trade-off game (Cohen, 1997; Weale, 1992).

These three waves describe the theoretical and empirical evolution of environmental politics in Western European countries, which means that EM has a limited scope of applicability in the developing world (Frijns et al., 2000). Ecological institutionalisation is highly reliant on modernity that is reflected through specific national conditions ranging from political economy to socio-cultural beliefs, right through to techno-industrial capability (Beck, 1992; Beck et al., 1994; Berger & Flynn, 2001; Obach, 2007). It therefore seems that investigating the footprint of EM in Brazil in this research is quite ambitious, challenging the key criticism of EM in the 1990s as to its inaccessibility to developing countries. Nonetheless, taking into account that tropical countries possess valuable Indigenous resources and play a decisive role in climate change mitigation and ecological conservation and restoration, being ecologically modernised seems necessary for tropical countries. Additionally, the global mainstream voices and institutions challenging the tropical rural areas are also usually driven by European countries; applying the EM discourse can better link the global-local political interconnections.

Moreover, applying this theory in Brazil is still rational considering the advanced regulation, technology progress, increased living standards and social forces in this typical tropical country. This forms favourable conditions, accelerating a nation's speed and probability of implementing environmental governance. The significant relevance of the material, institutional and sociocultural capacity of a country is in the outcome of the environmental policy (Jänicke, 1992). EM can not only explain the institutional changes that have already taken place but can also be applied as a guide for reforms (Berger & Flynn, 2001). Consequently, the next section is going to theoretically embed the discourse of EM into the concept of tropical rural resilience.

5 Introducing the Analytical Framework

Although the resilience theory faces a common criticism of yet theorising its interrelationship with political dynamics as one aspect of pluralistic globalisation processes intervening in the daily life of communities (Machado, 2019), the subject of rural studies has made certain inroads into politics, which has been hybridised in the theoretical frameworks of political economy and political ecology. The former examines governmental rural development agenda, probing into the incubation of capitalist production and privatisation in (sub-)national agricultural and rural regulation and policy in the background of global political economy (Newby, 1985; Cloke, 1987). It provides a lens into the government's identity, rhetoric and plans espousing a transition towards deregulated, market-oriented rural economies (Marsden et al., 1986; Cloke, 1987) and post-productivist hedonic and service functionalisation (Mormont, 1990; Munton, 1995). The latter expounds how neoliberal globalisation, capitalist penetration and multilateral institutions accelerate local inequality and oppression, uneven power relations and eventually conflicts over ecological and raw materials (Hollander, 2005; Bryant & Bailey, 1997). There is also a growing interest in demonstrating how micro-scale bottom-up grassroots movements, especially in the Global South, politically solve the local social-ecological problems (Rainer, 2016).

Consequently, both theories extend the technocratic rural management and provide insights to power relations at multiple levels. The political-economy framework elaborates on the process of decentralised decision-making within an overarching policy network for rural policy reform (Marsden et al., 1993; Murdoch & Marsden, 1994). Nonetheless, the global context of political economy appears to be static snapshots, making it struggle to portray the dynamics of international relations and global-local interconnections that construct the discourse of the environment. The political ecologists, comparatively, shed light on the supra-national political forces jeopardising while also addressing community-level social ecology (Bryant & Bailey, 1997; Woods, 2007).

However, the conceptualisation of politics in both rural-study theories shoulders the limitation of serving for the capitalist spreading. They lean towards the political endeavour of restructuring capitalist territorial policy, neoliberal agro-industrial policy, multifunctional landscape programme and the impacts of these regulations on grassroots decision-making, values, institutional arrangement and green governance (Watts, 1983; Pretty, 1995). Their narratives of evaluating rural development shrink down to industry maturity, market architecture and categorising socio-economic status (Marsden, 1999; Kelly et al., 2015). From the subject of rural studies, its political angle can hardly be comprehensive enough to analyse and evaluate tropical rural resilience, since it, to a large degree, ignores the main ecological problems in the tropical rural areas, i.e. deforestation and habitat conversion, together with the issue of tropical-social integrated resilience systems.

Through streamlining the theories of rural resilience, political economy and political ecology discussed hereinbefore, it seems indispensable to innovatively build up an analytical framework that subordinates the environmental perspective of political studies to tropical rural resilience. This helps testify to the hypothesis that, although the ecologically political dimension is largely missing from the notion of tropical rural resilience, it can be, firstly, witnessed in tropical countries when fostering and sustaining domestic rural resilience and, secondly, verified as an effective approach to facilitate and reconcile ecologically, socially and economically viable resilience in tropical countries.

Following this logic, the analytical framework is sketched below in Table 3. The four elements of the advanced environmental political discourse—EM, i.e. modern science and technology, ecology economisation, ecology politicisation and societal power—are supposed to be an integrated part of the tropical rural resilience that characterises and defines a typical tropical country. The relationship between the left column, i.e. tropical-political integrated

Table 3 Analytical framework of political resilience in tropical rural

Tropical-political integrated systems	Relationships	Tropical rural resilience
1. Modern science and technology (from criticism in simple modernisation to central, prime institutions) 2. Economisation of ecology (from antagonism to the marriage of economy and ecology) 3. Politicisation of ecology (from controlling to decentralising power, repairing institutions) 4. Societal power (social movements and changes by entrepreneurs, producers and consumers)	Interrelated & Reinforcing	1. Tropical resilience capacity 2. Tropical embeddedness 3. Tropical agroecosystems 4. Tropical-social integrated systems 5. Tropical-economic integrated systems

(Source: Authors' own)

systems, and the right column i.e. tropical rural resilience, is thus interrelated. It is also assumed that well-cultivated tropical-political integrated systems are, simultaneously, consolidating and reinforcing the six features generated in the concept of tropical rural resilience. That is, being politically sensitive helps foster, improve and sustain rural resilience in a tropical country.

Consequently, this research situates the analytical framework within the theories of both EM and resilience to conceptualise potent and logical inferences about how environmental politics challenges, maintains and/or improves the agroecosystem resilience in rural farming spaces. Their combination represents an innovative attempt at transferring political behaviour and rhetoric as one category of externalities in resilience study. Applying this analytical framework means that, through anatomising the appearance of environmental politics, i.e. the four EM characteristics, in the tropical-political integrated systems in Brazil, the level of political resilience is visualised in Brazil's rural resilience. This, therefore, answers the research question of whether or not Brazil is gaining its political dimension of resilience while being highly involved in international environmental politics. It argues that the rural in Brazil is not politically vulnerable provided its development is scientific-technologically embedded, ecologically economised and politicised and societally empowered. By achieving a politically resilient rural environment, we could further declaim that Brazil's rural areas achieve a comprehensive degree of resilience in various dimensions, including the ecological, economic, social and political perspectives.

6 Applying Political Resilience in Rural Brazil

Agriculture is a pivotal pillar of Brazil's national economy; approximately 85 billion US dollars in value were added by the sector to Brazil's gross domestic product (GDP) per year in the past decade, accounting for 7% of the GDP (Statista, 2023). It has continued to be a global force in agricultural supply by weaving international trade networks and structuring different export cycles, from sugar in the eighteenth century, through cotton and coffee in the nineteenth and twentieth centuries, respectively, to the period of agricultural modernisation in the twenty-first century characterised by agricultural diversification and intensification. That is, today, it supplies about 40 kinds of soft commodities comprising soybeans, sugar, beef, corn, etc., to more than 200 countries around the world. It took a significant share of 5.2% of total global agricultural exports in 2020, ranking third in the world (Klein & Luna, 2018; FAO, 2018). With the doubled gross value of Brazil's agricultural products in parallel with doubled meat production and tripled grain production in the past two decades (Zhou et al., 2022), Brazil's farming system is well structured and greatly extended. It can be categorised into four establishments—rural settlements (2%), smallholder farms (19%), middle-scale farms (34%) and large-scale farms (44%). Each of them echoes specific size, geography, number of producers and even types of agri-food production. With regards to agricultural land use, pastures account for 49% of Brazil's agricultural land, bushes and forests take 29% and crop production occupies 22% (Sette & Ekboir, 2013).

The organised farming system, as well as the strength of farmers' adaptability in enabling the sustainability and modernisation of agriculture across centuries in Brazil, are contributed by both the biologically abundant natural environment and national rural-relevant strategy, regulation and programmes. They are also largely shaped by national and international environmental politics. This section, correspondingly, employs the analytical framework established in Sect. 6.5 and empirically articulates the process of formulating the tropical-political integrated systems in Brazil as revealed by the four identified EM characteristics through a combination of policy document analysis and secondary information and data. It critically analyses the extent to which the level of environmental politics incorporated in Brazil's rural agriculture reinforces the achievement of overall multi-dimensional resilience.

6.1 Scientifically Technological Embedded Tropical Rural Resilience

Science and technology were paramount means of increasing agricultural productivity and generated the so-called green revolution in Brazil throughout the two decades from 1950 to 1970, through agrochemicals, including fertilisers and pesticides, genetically modified high-yield and tropical-resilient crop varieties, mechanisation and irrigation infrastructure (Zhou et al., 2022). In addition to concentrating on yield, science and technology help policymakers and agricultural practitioners, especially those new immigrants to the Amazon rainforests, comprehend and preserve while also smartly utilising tropical agroecosystems to accelerate the elimination and neutralisation of carbon emissions, tackle illegal land clearance and ultimately achieve net zero deforestation. To foster tropical rural resilience, this research finds that science and technology are of primary association with four elements in strengthening the tropical rural resilience systems, i.e. *tropical resilience capacity*, *tropical embeddedness*, *tropical agroecosystems* and *tropical-economic integrated systems*.

Science and technology in Brazil are closely linked with understanding and monitoring the transformation of *tropical agroecosystems* in rural resilience systems. Back to 1988, the National Institute for Space Research (INPE) developed *the Measurement of Deforestation by Remote Sensing (PRODES)* project to generate annual clear-cut deforestation rates in the Legal Amazon and inform the government to make evidence-based ecology-informed protection policy. *The Amazon Protection and Vigilance Service* was further launched in 1996 that established a surveillance system for investigating illegal logging. A *Real-Time Deforestation Detection System (DETER)* was created by the INPE in 2004 to issue daily deforestation alerts to support environmental inspection and prosecution agencies. Also since 2004, the government has been implementing and upgrading a cross-ministerial *Action Plan for Deforestation Prevention and Control in the Legal Amazon (PPCDAm)*. Including the recent one being activated in 2023, *the PPCDAm* articulates how advanced technologies and intelligence mechanisms, including the 100% Brazilian satellite system can identify irregularities and anticipate manmade and natural devastation (Bourscheit, 2022; IMF, 2023). Information on deforestation and degradation can also be received by landowners to self-organise and adapt into surrounding transforming agroecosystems and thus improve their tropical resilience capacity.

The institutionalisation of research and development (R&D) to control and contain environmental crimes is another lynchpin in Brazil's scientific-technological resilience building. The Brazilian Institute of Environment and Renewable Natural Resources (IBAMA), affiliated to the Ministry of the Environment, was established in 1989 to provide data analysis and technical advice on biodiversity, emissions and waste, environmental supervision and protection, licensing and chemicals (IBAMA, 2023). The Brazilian Agricultural Research Corporation (EMBRAPA) attached to the Ministry of Agriculture, Livestock and Supply (MAPA), founded in 1973, is another outstanding nation-wide R&D institution. It concentrates on technological solutions addressing environmental sustainability, together with tropical agricultural innovation and food security e.g. through generating technologies, methodology, software, applications and equipment (EMBRAPA, 2023). Its scientific interest in tropical agricultural productivity underscores the fusion of exogenous knowledge and endogenous tropical farming to produce multifunctional public goods and common gains ranging from food security to biodiversity conservation.

In the *tropical-economic integrated resilience systems*, technological and technical innovation and breakthroughs in Brazil are integrated into the subject of tropical agriculture leading to forest-friendly, low-emission and high-yield agricultural practices in tropical pastures and arable lands (EMBRAPA, 2021). Brazil perceives the adoption of agricultural technology in rural communities as a way to strengthen local *tropical resilience capacity*. It is harnessed by agricultural practitioners to possess productive but also climate-resilient agriculture, augment flexibility in international ecologically embodied soft commodity chains and consolidate farming infrastructure such as precision irrigation systems and soil conditions (IMF, 2023). Brazil supports a resilient transition of rural agriculture towards a restructured, compatible tropical-economic system. The resilient transition is strategically implemented through two approaches: i) alternative climate-smart forms of land use and ii) yields improvement on existing farmland.

Both can barely be separated from technological and technical innovation, which is devised and demonstrated through a series of climate- and conservation-informed technological solutions developed mainly by the EMBRAPA as a contribution to policy instruments such as *the Low Carbon Agriculture Plan 2010–2020 (ABC Plan)*, the second phase of it—*the ABC+ Plan 2020–2030*, *PRS*, etc. Some leading agricultural technologies include the integrated silvopastoral (livestock-forestry, ILF) and agrosilvopastoral (crop-livestock-forestry, ICLF) production systems, crop rotation or double cropping, low-carbon and carbon-neutral agriculture, green manure, no-till cultivation systems, soil correction, inoculation with Rhizobium, degraded pasture recovery, etc. (PRS, 2023; Zhou et al., 2022; Salton et al., 2014; WBG, 2023; Alves et al., 2015; Oliveira, 2014). This highlights that decarbonisation- and ecology-informed technologies and techniques in agriculture are stimulating the industrial competitiveness of Brazilian agriculture and are prone to reverse the industrial climate and ecological vulnerability. They are also facilitating the improvement of productivity and the standing of the tropical biomes (EMBRAPA, 2021).

6.2 Ecological Economisation and Tropical Rural Resilience

Exercising environmental superpower and mobilising environmental-related market-based and financial mechanisms on the global arena has efficiently echoed Brazil's internal economisation of its biologically diversified biomes. The enrichment of Brazil's environmental policy through the angle of ecological economisation can be traced back to 1995 in its *National Integrated Policy for the Legal Amazon*, which formalised the instrument of ecologically economic zones for achieving the economic implication of preserving native ecosystems surrounding the Amazon (Mahar, 2000). Brazil's Ministry of Environment (MMA) and the Ministry of Economy (ME) jointly developed the initiative—*the National Green Growth Program (PNCV)*—to realise the translation of conservation and decarbonisation into green jobs and a green economy via embedding the characteristics of each region and biome when incubating green criteria (GOV, 2022). By appreciating and protecting the tropical agroecosystems and complying with *the Forest Code* Constitution and national climate and deforestation pledges, Brazil's government has been tailoring 80 *Payment for Ecosystem Services (PES)* schemes so far, such as *the Proambiente,* to introduce economic activities for rural farming communities to reap fiscal returns and therefore reconcile environmental protection with agricultural production (Mamedes et al., 2023).

Specific activities to conserve and restore tropical ecological systems and enhance land use include the establishment of the agroforestry business in the understory of legal reserves, capitalising and transacting non-timber forest products, reforesting fast-growing and short-rotating Eucalyptus trees, recovering degraded land, soil and water conservation, trading forest certificates and running ecological tourist programmes (WBG, 2023; Mamedes et al., 2023). Landowners whose protected areas exceed the minimum legal requirement set out in *the Forest Code* are eligible for financial rewards through carbon offsets for eliminating emissions and/or payments for giving up legal rights to clear those forests (Stickler et al., 2013). These activities do not only reduce the potential exposure of rural communities to environmental adversities but also capture the value of the tropical land, economise and diversify the action of conservation and encourage national and international capital flow to the biomes. Both the 2008 *Amazon Fund* and the 2015 *National REDD+ Strategy* are creative fiscal institutions formulated by Brazil for the country to receive worldwide compensation by protecting local agroecosystems (Zhou et al., 2022). They are, therefore, enhancing the attributes of *tropical agroecosystems* and *embeddedness* in the framework of tropical rural resilience.

Economic ecologisation via sectoral reform is also an essential attribute in the *tropical-economic integrated system building*. Environmental damage is viewed as an institutionalised problem that is internalised in economic activities in local communities. As tropical rural resilience is largely bolstered by the resilience of tropical agri-business to climate and environmental uncertainties, Brazil is fostering its rural resilience through initiating a variety of sector-level interventions in agriculture, forestry and livestock that improve their adaptation capacity and tremendously accelerate the internalisation of environmental externalities in sectoral innovation and market solutions. These intervention measures are often government-invested financial stimuli and technical assistance, which enable resource-dependent industries and ecological construction to move from confrontation and trade-offs to balanced and win-win development.

The examples of these measures comprise the *Plano Safra programme* providing rural credits to subsidise low-emission and/or zero-illegal deforested agricultural operations; sustainability certifications on low-carbon, carbon-neutral, organic agricultural plantations and cattle ranching such as the Viva brand; climate-resilient rural livelihoods schemes such as *the PRS*, *Adapta Sertao*, *MAIS Programme* and *ABC Cerrado*; community safety nets and climate-sensitive insurance products guaranteeing first-mover returns from adopting climate-informed transitions, etc. (PRS, 2023; Zhou et al., 2022; WBG, 2023; IMF, 2023). The inclusion of the private sector and ecologisation of the food system transition in Brazil to a sustainable soft-commodity supplier and satisfy the increasing ecologically sound demand from the consumption markets. This reconciles the environmental problems brought about by the capitalist political economy and makes Brazil a combined agricultural and environmental power. This also steers the awareness and behaviour change of the business stakeholders operating in Brazil, which will be expounded in the societal power bit of this section. Finally, these measures also improve the *tropical agroecosystems* and *tropical embeddedness* in tropical rural resilience.

6.3 Ecological Politicisation and Tropical Rural Resilience

Brazil's competence in building a legal foundation in conserving the six incredibly biodiversified biomes occupied in its territory was revealed in its very first Constitution—*the Forest Code* of 1965, which was an unprecedented, stringent constitutional innovation that categorised the *Permanent Preservation Areas (APPs)* and *Legal Reserves (RLs)* in Brazil. The latter ruled rural properties to mandatorily protect a certain proportion of native vegetation on their lands as the minimum requirement (McDonald, 2010; Stickler et al., 2013). This dovetailed with

Jänicke's (1992) elaboration of the consolidation of legal constitutions that represented the commencement of EM and an almost leapfrogging progress towards the second wave of EM. It recognised the rationality of constructing a biosphere in Brazil based on its rich biomes, which was in parallel with and could also be exploited in a legal manner to meet the development interests in expanding social and economic spheres. *The Forest Code* was modified in 1988 and 2012, the former of which enlarged the conserved forests in private properties by 30% (Hall, 1997). The latter delivered both the management tools of the *Rural Environmental Registry (CAR)* and the *Environmental Regularisation Program (PRA)* for better executing *the Forest Code*. The *CAR* is devised for reducing land conversion through monitoring landholders' legally binding conservation requirements and punishing those *CAR*-qualified but noncompliant farms (Stickler et al., 2013). The *PRA* is for reforesting the deficits in *APPs* and *RLs* (Plenamata, 2023).

At the ministerial level, the Ministry of Agrarian Development (MDA) combines the interests of agri-business and environmental protection from the MAPA and MME, respectively, by focusing on agroecological rural development. The Federal Government also instituted the agricultural agency, EMBRAPA, in 1973 and the forestry agency, IBAMA, in 1989 to identify and analyse opportunities to augment productivity while appropriately using bio diversified natural resources in the Amazon (Kolk, 1996; Hall, 1997). With the pressure from international institutions and developed countries, *the Committee of Expert* was organised to revise *the Calha Norte Project* and included the Amazon as part of its national identity (Albert, 1992), which designated 6% and 17% of the Amazon as natural parks and Indigenous territories respectively in 1990 (Kolk, 1996). This doubtlessly demonstrates the overwhelming interconnectedness between global and national politics in tropical countries. *The Permanent Preservation Areas (APPs)*, including the parks and Indian reserves, together with the *CAR*-licensed *Legal Reserves (LRs)* in individual farms, are still assumed to increase until 2030 by following the legislation of *PRA*. The structured policy of *PPCDAm* in 2004, orchestrated by 13 federal agencies, as an effective command-and-control environmental instrument, is playing a key role in enlarging the protected areas in Brazil (Nepstad et al., 2014).

The establishment of *the Inter-ministerial Committee on Climate Change* in 2007, which is now the *Inter-ministerial Committee on Climate Change and Green Growth (CIMV)*, as a federal-level decision-making body, marked that Brazil had finally embarked on the path to domestic climate governance. The *CIMV* published Brazil's very own *National Climate Change Policy (Plano Clima)* in 2009 and emphasised the necessity to fully integrate it into all domains of national agenda-setting and sectoral programmes. It also acts as the single governing entity of the national green growth program—*the PNCV* (GOV, 2022). There is also ecology-embodied investigation and litigation in the legislative system in Brazil for the sake of reducing illegal logging and climate governance, e.g. the 2019 *National Observatory on Environmental, Economic and Social Issues of High Complexity and Great Impact and Repercussion*, *Panel on Environmental Actions for the Amazon Region* in 2020, and *Panel on Environmental and Inter-institutional Data (SireneJud)* in 2021 (WBG, 2023).

Accordingly, ecological politicisation in Brazil, at the national level, has been organised and planned quite comprehensively through establishing environmental law, policy, bureau, agency and committee. It has considerably improved the attribute of *tropical agroecosystems* in tropical rural resilience through normalising agricultural operation and land use based on appropriate legal deforestation. This was especially true after the introduction of the *PPCDAm* in 2004, which significantly reduced deforestation in the Amazon rainforests and deforestation-induced carbon emissions there by 80% and 65%, respectively, by 2012 (West & Fearnside, 2021). Simultaneously, during this period, Brazil also witnessed an increase in agricultural outputs and the involved international trade in soft commodities became closer, which boosted the tropical rural resilience if we perceive this through the angle of *integrating tropical-economic systems*.

6.4 Societal Power in Tropical Rural Resilience

Brazil is strategic in recognising, assembling and internalising societal power and institutional capacity from several different social resources through mechanisms and programmes to strengthen the tropical-social integrated systems in Brazil. The government quite often invites multiple stakeholders in the consultation process of decision-making to exchange views and facilitate respective tropical resilience capacity building. The very first national climate change policy in 2009—*Plano Clima*—involves the participation of researchers and sub-national government agencies. *The National Agricultural Research System (SNPA)* and its platform—*the Alliance for Innovation*—also include the state agricultural research organisations (OEPAS), universities, non-profit organisations and private agencies. National community programmes such as *PRS* promote the alleged 'social technologies', i.e. exchanges with rural producers and social organisations (PRS, 2023). Most importantly, given the persuasive power of the business sector in Brazil, which could be perceived as a sort of lobbying effort on behalf of global capitalisation and to suppress other domains of politics, it is essential to note that industry associations in Brazil are strategically positioned to bridge the private sector through membership management and government entities in Brazil.

Industry associations in Brazil have been exerting their initiatives in actively responding to and acting upon the international environmental campaigns against agriculture-embodied deforestation in the Amazon. Back in 2006, *the Brazil's Soy Moratorium (BSM)*, as the first voluntary zero-deforestation movement in tropical countries, witnessed the milestone of accumulating and exercising societal power for problem solving in the world. It was portrayed and signed by two national-level industry associations in the grain sector in Brazil—the Brazilian Association of Vegetable Oil Industries (ABIOVE) and the National Association of Grain Exporters (ANEC), with the assistance of the non-governmental organisation (NGO)—the Greenpeace. Through *the BSM*, both associations innovatively adopt the market-based approach to address deforestation through the angle of supply chain governance. They set up the signatory date, the 24th of July in 2006, as the deforestation cut-off date in the Amazon in Brazil. To implement this alleged zero deforestation, they leveraged major business members affiliated and channelled them not to trade or finance soybeans from farms converted after July 2006 in Brazil's Amazon. All three first movers, the ABIOVE, ANEC and Greenpeace, formed an alliance, named *the Soy Working Group (GTS)*, to take responsibility for the mission (ABIOVE, 2023; Imaflora, 2016; Gibbs et al., 2015).

The *GTS* group was further expanded after 2 years of implementation in 2008 by having Brazil's government, INPE, Bank of Brazil and the European Soy Consumer Group involved in the alliance, which made the initiative a real government-adopted, multi-stakeholder governance model. This eye-catching step rewrote the history of societal power in the tropics. In addition to enabling economic agents to become social carriers and adapt into a new atmosphere of deforestation-mainstreamed global environmental politics and international relations, civil society organisations and economic-pertinent institutions embed their societal power and environmental awareness in decision-makers at the top level. Brazil's government then integrates this initiative in its administrative-regulatory decision-making process and connects with the 2012 *Forest Code Constitution, CAR, PRODES*, etc. The endorsement of Brazil's government eventually brought the financial actor, Bank of Brazil, and national aerospace research centre, INPE, and the European Soy Consumer Group into the *GTS* to champion, deliberate, monitor and evaluate the implementation of *BSM*. There are also third-party audits becoming a new category of social carriers who verify the compliance of the signatory companies with the *BSM*

(ABIOVE, 2023; Imaflora, 2016; Gibbs et al., 2015).

The joint efforts have generated a reduction of direct deforestation for soy plantations (Zhou et al., 2022). Merely 3% of soybeans can be traced back to land that was illegally cleared after the revised 2008 deforestation deadline (ABIOVE, 2023). Consequently, social forces in ecological modernisation discourse enhance the *tropical agroecosystems* coupled with *tropical-social integrated systems,* considering the efforts being derived from multiple civil society organisations and the private sector. Moreover, very similar strategic public-private partnerships can also be discovered in the freshly established social relationships between EMBRAPA and major business stakeholders—Marfrig and Bayer. To contribute to *the ABC+ Plan*, EMBRAPA partners with Marfrig to neutralise methane gas released by cattle through farming in integrated silvopastoral and agrosilvopastoral systems and thus innovatively produce *Carbon Neutral Beef (CNBB),* for instance (EMBRAPA, 2020). Additionally, they also inject the principles of good agricultural practices and proper pasture management to produce Low *Carbon Brazilian Beef (LCBB)*, which, in turn, increases productive efficiency, i.e. the kilogram per pasture, reduces pasture expansion via land conversion and increases farmer profitability (EMBRAPA, 2021).

7 Discussions and Conclusions

By applying the established analytical framework of understanding and assessing the relationship between environmental politics and tropical rural resilience in Brazil's scenario, this chapter articulates that Brazil is institutionalising and programming the issues of deforestation and climate change and, correspondingly, stimulating its resilience in rural agriculture. The empirical analysis upon Brazil unequivocally underpins the hypothesis that the development of environmental politics effectively strengthens the compatible interrelationship between ecology and social economy. Through the discourse of the environmental political theory and approach—ecological modernisation—the chapter critically analyses the embeddedness of modern science and technology, ecological economisation or economy ecologisation, ecological politicisation and societal power by Brazil's government in tropical rural resilience, specifically, tropical capacity building, tropical embeddedness, agroecosystems, tropical-social integrated systems and tropical-economic integrated systems. It believes that Brazil's tropical rural resilience is technologically, economically, politically and socially feasible.

We find out that the governmental behaviour of integrating science and technology in tropical rural resilience is competent in i) building the capacities of agricultural adaptability and decision-making in the transforming ecosystems and ecologically induced social-economic systems; ii) bridging exogenous scientific knowledge and endogenous agroecosystems to foster knowledge innovation and reach multifunctionality; iii) enabling the understanding, capitalising and cherishing the value of ecosystems to community, industrial development and the common public goods; and finally iv) reorganising land-based sectoral evolution and innovation. Nonetheless, comparatively, this finding mirrors the diminution of effects on tropical-social integrated systems of the Brazilian decision-makers' administrative-regulatory and pragmatic programming efforts. This can be illuminated from the perspective of a lack of instruments and mechanisms transferring many different technologies and techniques to a variety of social actors across sectors and scales to augment their scientific values and capacity. There are also limited support programmes and projects available upon experimenting with the distribution of technology to various stakeholders in rural communities in a just and fair diffusion manner.

The direct causal relationship can be figured out between ecological economisation and the cultivation of tropical-economic integrated systems in Brazil's environmental politics. It is quite apparent that policymakers are devising and innovating ecology-embodied, market-based and financial conservation policy and activities, including afforestation, ecotourism and agroforestry business. These land-based industries in Brazil, especially the agricultural sector that is crucial to the national economy and people's livelihood, have a good momentum of development. By reviewing and assessing the progress in sectoral ecologisation, a variety of intervening policy options are available to mediate the environmental problems arising from capitalist economic development. These touch upon the fields of alternative livelihoods, sustainability certification, insurance and subsidies. Accordingly, policymaking surrounding ecological economisation and economic ecologisation in Brazil also augments the attributes of tropical agroecosystems and tropical embeddedness.

However, considering the policy development in the agriculture industry, agricultural economy and international trade, the dimension of sustainability policy championing economic ecologisation may still be a small segment affecting rural agrarian communities. The World Bank Group (WBG, 2023) highlighted that investments for addressing illegal deforestation and improving agricultural productivity are relatively modest. There is a level of conflict of interest between ecological economisation and agricultural economisation, which means that it can be more cost-effective to convert land for industrial purposes than the value gained from maintaining the native land. This may, therefore, hinder the resilience building in Brazil. Moreover, the problem of sustainable programmes and good-practice schemes upon ecological economisation and economic ecologisation is that they are usually on a small scale, such as through a demonstration pilot in a couple of selected farms in a community. Reorganising sustainable agricultural products and the international marketisation and distribution of these products are missing elements in the tropical-economic integrated systems. This underscores that it is indispensable to structure innovative and solid fiscal and economic arrangements to expand and even industrialise those tried experiments and practices.

The process of ecological politicisation in Brazil has consolidated its rural resilience building. The institutionalisation of environmental law—*Forest Code*—coupled with climate change policy and programmes sketched the biosphere in Brazil and significantly improved the position of tropical agroecosystems in communities. The government is also shifting its position from a controller to a facilitator through organising and repairing ecologically pertinent institutions, agencies, panels and committees and decentralising its exclusive governing power to these entities, which reinforces the tropical resilience capacity by shortening the distance between decision makers and tropical communities. However, we could argue that the shifting governing power happened too early in Brazil when the central government had not yet formulated any solid strategy or working plans to deliver zero deforestation pledges and/or climate commitments (WBG, 2023). Moreover, the problem that restricts the contribution of ecological politicisation to rural resilience in Brazil is at the implementation level. Ecology is not fully politicised owing to the limited enforcement capacity of the *Forest Code,* e.g. it takes time to validate and issue the *CAR* license to farmers. It is also hard for the governing bodies to deliver creative incentives when they confront deducted budgets, staff cutting down, business lobby groups interfering in agenda setting, etc. Furthermore, the protection agenda itself remains frustrating to execute, as Brazilian Amazon states still set priorities over poverty eradication and economic reform (Consorcio dos Governadores da Amazonia, 2021).

This may, to a certain degree, lead to their engagement with actors in civil society. The inference that societal power reinforces the tropical-social integrated systems is plausible in Brazil. The government tends to consult and build partnerships with multiple stakeholders, especially researchers from think tanks and universities, for policymaking. It also pays attention to the initiatives and performance of industry associations and can, strategically, incorporate their actions as part of the governmental behaviour and/or good governance, which enhances the involvement of the private sector in activating innovation and entrepreneurship. It becomes mind-boggling considering how cutting-edge, avant-garde and bold some of these initiatives are. The government can also provide supports and manoeuvre a wider group of stakeholders ranging from economic agents to financial institutions to change their behaviours and become new social carriers. These new actors, therefore, create ecologically relevant social relations, values and capacities, and strengthen the integrated tropical-social resilience systems in Brazil. Nevertheless, there can be more complicated networks and social relationships formed across not only horizontal sectors and scales but also vertical layers to better amplify and rationalise a social-ecological foundation.

Based on our analysis, this chapter concludes that Brazil's agrarian communities are gaining rural resilience while the government focuses on rooting and exercising environmental politics. Brazil develops greening and developing policies simultaneously and there are certain good governance and best practices identified as means of driving social, productive and environmental rationale destined to rural resilience. This further implies the necessity to add the politically driven rural resilience when narrating the development-conservation compatible resilience building in tropical countries. The empirical analysis, therefore, confirmed the rationality and innovation of displaying a level of reflexivity with regard to the traditional resilience approach by embedding environmental politics in resilience study, theoretically and then empirically examined. To offer a holistic picture of environmental politics reinforcing tropical resilience, this research suggests that probing into sub-national policy and programmes is a future study pathway to gain sufficient mechanistic evidence. Moreover, it is necessary to apply the established analytical framework in other tropical countries to provide more evidence-based suggestions to tropical countries for them to better undertake the important roles of both conserving the biomes as a tropical solution to combat climate change and securing the international sustainable agenda and food provision.

References

ABIOVE. (2023). "Sustainability", Brazilian Association of Vegetable Oil Industries, [Online], Retrieved 20 December 2023, from https://abiove.org.br/en/sustainability/.

ACTO. (2023). "Declaration of Belem", Amazon Cooperation Treaty Organization, [Online], Retrieved 25 October 2023, from https://otca.org/en/wp-content/uploads/2023/10/Declaration-of-Belem.pdf.

Aggarwal, R. M. (2006). Globalisation, local ecosystems, and the rural poor. *World Development, 34*(8), 1405–1418.

Albert, B. (1992). Indian lands, environmental policy and military geopolitics in the development of the Brazilian Amazon: The case of the Yanomami. *Development and Change, 23*(1), 55.

Alves, F. V., Almeida, R. G., & Laura, V. A. (2015). *Carne Carbono Neutro: um novo conceito para carne sustentável produzida nos trópicos*. Embrapa.

Anderson, M. S. (2002). Ecological modernization or subversion: The effect of Europeanization on Eastern Europe. *American Behavoral Scientists, 45*, 1394–1416.

Arezecki, R., Deininger, K., & Selod, H. (2012). The global land rush. *Finance and Development*, 46–49.

Bahro, R. J. (1984). *From red to green*. Verso.

Baird, J., Plummer, R., Haug, C., & Huitema, D. (2014). Learning effects of interactive decision-making processes for climate change adaptation. *Global Environmental Change, 27*, 51–63.

Barbosa, L. C. (2000). *The Brazilian Amazon rainforest: Global ecopolitics, development, and democracy*. University Press of America.

Barnes, T. J. (1997). Theories of accumulation and regulation: Bringing life back into economic geography—introducing the section three. In R. Lee & J. Wills (Eds.), *Geographies of economies*. Edward Arnold.

Barthel, S., Parker, J., & Ernstson, H. (2015). Food and green space in cities: A resilience lens on gardens and urban environmental movements. *Urban Studies, 52*(7), 1321–1338.

Baylis, J., Smith, S., & Owens, P. (2008). *The globalisation of world politics: An introduction to international relations*. Oxford University Press.

Beck, U. (1992). *Risk society: Towards a new modernity*. Sage Publishers.

Beck, U., Giddens, A., & Lash, S. (1994). *Reflexive modernization: Politics, tradition and aesthetics in the modern social order*. Polity Press.

Bell, M. (1998). *An invitation to environmental sociology*. Pine Forge Books.

Berger, G., & Flynn, A. (2001). Ecological modernization as a basis for environmental policy: Current environmental discourse and policy and the implications on environmental supply chain management. *European Journal of Social Science Research, 14*(1), 55–72.

Berkes, F., & Ross, H. (2013). Community resilience: Toward an integrated approach. *Society & Natural resource, 26*(1), 5–20.

Biggs, R., Westley, F. R., & Carpenter, S. (2010). Navigating the Back loop: Fostering social innovation and transformation in ecosystem management. *Ecology and Society, 15*(2).

Binswanger, H. (1991). Brazilian policies that encourage deforestation in the Amazon. *World Development, 19*, 821–829.

Black, J. (2002). Critical reflections on regulation. *Australian Journal of Legal Philosophy, 27*(1), 1–35.

Bourscheit, A. (2022). Prodes and deter: Get to know these strategic systems in the fight against deforestation in the Amazon, Infoamazonia, [Online], Retrieved 10 December 2023, from https://infoamazonia.org/en/2022/02/15/prodes-and-deter-systems-against-deforestation-amazon/.

Brown, K. (2016). *Resilience, development and global change*. Routledge.

Brown, K., & Westaway, E. (2011). Agency, capacity, and resilience to environmental change: Lessons from human development, well-being, and disasters. *Annual Review of Environment and Resources, 36*(1), 321–342.

Bryant, R., & Bailey, S. (1997). *Third world political ecology*. Routledge.

Bryceson, D. F. (2002). Multiplex livelihoods in rural Africa: Recasting the terms and conditions of gainful employment. *Journal of Modern African Studies, 40*(1), 1–28.

Bulkeley, H., & Newell, P. (2010). *Governing climate change*. Routledge.

Buttel, F. H. (2000). Ecological modernization as social theory. *Geoforum, 31*, 57–65.

Byrne, J. (2023). COP28: Brazil Launches 250 Billion USD global fund to safeguard forests across 80 Nations, FeedNavigator, [Online], Retrieved 28 October 2023, from https://www.feednavigator.com/Article/2023/12/05/COP28-Brazil-launches-250bn-global-fund-to-safeguard-forests-across-80-nations.

Carter, N. (2007). *The politics of the environment*. Cambridge University Press.

Christoff, P. (1996). Ecological modernization, ecological modernities. *Environmental Politics, 5*, 476–500.

Cloke, P. (1987). *Rural planning: Policy into action?* Harper and Row.

Cohen, M. (1997). Risk society and ecological modernization: Alternative visions for postindustrial nations. *Futures, 29*(2), 105–119.

Cutter, S. L., Barnes, L., Berry, M., Burton, C., Evans, E., Tate, E., & Webb, J. (2008). A place-based model for understanding community resilience to natural disasters. *Global Environmental Change, 18*, 598–606.

Darnhofer, I., Lamine, C., Strauss, A., & Navarrete, M. (2016). The resilience of family farms: Towards a relational approach. *Journal of Rural Studies, 44*, 111–122.

Dryzek, J. S. (1987). *Rational ecology, environment and political economy*. Blackwell.

Dummett, C., & Calder, E. (2023). Rainforests explained: Where they are, what they do, and why they are crucial for our planet, Global Witness, [Online], Retrieved 20 October 2023, from https://www.globalwitness.org/en/blog/rainforests-explained-where-they-are-what-they-do-and-why-theyre-crucial-for-our-planet/?gclid=EAIaIQobChMIjea90LyBgwMVw1xHAR1oDwwvEAAYAiAAEgKSCPD_BwE.

E3G. (2023). Brazil and its upcoming G20 Presidency: empowering emerging giants, E3G, [Online], Retrieved 30 October 2023, from https://www.e3g.org/news/brazil-and-its-upcoming-g20-presidency-empowering-emerging-giants/.

EMBRAPA. (2020). Integrated crop-livestock-forestry systems, EMBRAPA, [Online], Retrieved 17 December 2023, from https://www.embrapa.br/en/tema-integracao-lavoura-pecuaria-floresta-ilpf/busca-de-noticias/-/noticia/55338720/marfrig-launches-carbon-neutral-beef-line-with-embrapa.

EMBRAPA. (2021). New low carbon Brazilian beef protocol allows increased stocking rates with sustainability, EMBRAPA, [Online], Retrieved 3 November 2023, from https://www.embrapa.br/en/busca-de-noticias/-/noticia/61446348/new-low-carbon-brazilian-beef-protocol-allows-increased-stocking-rates-with-sustainability.

EMBRAPA. (2023). "About Embrapa", EMBRAPA, [Online], Retrieved 17 November 2023, from https://www.embrapa.br/en/sobre-a-embrapa.

Emirbayer, M., & Sheller, M. (1999). Publics in history. *Theory and Society, 28*(1), 143–197.

FAO. (2018). *World food and agriculture—statistical pocketbook*. FAO.

Filho, J. P. O. (1990). Frontier security and the new indigenism: Nature and the origins of the Calha Norte project. In D. Goodman & A. Hall (Eds.), *The future of Amazonia: Destruction or sustainable development*. Macmillan.

Fisher, D., & Freudenburg, W. (2001). Ecological modernization and its critics: Assessing the past and looking toward the future. *Society Natural Resources, 14*, 701–709.

Folke, C. (2006). Resilience: The emergence of a perspective for social-ecological systems analyses. *Global Environmental Change, 16*(3), 253–267.

Folke, C., Biggs, R., Norstrom, A. V., Reyers, B., & Rockstrom, J. (2016). Social-ecological resilience and biosphere-based sustainability science. *Ecology and Society, 21*(3).

Folke, C., Carpenter, S. R., Walker, B., Scheffer, M., Chapin, T., & Rockstrom, J. (2010). Resilience thinking: Integrating resilience, adaptability and transformability. *Ecology and Society, 15*(4), 20.

Frijns, J., Phuong, P., & Mol, A. (2000). Ecological modernisation theory and industrialising economies: The case of Viet Nam. *Environmental Politics, 9*(1), 257–292.

Gibbs, D. (2000). Ecological modernisation, regional development and regional development agencies. *Geoforum, 31*(1), 9–19.

Gibbs, H. K., Rausch, L., Munger, J., Schelly, I., Morton, D. C., Noojipady, P., Soares-Filho, B., Barreto, P., Micol, L., & Walker, N. F. (2015). Brazil's Soy moratorium. *Science, 347*(6220), 377–378.

Giddens, A. (2009). *Sociology* (6th ed.). Polity Press.

Glasson, B. (2012). Gentrifying climate change: Ecological modernisation and the cultural politics of definition. *M/C Journal, 15*(3).

Goldstein, B. E., Wessells, A. T., Lejano, R., & Butler, W. (2015). Narrating resilience: Transforming urban systems through collaborative storytelling. *Urban Studies, 52*(7), 1285–1303.

Goodman, D., & Hall, A. (1990). Introduction. In D. Goodman & A. Hall (Eds.), *The future of Amazonia: Destruction or sustainable development*. Macmillan.

Gorz, A. (1989). *Critique of economic reason*. Verso.

Gouldson, A., & Bebbington, J. (2007). Corporations and the governance of environmental risks. *Environment and Planning, 25*(1), 4–20.

GOV. (2022). Brazil creates the National Green Growth Program to consolidate the country as the world's largest environmental superpower. Brazil Government, [Online], Retrieved 14 November 2023, from https://www.gov.br/en/government-of-brazil/latest-news/brazil-creates-the-national-green-growth-program-to-consolidate-the-country-as-the-worlds-largest-environmental-superpower.

GOV. (2023). "Brazil at Rio+20", Brazil Government, [Online], Retrieved 25 October 2023, from http://www.rio20.gov.br/brazil.html.

Gray, J. (2002). *False Dawn: The delusions of global capitalism*. Granta Books.

Hageman, H. (1996). *Bancos, incendiaries, e Florestas Tropicais (banks, fires and tropical forests)*. FASE/IBASE/Instituto Socioambiental.

Hajer, M. (1995). The politics of environmental discourse: Ecological modernisation and the regulation of acid rain. [Online], Retrieved 29 November 2023, from http://www.oxfordscholarship.com https://doi.org/10.1093/019829333X.001.0001.

Hall, A. (1989). *Developing Amazonia: Deforestation and social conflict in Brazil's Carojas Programme*. Manchester University Press.

Hall, A. (1997). *Sustaining Amazonia: Grassroots action for productive conservation*. Manchester University Press.

Hollander, G. (2005). Securing sugar: National security discourse and the establishment of Florida's sugar-producing region. *Economic Geography, 81*(4), 339–358.

Huber, J. (1982). *Die verlorene Unschuld der Ökologie. Neue Technologien und superindustrielle Entwicklung*. Fisher Verlag.

IBAMA. (2023). "Matters", Brazil Government, [Online], Retrieved 7 October 2023, from https://www.gov.br/ibama/pt-br/assuntos.

IEA (International Energy Agency). (2017). Carbon capture, utilisation and storage: A critical tool in the climate energy toolbox, [Online], Retrieved 12 June 2018, from https://www.iea.org/topics/carbon-capture-and-storage/industry/.

Imaflora. (2016). 10-Year of Soy moratorium in the Amazon: History, impacts and expansion into Cerrado Areas. Imaflora, [Online], Retrieved 16 December 2023, from https://www.imaflora.org/public/media/biblioteca/IMF-10-years-of-soy-moratorium-WB.pdf.

IMF (2023). Policy options to address climate vulnerabilities and green the economy. International Monetary Fund, [Online], Retrieved 10 December 2023, from https://www.elibrary.imf.org/view/journals/002/2023/289/article-A005-en.xml.

InfoHub. (2022). Info Hub Brazil, [Online], Retrieved 25 October 2023, from https://infohubbrasil.mma.gov.br/en.

Inglehart, R. (1990). *Culture shift in advanced industrial society*. Princeton University Press.

Jänicke, M. (1990). *State failure*. Pennsylvania State University Press.

Jänicke, M. (1992). Conditions for environmental policy success: An international comparison. *Environmentalist, 12*(1), 47–58.

Jänicke, M. (1995). *The political system's capacity for environmental policy*. Freie Universität.

Jänicke, M. (2005). Trend-setters in environmental policy: The character and role of Pioneer countries. *European Environment, 15*(2), 129–142.

Jessop, B. (2000). The crisis of the national spatio-temporal fix and the tendential ecological dominance of globalizing capitalism. *International Journal of Urban and Regional Research, 24*, 323–360.

Johannisson, B., & Olaison, L. (2007). The moment of truth—reconstructing entrepreneurship and social capital in the eye of the storm. *Review of Social Economy, 65*(1), 55–78.

Jones, R. W., & Diepart, J.-C. (2015). Learning towards resilience. In J.-C. Diepart (Ed.), *Learning for resilience: Insights from Cambodia's rural communities*. The Learning Institute.

Kelly, C., Ferrara, A., Wilson, G. A., Ripullone, F., Nole, A., Harmer, N., & Salvati, L. (2015). Community resilience and land degradation in forest and shrub land socio-ecological systems: A case study in Gorgoglione, Basilicata region, Italy. *Land Use Policy, 46*, 11–20.

Klein, H., & Luna, F. V. (2018). *Feeding the world: Brazil's transformation into a modern agricultural economy*. Cambridge University Press.

Kolk, A. (1996). Forests in international environmental politics. International.

Korhonen, J., & Seager, T. P. (2008). Beyond eco-efficiency: A resilience perspective. *Business Strategy and the Environment, 17*(7), 411–419.

Korsgaard, S., Anderson, A., & Gaddefors, J. (2016). Entrepreneurship as resourcing: Towards a new image of entrepreneurship in a time of financial, economic and socio-spatial crisis. *Journal of Enterprising Communities: People and Places in the Global Economy, 10*(2), 178–202.

Laferriere, E. (1994). Environmentalism and the global divide. *Environmental Politics, 3*(1), 91–113.

Langhelle, O. (2000). Why ecological modernization and sustainable development should not be conflated. *Journal of Environmental Policy and Planning, 2*, 303–322.

Light, I., & Dana, L. P. (2013). Boundaries of social capital in entrepreneurship. *Entrepreneurship: Theory and Practice, 37*(3), 603–624.

Machado, F. S. (2019). Relational rural geographies, resilience, and narratives of small-scale fruit farming in the Metropolitan Countryside of Rio de Janeiro, Brazil. PhD thesis.

Mahar, D. J. (1989). *Government policies and deforestation in Brazil's Amazon region*. World Bank.

Mahar, D. J. (2000). Agro-ecological zoning in Rondonia, Brazil: What are the lessons? In A. Hall (Ed.), *Amazonia at the crossroads*. University of London Press.

Mair, J., & Marti, I. (2006). Social entrepreneurship Eesearch: A source of explanation, prediction and delight. *Journal of World Business, 41*(1), 36–44.

Mamedes, I., Guerra, A., Rodrigues, D. B. B., Garcia, L. C., Godoi, R. F., & Oliveira, P. T. (2023). Brazilian payment for environmental services programs emphasize water-related services. *International Soil and Water Conservation Research, 11*(2), 276–289.

Marsden, T. (1996). Rural geography trend report: The social and political bases of rural restructuring. *Progress in Human Geography, 20*, 246–258.

Marsden, T. (1999). Rural futures: The consumption countryside and its regulation. *Sociologia Ruralis, 39*, 501–520.

Marsden, T., Munton, R., Whatmore, S., & Little, J. (1986). Towards a political economy of capitalist agriculture: A British perspective. *International Journal of Urban and Regional Research, 10*, 489–521.

Marsden, T., Murdoch, J., Lowe, P., Munton, R., & Flynn, A. (1993). *Constructing the countryside*. UCL Press.

McCleary, R. (1991). The international community's claim to rights in Brazilian Amazonia. *Political Studies, 39*(3), 692.

McDonald, M. (2010). Environment and security: Global eco-politics and Brazilian deforestation. *Contemporary Security Policy, 24*(2), 69–94.

McNaughton, R. B., & Gray, B. (2017). Entrepreneurship and resilient communities—introduction to the special issue. *Journal of Enterprising Communities: People and Places in the Global Economy, 11*(1), 2–19.

MFA. (2023). Letter from Developing Countries to European Authorities regarding the Entry into Force of the European Union's So-called 'Anti-Deforestation Law, Brazil Government, [Online], Retrieved 30 October 2023, from https://www.gov.br/mre/pt-br/canais_atendimento/imprensa/notas-a-imprensa/carta-de-paises-em-desenvolvimento-a-autoridades-europeias-sobre-a-entrada-em-vigor-da-chamada-201clei-antidesmatamento201d-da-uniao-europeia.

Mol, A. (1995). *The refinement of production. Ecological modernization theory and the chemical industry*. Jan van Arkel/International Books.

Mol, A. (1999). Ecological modernization and the environmental transformation of Europe: Between national variations and common denominators. *Journal of Environmental Policy and Planning, 1*, 167–181.

Mol, A. (2000). The environmental movement in an era of ecological modernization. *Geoforum, 31*, 45–56.

Mol, A. (2002). Ecological modernization and the global economy. *Global Environmental Politics, 2*(2), 92–115.

Mol, A. (2006). From environmental sociologies to environmental sociology? A comparison of US and European Environmental Sociology. *Organisation and Environment, 19*(1), 5–27.

Mol, A., & Spaargaren, G. (1993). Environment, modernity and the risk society: The apocalyptic horizon of environmental reform. *International Sociology, 8*, 431–459.

Mol, A., & Spaargaren, G. (2000). Ecological modernisation theory in debate: A review. *Environmental Politics, 9*(1), 17–49.

Mol, A., Sonnenfeld, D., & Spaargaren, G. (2009). *The ecological modernization reader: Environmental reform in theory and practice*. Routledge.

Mormont, M. (1990). Who is rural? Or, how to be rural: Towards a sociology of the rural. In T. Marsden, P. Lowe, & S. Whatmore (Eds.), *Rural restructuring: Global processes and their response*. David Futon.

Morris, C., & Kirwan, J. (2011). Ecological embeddedness: An interrogation and refinement of the concept within the context of alternative food networks in the UK. *Journal of Rural Studies, 27*(3), 322–330.

Morrison, C., Ramsey, E., & Bond, D. (2017). The role of social entrepreneurs in developing community resilience in remote areas. *Journal of Enterprising Communities: People and Places in the Global Economy, 11*(1), 95–112.

Munoz, P., & Cohen, B. (2017). Towards a social-ecological understanding of sustainable venturing. *Journal of Business Venturing Insights, 7*, 1–8.

Munton, R. (1995). Regulating rural change: Property rights, economy and environment—a case study from Cumbria, UK. *Journal of Rural Studies, 11*, 269–284.

Murdoch, J., & Marsden, T. (1994). *Reconstituting rurality*. UCL Press.

Nepstad, D., McGrath, D., Stickler, C., Alencar, A., Azevedo, A., Swette, B., Bezerra, T., DiGiano, M., Shimada, J., da Motta, R. S., & Armijo, E. (2014). Slowing Amazon deforestation through public policy and interventions in Beef and Soy supply chains. *Science, 344*(6188), 1118–1123.

Newby, H. (1985). *Green and pleasant land? Social change in rural England*. Hutchinson.

Obach, B. (2007). Theoretical interpretations of the growth in organic agriculture: Agricultural modernization or an organic treadmill. *Society & Natural Resources: An International Journal, 20*(3), 231.

O'Brien, K., Hayward, B., & Berkes, F. (2009). Rethinking social contracts: Building resilience in a changing climate. *Ecology and Society, 14*(2), 12.
O'Brien, R., & Williams, M. (2016). *Global political economy: Evolution and dynamics* (5th ed.). Palgrave.
OC. (2021). "Laboratório do Observatório do Clima v. Minister of Environment and Brazil", Laboratório do Observatório, [Online], Retrieved 28 October 2023, from https://climatecasechart.com/non-us-case/laboratorio-do-observatorio-do-clima-v-environmental-ministry-and-brazil/.
Oliveira, P. P. A. (2014). *Protocolo para Quantificação dos Estoques de Carbono do Solo da Rede de Pesquisa Pecus*. Embrapa Pecuária Sudeste.
O'Neill, K. (1998). Global to local, public to private: Influences on environmental policy change in industrialized countries. Paper presented at the 14th world congress of sociology, Montreal, July. Montreal: CSA.
Paehlke, R., & Torgerson, D. (1990). *Managing leviathan: Environmental politics and the administrative state*. Broadview Press.
Pepper, D. (1984). *The roots of modern environmentalism*. Croom Helm.
Perz, S. G. (2002). The changing social contexts of deforestation in the Brazilian Amazon. *Social Science Quarterly, 83*(1), 35–43.
Pieterse, J. N. (1995). Globalisation as hybridisation. In M. Featherstone, S. Lash, & R. Robertson (Eds.), *Global modernities*. Sage.
Pilgrim, S. E., Cullen, L. C., Smith, D. I., & Pretty, J. (2008). Ecological knowledge is lost in wealthier communities and countries. *Environmental Science & Technology, 42*(4), 1004–1009.
Plenamata. (2023). Environmental Regularization Program (PRA), Plenamata, [Online], Retrieved 26 October 2023, from https://plenamata.eco/en/verbete/programa-de-regularizacao-ambiental-pra/.
Pretty, J. N. (1995). *Regenerating agriculture: Policies and practice for sustainability and self-reliance*. Earthscan.
PRS. (2023). "Sustainable rural program—strengthening low-carbon agriculture", Sustainable Rural Program, [Online], Retrieved 12 October 2023, from https://programaruralsustentavel.org.br/.
Rainer, G. (2016). Constructing globalised spaces of tourism and leisure: Political ecologies of the Salta Wine Route (NW-Argentina). *Journal of Rural Studies, 43*, 104–117.
Rofe, M. W. (2009). Globalisation, gentrification and spatial hierarchies in and beyond New South Wales: The local/global nexus. *Geographical Research, 47*(3), 292–305.
Schnaiberg, A. (1980). *The environment*. Oxford University Press.
Salton, J. C., Mercante, F. M., Tomazi, M., Zanatta, J. A., Concenço, G., Silva, W. M., & Retore, M. (2014). Integrated crop-livestock system in tropical Brazil: Toward a sustainable production system. *Agriculture, Ecosystems & Environment, 190*, 70–79.
Sette, C., & Ekboir, J. (2013). *An overview of rural extension in Brazil: The current situation*. Rome: Institutional learning and change working paper 14.
Seyfang, G., & Smith, A. (2007). Grassroots innovations for sustainable development: Towards a new research and policy agenda. *Environmental Politics, 16*(4), 584–603.
Shepherd, D. A., & Patzelt, H. (2011). The new field of sustainable entrepreneurship: Towards a new research and policy agenda. *Environmental Politics, 16*(4), 137–163.
Shrivastava, P., & Kennelly, J. J. (2013). Sustainability and place-based enterprise. *Organisation & Environment, 26*(1), 83–101.
Simonis, U. (1989). Ecological modernization of industrial society: Three strategic elements. *International Social Science Journal, 121*, 347–361.
Smith, N. (1990). *Uneven development: Nature, capital and the production of space* (2nd ed.). Blackwell.
Sonnino, R., & Griggs-Trevarthen, C. (2013). A resilient social economy? Insights from the community food sector in the UK. *Entrepreneurship & Regional Development, 25*(3/4), 272–292.
Spaargaren, G. (1997). The ecological modernization of production and consumption: Essays in environmental sociology. Wageningen University. PhD thesis.
Spaargaren, G., & Mol, A. (1992). Sociology, environment, and modernity: Ecological modernization as a theory of social change. *Society & Natural Resources, 5*(4), 323–344.
Staggenborg, S., & Ogrodnik, C. (2015). New environmentalism and transition Pittsburgh. *Environmental Politics, 24*(5), 723–741.
Statista. (2023). Value added by the agriculture, forestry, and fishing sector to the gross domestic product (GDP) in Brazil from 2013 to 2022, Statista, [Online], Retrieved 23 November 2023, from https://www.statista.com/statistics/1075048/brazil-agriculture-value-added-gdp/.
Stern, N., Peters, S., Bakhshi, V., Bowen, A., Cameron, C., Catovsky, S., Crane, D., Cruickshank, S., Dietz, S., Edmonson, N., Garbett, L., Hamid, L., Hoffman, G., Ingram, D., Jones, B., Patmore, N., Radcliffe, H., Sathiyarajah, R., Stock, M., Taylor, C., Vernon, T., Wanjie, H., & Zenghelis, D. (2006). *Stern review: The economics of climate change*. HM Treasury.
Stickler, C. M., Nepstad, D., Azevedo, A. A., & McGrath, D. G. (2013). Defending public interests in private lands: Compliance, costs and potential environmental consequences of the Brazilian Forest code in Mato Grosso. *Philosophical Transactions Royal Society B-Biological Sciences, 363*(1619), 20120160.
Storr, V. H., Haeffele-Balch, S., & Grube, L. E. (2015). *Community revival in the wake of disaster—lessons in local entrepreneurship*. Palgrave Macmillan.
UN. (2023). United Nations Conference on Sustainable Development, 20–22 June 2012, Rio de Janeiro. UN, [Online], Retrieved 28 October 2023, from https://www.un.org/en/conferences/environment/rio2012.

UNFCCC. (2023). Federative Republic of Brazil Nationally Determined Contribution (NDC) to the Paris Agreement under the UNFCCC, UNFCCC, [Online], Retrieved 28 October 2023, from https://unfccc.int/sites/default/files/NDC/2023-11/Brazil%20First%20NDC%202023%20adjustment.pdf.

Valdes, C. (2022). Brazil's momentum as a global agricultural supplier faces headwinds. USDA, [Online], Retrieved 10 October 2023, from https://www.ers.usda.gov/amber-waves/2022/september/brazil-s-momentum-as-a-global-agricultural-supplier-faces-headwinds/.

Vlasov, M., Bonnedahl, K. J., & Vincze, Z. (2018). Entrepreneurship for resilience: Embeddedness in place and in trans-local grassroots networks. *Journal of Enterprising Communities: People and Places in the Global Economy, 12*(3), 374–394.

Walker, B., & Salt, D. (2006). *Resilience thinking: Sustaining ecosystems and people in a changing world.* Island Press.

Walter, G. R. (2002). Economic, ecology-based communities, and sustainability. *Ecological Economics, 42*(1/2), 81–87.

Watson, C., & Schalatek, L. (2020). Climate finance thematic briefing: REDD+ Finance, Heinrich Boll Stiftung, [Online], Retrieved 12 October 2023, from https://climatefundsupdate.org/wp-content/uploads/2020/03/CFF5-2019-ENG-DIGITAL.pdf.

Watts, J. (2023). Lula's bid to style himself climate leader at Cop28 undermined by opec move, the Guardian, [Online], Retrieved 21 October 2023, from https://www.theguardian.com/environment/2023/dec/02/lula-climate-leader-cop28-brazil-undermined-by-opec-move.

Watts, M. (1983). *Silent violence: Food, famine and peasantry in northern Nigeria.* University of California Press.

WBG. (2023). Brazil country climate and development report. World Bank Group, , [Online], Retrieved 12 December 2023, from https://www.worldbank.org/en/news/infographic/2023/05/08/brazil-country-climate-and-development-report.

Weale, A. (1992). *The new politics of pollution.* Manchester University Press.

West, T. A. P., & Fearnside, P. M. (2021). Brazil's conservation reform and the reduction of deforestation in Amazonia. *Land Use Policy, 100*, 105072.

Whiteman, G., & Cooper, W. H. (2011). Ecological Sensemaking. *Academy of Management Journal, 54*(5), 889–911.

Wilson, G. A. (2007). *Multifunctional agriculture: A transition theory perspective.* CABI.

Wilson, G. A. (2012). *Community resilience and environmental transitions.* Earthscan.

Woods, M. (2005). *Rural geography: Processes, responses and experiences in rural restructuring.* Sage.

Woods, M. (2007). Engaging the global countryside: Globalisation, hybridity and the reconstitution of rural place. *Progress in Human Geography, 31*, 485–507.

York, R., & Rosa, E. (2003). Key challenges to ecological modernization theory: Institutional efficacy, case study evidence, units of analysis, and the pace of eco-efficiency. *Organization & Environment, 16*(3), 273–288.

Young, O. (1994). *International governance: Protecting the environment in a stateless society.* Cornell University Press.

Zhou, Z. W., Nepstad, D., Shimada, J., & Baxter, T.Y. (2022). Country specific study: Understanding Brazilian agriculture sector and trade and investment relationship with China. Global Environmental Institute, [Online], Retrieved 10 October 2023, from http://www.geichina.org/news/publications/.

Vegetation Rehabilitation for Rangeland Resilience in China: Comparative Effectiveness Between Grazing Control and Ecological Engineering

Xiangwei Zhang and Haibin Chen

Abstract

Vegetation rehabilitation is the key for recovering the ecological resilience of degraded rangelands, yet an important knowledge gap remains in understanding the comparative effectiveness of measures between grazing control and ecological engineering. Taking Qinghai Province of China as a case study, the chapter tries to narrow the gap by exploring the comparative efficiency of two vegetation rehabilitation programs, i.e., the Rangeland Ecological Protection Subsidy and Reward (REPSR) scheme, which focuses on grazing intensity management, and the Return Grazing Land to Rangeland (RGLR) project, which combines ecological engineering measures. Spatial panel modelling results confirm that both programs promoted the desired vegetation rehabilitation, while the marginal return of the REPSR was larger than that of the RGLR. Differential levels of negative spatial spillover problem in terms of grazing pressure shift are revealed in sub-regions. To improve the investment effectiveness and address the spatial spillover problem, locally adapted and spatially holistic intervention measures and a flexible compensation scheme are suggested.

Keywords

Ecological resilience · Vegetation rehabilitation · Rangeland degradation · Grazing restriction · Ecological engineering

1 Introduction

Vegetation constitutes an essential element of the terrestrial ecosystem (Zhang & Jin, 2021), and vegetation conservation is critical for sustaining a multiplicity of ecosystem services, such as soil retention (Liu et al., 2022), carbon sequestration (Shi et al., 2023), freshwater supply, and hydrological regulation (Lu et al., 2018). However, due to climate change and the ongoing encroachment by human activities, vegetation degradation is evident across the globe and has triggered a series of natural and social problems, such as the increased frequency, severity, and duration of droughts and floods, food insecurity, and land desertification (Kooch et al., 2022; Liu et al., 2019b; Maria Teixeira Lins et al., 2023). Reversing the degradation trend and rebuilding plant ecosystems are thus imperative and have been a common concern of international academia and conservation practitioners. In the Anthropocene, vegetation rehabilitation is also of

X. Zhang · H. Chen (✉)
College of Economics and Management, Northwest A&F University, Yangling, China
e-mail: zhang_xw@nwafu.edu.cn; hbchen@nwafu.edu.cn

Y. Li (ed.), *Global Perspectives on Building Resilience for Sustainable Rural Development*, Sustainable Development Goals Series, https://doi.org/10.1007/978-981-96-6149-7_7

significant value for enhancing resilience of social-ecological systems (SESs) to deal with the increasing uncertainties. In sustainability science, resilience emphasizes human adaptation to environmental change and explores alternative paths and transformations in the face of external pressures and shocks (Folke et al., 2010; Cumming & Peterson, 2017). Nature-based solutions (NBSs) are recognized as an effective approach for the conservation, sustainable use, and restoration of natural or modified ecosystems, providing benefits to both human well-being and biodiversity simultaneously (Bianciardi et al., 2023). Through the purposeful protection, management, or restoration of ecosystems, NBSs emphasize that people are not passive beneficiaries of nature but can enhance system resilience and address major challenges (Cohen-Shacham et al., 2019).

Ecological restoration programs (ERPs) are an important component of NBSs (Bustamante et al., 2019; Cohen-Shacham et al., 2019). A large number of ERPs have been rolled out globally (Liu et al., 2019b), in which enlarging vegetation coverage and improving biomass productivity are frequently among the priority goals and listed as primary evaluation criteria of programs' success (Jiang et al., 2020). For instance, it is reported that 16 nationwide ERPs have been implemented in China from 1983 to 2015 (Wang et al., 2021), and have remarkably improved habitat quality and vegetation productivity across the country (Chen et al., 2022b). Covering nearly 25% of the earth's land surface, rangeland contributes 20% of the world's carbon stock and supports the livelihoods of more than 1 billion people, mostly in the developing world (Chen et al., 2017; Liu et al., 2019b), thus also receives significant attention among the conservation community. Of the 16 ERPs in China, for example, the area of grass gain projects accounts for 47.50%, slightly behind tree gain projects (48.55%) (Chen et al., 2022b). Likewise, 539 LIFE Nature projects have been rolled out by the European Union between 1996 and 2020, with a core goal to restore European natural and semi-natural grasslands (Goret et al., 2021).

In China, rangeland means more than ecological services. Since it is home to more than 70% of the country's ethnic minority population, rangeland conservation is of vital value in sustaining the livelihoods of millions of ethnic minority herders and maintaining social stability in pastoral regions (Chen et al., 2022a; Li et al., 2021a). Still, about 90% of natural rangeland in China has been degraded to varying extents by the end of the twentieth century (Li et al., 2021a; Zhang et al., 2020). In order to restore the rangeland ecology, the government has executed a range of nature-based ERPs, in which the Return Grazing Land to Rangeland project (RGLR project), launched in 2003, and the Rangeland Ecological Protection Subsidy and Reward scheme (REPSR scheme), launched in 2011, are the two most important. The RGLR mainly applies a combination of grazing intensity management and ecological engineering measures to certain designated project areas, such as grazing ban, grazing rest, rotational grazing, supplementary grass seed sowing, etc. (Chen et al., 2017); the REPSR covers a much wider area and predominantly applies livestock management strategy, in which grazing ban and rangeland-livestock balance are the two principal approaches (Yin et al., 2019). Grazing ban largely aims at seriously degraded rangelands and ecological hotspots, such as water and biodiversity conservation areas, while rangeland-livestock balance is typically applied on rangelands with more benign soil and water conditions and without significant degradation (Hu et al., 2019). As of 2018, the RGLR project had invested a total amount of 29.57 billion Yuan.[1] As of 2021, the first two rounds of the REPSR scheme had invested over 150 billion Yuan, with a total implementation area of 3.82 billion mu (1 mu = 0.067 ha), of which the implementation area of grazing ban was 1.21 billion mu and the implementation area

[1]The State Council of the People's Republic of China, 2018. China has invested nearly 30 billion yuan on the Return Grazing Land to Rangeland project. Available online at. http://www.gov.cn/xinwen/2018-07/17/content_5307177.htm. (Accessed 20 March 2023).

of rangeland-livestock balance was 2.61 billion mu.[2]

The positive effect of Chinese rangeland ERPs in restoring vegetation coverage has been generally recognized by both government reports and academic literature (Chen et al., 2019; Ning et al., 2022). There is ample evidence that the implementation of these projects has led to an increase in vegetation coverage and biodiversity and an alleviation of soil erosion and water loss problems, therefore contributing to the virtuous cycle and resilience of rangeland ecosystems (Luo et al., 2019; Yang et al., 2018). Nonetheless, several studies argue that engineering-restored vegetation that normally features a monoculture structure displays a low survival rate and is unstable (He et al., 2021). Several studies also point out that although ERPs are designed to mitigate the impacts of climate change and other shocks, their effectiveness depends on local social-ecological contexts (Bustamante et al., 2019). Moreover, in the context of uncertainties, the design and implementation of ERPs need to strengthen the adaptability to changes rather than simply restore to the status quo ante (Simonson et al., 2021). Some raise concerns about the rigidity of the top-down implementation and the lack of local adaptability of the one-size-fits-all approach, pointing out that moderate grazing pressure can in fact promote the synergy of the supplies of various ecosystem services (Fan et al., 2019), whereas the negligence of the interconnectedness and integrity of rangeland ecosystems may lead to the increase in ecosystem vulnerability (Li et al., 2015; Lu et al., 2017). Previous field experiments on grazing intensity find that moderate-intensity grazing could help inhibit the growth of poisonous weeds and increase plant species diversity by selective foraging of animals (McNaughton, 1983), promote the activity of the soil microbial community and improve soil fertility by supplying animal excrements, etc. (Du et al., 2020; McNaughton, 1985), and meanwhile enhance the ecosystem resilience by continuous moderate disturbance (Li et al., 2021b). Grazing ban by building fences limits the animal mobility, which might contribute to vegetation restoration within the project area in the short run, but could inflict harm on the ecosystem integrity by eliminating compensatory interactions between plants and animals in the long run (Liu et al., 2020; Sun et al., 2020).

The negative impact on the livelihoods of some smallholders is also a major concern, since the indiscriminate area-based compensation scheme ignores the unevenness problem in household rangeland holding as well as the differential livelihood dependence on livestock production. As a result, illegal grazing activities are not uncommon, given the difficulty of detecting overgrazing in the vast rangeland regions and enforcing government rules in the pastoral countryside with strong bonding ties (Qiu et al., 2020; Wang & Lo, 2022). Grazing pressure may be transferred from project areas to non-project areas, causing a spatial spillover problem without reasonable accountability (Chen et al., 2017; Li & Bennett, 2019; Sun et al., 2020). More importantly, as a Payment for Ecosystem Service program, the compensation scheme is the main incentive for pastoralists to comply with the rules of grazing exclusion and maintain a reasonable stocking rate set by government. However, studies on the impacts of program investment have remained barely seen, speculatively due to a lack of disclosure of official statistics regarding investment amounts by programs across administrative regions. Thus, more empirical investigations are needed to better understand the differential impacts of investments on vegetation recovery under heterogeneous local natural and social contexts and through divergent intervention approaches promoted by different programs.

Taking Qinghai Province of China as a case study, the current study aims to narrow the knowledge gap regarding the comparative effectiveness of measures between grazing restriction and ecological engineering taken by the range-

[2]The State Council of the People's Republic of China, 2021. Over 12 million farmers and pastoralists have benefited from the implementation of the Rangeland Ecological Protection Subsidy and Reward Scheme in the past 10 years. Available online at. http://www.gov.cn/xinwen/2021-12/06/content_5656033.htm. (Accessed 20 March 2023).

land ERPs. Using a longitudinal dataset comprising MODIS NDVI, meteorological observation, demographic and economic statistics, as well as the investment statistics on the RGLR project and REPSR scheme, the study examines the spatial and temporal changes in vegetation coverage across Qinghai from 2000 to 2017 and evaluates the investment effectiveness and relative efficiency of the two programs on vegetation rehabilitation at the county level, while also exploring the potential spatial spillover problems across county boundaries. The study area is selected based on two considerations. First, with a wide distribution of alpine rangeland and with livestock farming as the primary livelihood source for a majority of the rural population, Qinghai is a key region that has implemented both RGLR project and REPSR scheme since their beginning. Still, compared with other well-studied areas, such as Inner Mongolia, the effectiveness of the ERPs in Qinghai has received less attention. Second, situated at the Qinghai-Tibet Plateau, climate in Qinghai is predominantly hypoxic, cold, and arid, with significant spatial and temporal variation. Likewise, landform is also very diverse and changeable. The high degree of heterogeneity in local climate, soil, and hydrological conditions may lead to noteworthy spatial variability in vegetation cover and may affect the local effectiveness of the ERPs. With empirical evidence and spatial modelling, the study contributes to informing future refinement of the RGLR project and REPSR scheme as well as other rangeland conservation interventions, in the hope of informing the development of more locally adapted, spatially holistic, investment-wise conservation measures and safeguarding the long-term resilience and sustainability of rangeland ecosystems in China and beyond.

2 Material and Methods

2.1 Study Area

Qinghai Province is located in northwest China (31°36′–39°19′N, 89°35′–103°04′E) and on the Qinghai-Tibetan Plateau, with an average elevation exceeding 3000 m. Because of high altitude, the climate is distinguished by hypoxia, coldness, and dryness, exhibiting an annual precipitation that averages between 50 and 550 mm, and an annual temperature that typically falls within the range of −5 to 8.5 °C (Teng et al., 2022). Due to the diversity and complexity of the terrain, the climate is also typified by high spatial and temporal variability. As one of the five major pastoral areas in China, Qinghai is rich in rangeland resource, covering about 96% of the province's land area. There is a total of 40.53 million ha of natural rangeland, in which the available rangeland accounts for 87% (35.26 million ha). Alpine meadow is the main type of rangeland and widely distributed. As of 2019, the province has a population of 6,078,200 people, mainly of ethnic minorities such as the Tibetan, and has a per capita GDP of 49,976 yuan (BS, 2022) (Fig. 1).

During the last two decades, Qinghai Province has executed a set of national rangeland ERPs. Among them, the RGLR project began in 2003 and is renewed every 5 years. In the first round, 12 counties from Yushu and Guoluo prefectures were involved; since then, the implementation scope has been expanded.[3] By the end of 2017, Qinghai Province had invested 5.398 billion yuan in the RGLR project, improving about 38.026 million mu of grassland, constructing nearly 280,000 mu of artificial forage land, restoring 400,000 mu of 'black beach' degraded rangeland, and building about 18,782.16 mu of fencing. The subsidy offered is 30 yuan/mu for fence construction, 60 yuan/mu for sowing improved grass seeds, 200 yuan/mu for artificial forage planting, and 180 yuan/mu for the 'black beach' degraded rangeland restoration.[4] The REPSR scheme was launched in 2011, covering a wider

[3]National Food and Strategic Reserves Administration. 2003. Qinghai Province Launches the Return Grazing Land to Rangeland project. Available online at. http://www.lswz.gov.cn/html/gzdt2007/2018-06/13/content_221058.shtml. (Accessed 20 March 2023).

[4]People's Government of Xinghe County. 2016. The subsidy standard of the Return Grazing Land to Rangeland project has been increased. Available online at. https://www.xinghe.gov.cn/information/xinghe11356/msg2323556784012.html. (Accessed 20 March 2023).

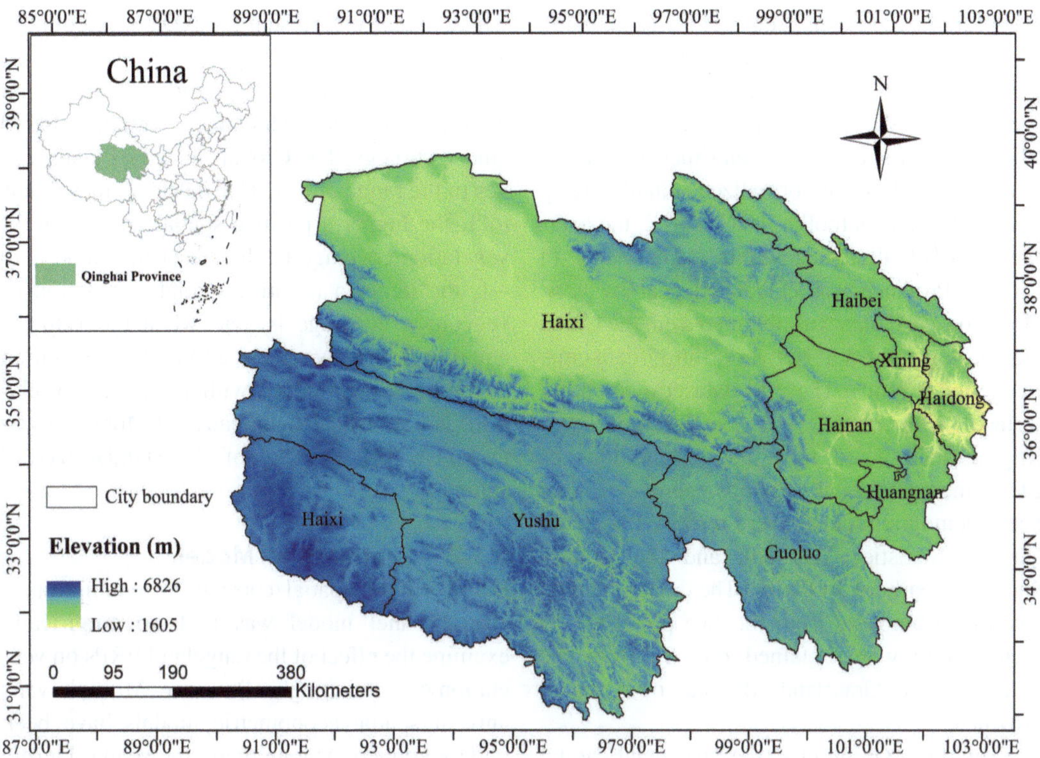

Fig. 1 Study area

area and making more investment. The first round (2011–2015) implemented in Qinghai Province enrolled 24,547 mu for grazing ban, 22,888 mu for rangeland-livestock balance, with an annual total investment of 1.816 billion yuan; the total enrollment areas under both grazing ban and rangeland-livestock balance remained basically unchanged in the second round (2016–2020), but the subsidy standards were raised, and the annual investment reached 2.413 billion yuan for the whole province.[5] At present, the subsidy for grazing ban and the reward for rangeland-livestock balance offered by the central government are 7.5 and 2.5 yuan/mu respectively, while local governments are encouraged to make matching investment.[6]

2.2 Data Preparation

The NDVI data from 2000 to 2017 were retrieved from the MODIS MOD13Q1 product produced by NASA. Considering that the vegetation index is significantly affected by snow reflection during the non-growth season, this study selects to only use the data from the growth season (May to October). A total of 216 remote sensing images were obtained. First, the MODIS Reprojection Tool (MRT) was used to perform image merge,

[5]The General Office of the People's Government of Qinghai Province. 2016. The Implementation Plan for the New Round of the Rangeland Ecological Protection Subsidy and Reward scheme (2016-2020). Available online at. http://www.qinghai.gov.cn/xxgk/xxgk/fd/zfwj/201712/t20171222_20715.html. (Accessed 20 March 2023).

[6]The General Office of the People's Government of Qinghai Province, 2021. The Implementation Plan for the Third Round of the Rangeland Ecological Protection Subsidy and Reward scheme in Qinghai Province (2021–2025). Available online at. http://www.qinghai.gov.cn/xxgk/xxgk/fd/zfwj/202201/t20220113_188910.html. (Accessed 20 March 2023).

reprojection, and format conversion. Then, using the Band Math tool in *ENVI 5.2,* the images of each month were processed by the Average Value Composite (AVC) method to obtain the average NDVI value for each year, and then the NDVI values of pixels were normalized. Finally, using the zonal statistics tool in *ArcGIS 10.2*, the average value of all pixels within each county in Qinghai Province was calculated year by year. Meteorological data was obtained from the China Meteorological Data Network (http://data.cma.cn/site/index.html). The grid dataset of monthly surface temperature and precipitation was Kriging-interpolated from the observations of 2472 meteorological stations across China. Socio-demographic data were retrieved from the Qinghai Statistics Yearbook and the County Statistics Yearbook of China. The data on annual investment amounts from the two programs by each county were obtained directly from the Forestry and Grassland Bureau of Qinghai Province.

Excluding Xining city (the provincial capital with little rangeland) and Golmud city, Dachaidan Administrative Committee and Mangya city in the northwest (due to a lack of statistical data and changes in administrative boundaries), the study takes 38 county-level administrative regions as analysis units. All above indicators were calculated for each county on an annual basis.

2.3 Methods

2.3.1 Global Moran's I

The global Moran's I index was used to evaluate the spatial autocorrelation of vegetation coverage in Qinghai Province (Fu et al., 2014). Its calculation formulas are as follows:

$$\text{Global Moran's I} = \frac{\sum_{i=1}^{n}\sum_{j=1}^{n} W_{ij}\left(Y_i - \bar{Y}\right)\left(Y_j - \bar{Y}\right)}{S^2 \sum_{i=1}^{n}\sum_{j=1}^{n} W_{ij}} \tag{1}$$

$$S^2 = \frac{1}{n}\sum_{i=1}^{n}\left(Y_i - \bar{Y}\right)^2 \tag{2}$$

where y_i and y_j represent the NDVIs of county i and j at a given year, respectively. n is the total number of counties. W is the spatial weight matrix generated using the contiguity-based Queen strategy. The Global Moran's I takes values in the range of $[-1, 1]$. The closer its value is to 1, the higher a positive spatial correlation of vegetation coverage is; the closer the value is to -1, the larger a negative spatial correlation of vegetation coverage is; and when the value is equal to 0, it is hypothesized that the vegetation coverage is randomized without spatial correlation. The statistical significance of Moran's I was tested by the P value of the standardized Z statistic.

2.3.2 Spatial Panel Model

Based on the spatial correlation assumption, a spatial panel model was built to empirically examine the effect of the rangeland ERPs on vegetation cover in Qinghai Province. Multiple variants of spatial econometric models have been widely applied. Among them, the spatial Durbin model (SDM) is the most general form and takes the spatial correlation of both independent and dependent variables into account and can be simplified to SAR or SEM under certain conditions (Espoir & Sunge, 2021). Thus, this study first constructs an SDM model and then applies the Lagrange Multiplier test, Likelihood Ratio (LR) test, and Wald test to determine whether the SDM should be simplified. The SDM model is specified as follows:

$$Y_{it} = \rho\sum_{j=1}^{n} W_{ij}Y_{jt} + \beta X_{it} + \theta\sum_{j=1}^{n} W_{ij}X_{it} + \mu_i + \nu_t + \varepsilon_{it} \tag{3}$$

where Y_{it} is the explained variable, representing the NDVI value of county i at year t, X_{it} is a set of explanatory variables, and W_{ij} is the spatial weight of county i and j. ρ and θ are the to-be-estimated parameters of spatial dependency; β is a group of parameters for the explanatory variables; μ_i and ν_t represent the spatial fixed effect and the time fixed effect, respectively; ε_{it} is a random error that is assumed to follow an independent identical distribution.

2.3.3 Variable Selection and Definition

Referring to previous studies, the current chapter takes the annual mean NDVI at the county level as the explained variable. Vegetation coverage is a key indicator of ecological quality, and its change is driven by both natural and human factors (Gu et al., 2022). As for the former, climate, topography, and soil factors are frequently considered (He et al., 2021; Zhang & Jin, 2021), whereas the latter mainly involves human activities that may affect vegetation dynamics directly or indirectly, which in the current study encompass livestock grazing, infrastructure construction, mineral extraction, tourism development, as well as the implementation of the rangeland ERPs (Gu et al., 2022; Liu et al., 2019b; Zhang & Jin, 2021). Accordingly, this study chooses the accumulated precipitation and mean temperature during the growth season (May to October), grazing intensity, economic intensity, population density, and investment intensity of rangeland ERPs as explanatory variables. The definition and descriptive statistics of variables are reported in Tables 1 and 2, respectively.

3 Results

3.1 NDVI Change Across Time and Space

The vegetation coverage in Qinghai displayed an obvious gradient decreasing from southeast to northwest. Five segments of multi-year mean NDVI accounted for 31.56% (0–0.15), 20.03% (0.15–0.3), 16.04% (0.3–0.5), 16.99% (0.5–0.7), and 15.38% (0.7–1) of the total area of Qinghai Province, respectively (Fig. 2). Among them, the lowest-value segment (0–0.15) covered the largest area and was mainly found in the northwest, where the Tsaidam Desert is situated, characterized by low precipitation, high evaporation, and poor soil quality. The highest-value segment (0.7–1) covered the least area and was largely found around Qinghai Lake, and at the foot of the Qilian and Kunlun Mountains. Despite low rainfall, water flow melting from alpine snow and glaciers, as well as water convergence in the lowland of the Qinghai Lake, provide a beneficial condition for vegetation growth.

From 2000 to 2017, the mean NDVI in Qinghai Province exhibited an overall increasing

Table 1 Variable definitions

Type	Notation	Definition	Unit	Exp. sign
Explained variable	$NDVI_{it}$	Mean NDVI value of all pixels within county *i* during the growth season at year *t*	na	–
Explanatory variables	$RGLR_{it}$	Investment amount of the RGLR project at year *t* in county *i* divided by its total land area	10,000 Yuan/km^2	+
	$REPSR_{it}$	Investment amount of the REPSR scheme at year *t* in county *i* divided by its total land area	10,000 Yuan/km^2	+
	$prec_{it}$	Total precipitation during the growth season of year *t* in county *i*	mm	+
	$temp_{it}$	Mean temperature during the growth season of year *t* in county *i*	°C	+/–
	$stock_{it}$	Total meat production of county *i* at year *t* divided by its total land area	Ton/km^2	–
	pop_{it}	Total human population of county *i* at year *t* divided by its total land area	Person/km^2	–
	gdp_{it}	CPI-adjusted GDP of county *i* at year *t* divided by its total land area (Year 2000 as the base)	10,000 Yuan/km^2	–

Table 2 Descriptive statistics for variables

Variable	Unit	Mean	Std. dev.	Sample size
$NDVI_{it}$	na	0.5736	0.1579	684
$RGLR_{it}$	10,000 Yuan/km^2	0.6421	1.5143	684
$REPSR_{it}$	10,000 Yuan/km^2	1.8318	3.1909	684
$prec_{it}$	mm	719.7728	129.7706	684
$temp_{it}$	°C	6.1283	1.8105	684
$stock_{it}$	Ton/km^2	2.2895	3.1083	684
pop_{it}	Person/km^2	37.0167	57.0722	684
gdp_{it}	10,000 Yuan/km^2	361.5610	758.0698	684

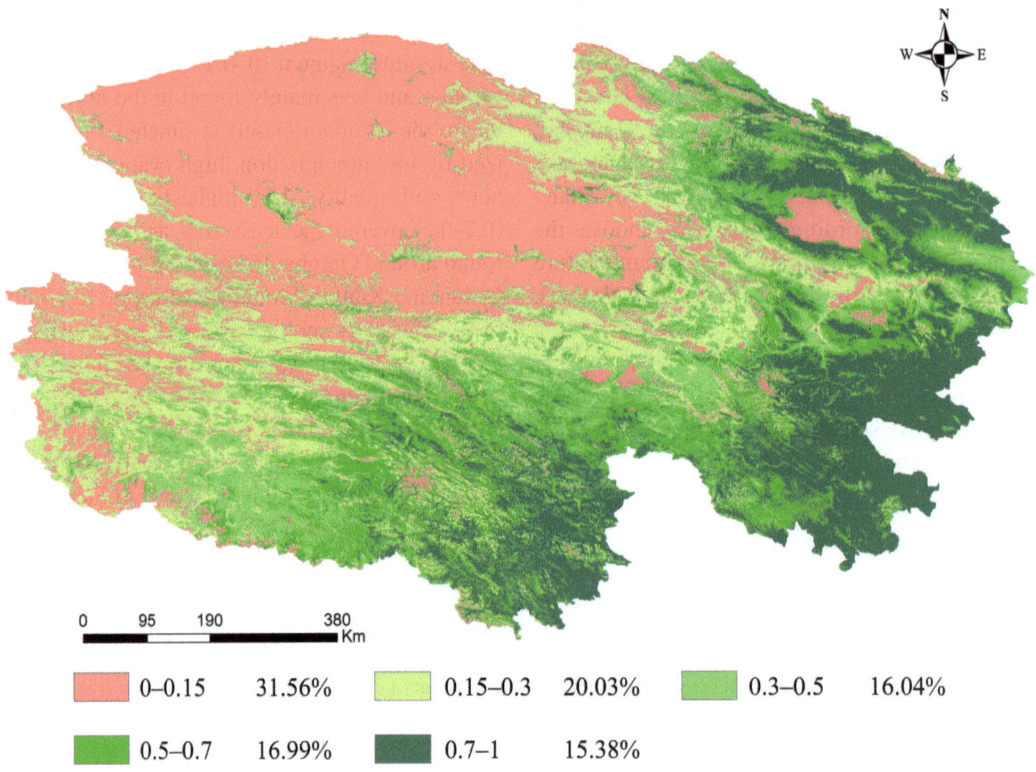

Fig. 2 Multi-year average of NDVI across Qinghai Province

trend at a growth rate of 0.0021 annually (R^2=0.3592, Fig. 3). Three distinct phases can be identified, with a steady greening phase at the beginning (2000–2005), followed by a fluctuating greening phase (2005–2012), and finally a browning phase (2012–2017). Across space, the areas experiencing an improvement in vegetation coverage comprised 75.30% of the total, of which 11.02% was significantly improved (annual mean NDVI increase larger than 0.1) and 64.28% slightly improved (annual mean NDVI increase larger than 0 but lower than 0.1). Vegetation greening was found across the province, but a majority of the significantly improved areas were distributed in the northern and western parts. Still, the vegetation coverage in 24.70% of the province exhibited a reduction in NDVI, although most of them were only slightly degraded (22.45%, annual mean NDVI decrease larger than 0.1). Vegetation browning was largely found in the southeast of the province and sporadically in the north (Fig. 4d). Both

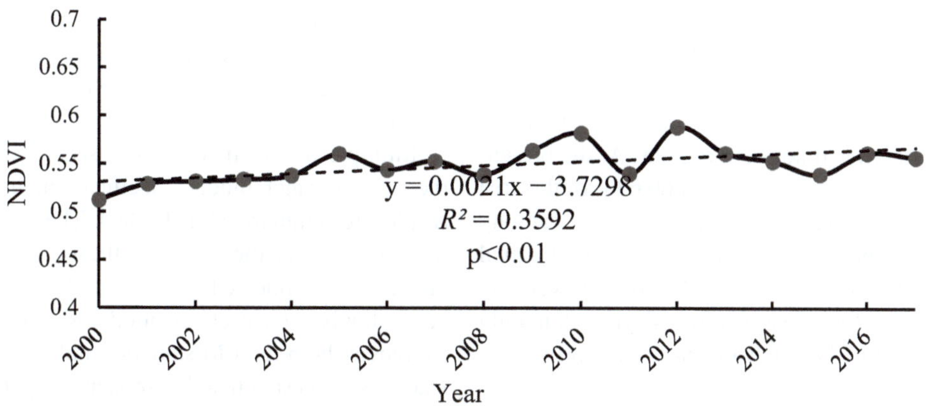

Fig. 3 Interannual variation of mean NDVI in Qinghai Province, 2000–2017

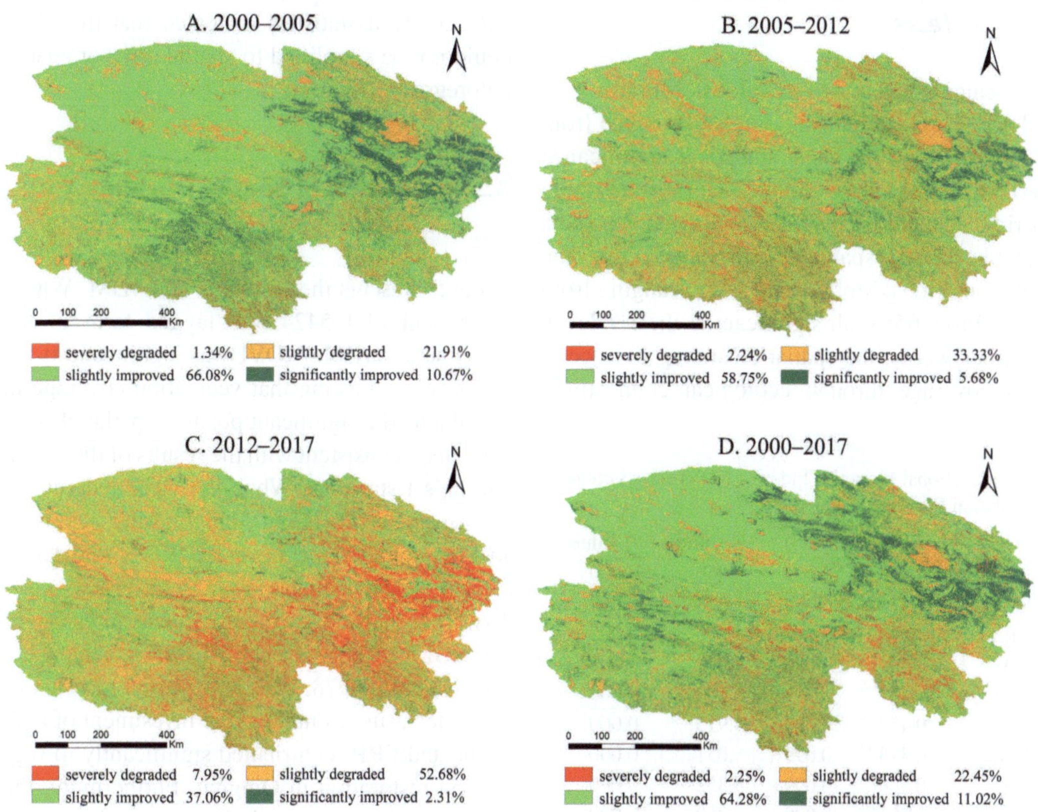

Fig. 4 Spatial distribution of NDVI change in Qinghai Province at periods of 2000–2005 (**a**), 2005–2012 (**b**), 2012–2017 (**c**), and 2000–2017 (**d**). Red represents areas severely degraded (−0.987 to −0.1), yellow slightly degraded (−0.1 to 0), light green slightly improved (0 to 0.1), and deep green significantly improved (0.1 to 1)

areas are mainly mountain-dominated terrain at higher altitudes.

At the steady greening phase (2000–2005), improvement in vegetation coverage was remarkable, occurring in 76.75% of the area, although a majority was slightly greened (66.08%). Still, a few spots of medium-scale contiguous degeneration could be found in the northern and western regions. At the fluctuating greening phase (2005–2012), about 64.43% of the province showed an improvement in vegetation cover, whereas 35.57% was degraded, largely found in the southern and northern regions. At the browning phase (2012–2017), about 60.63% of the area experienced vegetation degradation, among which 52.68% was slightly degraded (annual mean NDVI decrease larger than 0 but lower than 0.1), while 7.95% was severely degraded, mainly distributed in the south and east of the province.

3.2 Results of Global Moran's I and Model Specification Tests

The calculation results of global Moran's I of NDVI across counties in Qinghai Province from 2000 to 2017 and their statistical significance tests are reported in Table 3. During the study period, vegetation coverage in Qinghai showed a notable positive spatial autocorrelation, with the values of global Moran's I of NDVI ranging from 0.5989 to 0.6558, all significant at the 1% level. In other words, the spatial clustering of vegetation coverage through ecological connectivity and/or human-induced spillover across neighboring counties was evident and should be taken into account when investigating the effects of rangeland ERPs.

Table 3 Global Moran's I Index of NDVI across counties in Qinghai Province, 2000–2017

Year	Moran's I	SD(I)	Z-value	*P*-value
2000	0.476***	0.099	5.069	0.000
2001	0.498***	0.099	5.328	0.000
2002	0.448***	0.099	4.806	0.000
2003	0.494***	0.099	5.262	0.000
2004	0.487***	0.099	5.200	0.000
2005	0.457***	0.099	4.915	0.000
2006	0.473***	0.099	5.051	0.000
2007	0.500***	0.099	5.352	0.000
2008	0.501***	0.099	5.359	0.000
2009	0.487***	0.098	5.222	0.000
2010	0.489***	0.098	5.252	0.000
2011	0.471***	0.099	5.043	0.000
2012	0.493***	0.098	5.303	0.000
2013	0.522***	0.098	5.579	0.000
2014	0.499***	0.098	5.344	0.000
2015	0.483***	0.099	5.157	0.000
2016	0.509***	0.098	5.450	0.000
2017	0.480***	0.099	5.142	0.000

Note: *** indicates significance at the 1% levels

Multiple test methods were used to validate the spatial panel model specification, and the results are summarized in Table 4. The Hausman test showed that the null hypothesis of random effects was rejected (Liu & Nie, 2022). Accordingly, a fixed-effects model was specified. Moreover, both the LM statistics and robust LM statistics suggested that the impacts of spatial lag and spatial error should be considered when building models. Hence, the SDM was proven to be appropriate in the current case. Further, all the LR and Wald statistics indicated that the SDM could not be simplified to spatial error or spatial autoregressive models.

3.3 Results of Spatial Durbin Model

Table 5 presents the results of the SDM. With a coefficient of 0.5424, the lagged term of the dependent variable NDVI was significant at the 1% level, indicating that vegetation coverage in Qinghai had a significant positive spatial clustering effect, consistent with the results of the global Moran's I statistics. When the weighted sum of the mean NDVIs of neighboring counties increased by 1, the mean NDVI of the target county would increase by 0.5424, *ceteris paribus*. The coefficients of investment intensity of the RGLR project and the REPSR scheme were 0.0610 and 0.0978, and both passed the significance test, suggesting that the investment of both rangeland ERPs contributed significantly to vegetation restoration in Qinghai. Furthermore, the coefficient of the spatially weighted sum of *REPSR* was negative and significant ($P<0.01$), meaning that there might be a negative spatial spillover effect in the investment of the REPSR scheme. Regarding the control variables, grazing intensity (*stock*), economic intensity (*GDP*), population density (*pop*), precipitation (*prec*), and temperature (*temp*) were all statistically significant, although the coefficient signs of *stock* and *pop* did not meet the expectation.

Table 4 Results of tests on spatial panel model specification

Test method	Statistic	Estimate	*P*
Lagrange multiplier	LM-lag	156.00	0.000
	Robust LM-lag	14.12	0.000
	LM-Error	273.85	0.000
	Robust LM-Error	131.97	0.000
Hausman	Hausman statistic	49.59	0.000
Wald	Wald spatial lag	15.71	0.027
	Wald spatial error	18.39	0.010
LR	LR spatial lag	156.46	0.000
	LR spatial error	86.28	0.000

Table 5 Results of spatial Durbin model

Variable	Coefficient	*t*	*P*
RGLR	0.0610**	2.07	0.039
REPSR	0.0978***	4.49	0.000
stock	0.0122***	6.22	0.000
GDP	−0.0002**	−3.29	0.001
pop	0.0003**	2.54	0.011
prec	0.0012***	19.75	0.000
temp	−0.0104***	−3.20	0.001
W×RGLR	−0.0737	−1.53	0.126
W×REPSR	−0.1092***	−2.63	0.008
W×stock	0.0108**	2.64	0.008
W×GDP	0.0005***	3.03	0.002
W×pop	−0.0015***	−6.20	0.000
W×prec	−0.0007***	−8.83	0.000
W×temp	0.0299***	5.53	0.000
Spatial rho	0.5424***	13.22	0.000
R^2	0.5981		
Log-likelihood	784.6565		

Note: ***, **, and * indicate significance at the 1%, 5%, and 10% levels, respectively

Table 6 Decomposition of the total marginal effect of each driver on NDVI

Variables	Coefficient	*t*	*P*
Direct marginal effect			
RGLR	0.0547*	1.87	0.061
REPSR	0.0874***	4.39	0.000
stock	0.0155***	7.19	0.000
GDP	−0.0002**	−2.48	0.013
pop	0.0001	0.75	0.451
prec	0.0012***	22.41	0.000
temp	−0.0061**	−1.96	0.050
Indirect marginal effect			
RGLR	−0.0845	−1.02	0.307
REPSR	−0.1081	−1.51	0.131
stock	0.0352***	4.12	0.000
GDP	0.0007**	2.41	0.016
pop	−0.0027***	−5.47	0.000
prec	−0.0002	−1.52	0.129
temp	0.0493***	5.42	0.000

Note: ***, **, and * indicate significance at the 1% 5%, and 10% levels, respectively

Because of the inclusion of spatial lagged terms of both explained and explanatory variables, the SDM is not capable of calculating the marginal effects of explanatory variables (Espoir & Sunge, 2021). As an alternative, the current study used the partial differential method to decompose the fitted values of the *NDVI* based on the SDM, so as to unpack the total marginal effect of each explanatory variable into one marginal effect on the mean NDVI of the local county (direct marginal effect) and another on bordering counties (indirect marginal effect, or spatial spillover effect). The results are presented in Table 6. Both direct marginal effects of the RGLR project and the REPSR scheme on NDVI were positive and significant, despite neither of their indirect marginal effects being statistically significant. An increase of 10,000 Yuan/km^2 in the investment of the RGLR project in a certain county will directly increase the county's mean NDVI by 0.05474, *ceteris paribus*, while the same rise from the REPSR scheme will lead to a larger direct marginal effect—boosting the mean NDVI by 0.0874.

With regard to the control variables, an increase of 1 mm in a county's annual accumulated precipitation during the growth season will result in a marginal improvement of its mean NDVI by 0.0012, without a statistically significant spatial externality effect. Such results are reasonable and aligned with the conclusions drawn by previous studies (Piao et al., 2014; Zhai et al., 2022). Conversely, a warming by 1 °C in a county's annual mean temperature during the growth season will lead to a marginal decline in its mean NDVI by 0.0061 but a marginal improvement of 0.0493 in neighboring counties. The impact of warming on biomass productivity is inconclusive and may vary with topography, precipitation, latitude, and longitude. Previous studies in arid and cold areas suggested that warming contributes to mitigating the harmful effects caused by low temperatures on vegetation growth (Dong et al., 2013), but it could accelerate water evaporation and exacerbate aridity (Piao et al., 2006; Zuo et al., 2022). On the Qinghai-Tibetan Plateau, our modeling results indicate that the inhibiting effect exceeds the facilitating effect. As for the positive spatial spillover effect, a tentative explanation may be that although weather warming may accelerate the respiration of plant communities and reduce local biomass productivity, it may attract herbivores from neighboring areas to forage for food in this warmer place, given the frigid climate on the alpine rangeland (Li et al., 2021b; Piao et al., 2014).

The direct and indirect marginal coefficients of grazing intensity were 0.0155 and 0.0352, respectively, and both were statistically significant. However, such results did not meet our expectation. Such results might only reflect the statistical correlation between variables, but not causality. While promoting grazing bans and rangeland-livestock balance through the REPs, local governments have also proactively promoted the transformation of livestock farming from a traditional, open grazing pattern to a concentrated, intensive production pattern by providing subsidies for human settlement, shed construction for confined animal feeding, and artificial forage planting to reduce herders' dependence on natural rangelands while maintaining their livelihood (Chen et al., 2017). As a result, the vegetation recovers due to mitigated grazing pressure, while stocking rate continues to rise because of production mode transformation. Therefore, with well-designed strategies, reducing grazing intensity and increasing herd size can be compatible (Miao et al., 2020).

A growth of annual economic output by 10,000 Yuan/km^2 will result in a marginal decline of local mean NDVI by approximately 0.0002 ($P<0.05$), meeting the expectation, but will indirectly boost the mean NDVI of neighboring counties by 0.0002 ($P<0.05$). A possible explanation for the spatial effect of economic growth is that counties with a better economic development level might siphon labor and capital from surrounding counties, therefore reducing rangeland use scale and intensity and enabling vegetation recovery in neighboring counties. A increase of population density by 1 person/km^2 at a certain county would negatively affect the mean NDVI of its neighboring counties by 0.0027 ($P<0.01$), but had no statistically significant direct marginal effect. Theoretically, population growth and subsequent human activity intensification usually lead to vegetation degradation. However, due to a lack of data on human migration and population urbanization at the county level, the current study simply used statistical data from the household population register. Human migration across counties and even outside the province, as well as the ongoing urbanization, might relieve the pressure of local population growth on vegetation recovery. As such, we refain from making a clear interpretation of the regression results of population density.

4 Discussion

4.1 Comparative Effectiveness of Grazing Control and Ecological Engineering

The vegetation conservation investment yielded the desired return. From 2000 to 2017, the vegetation coverage in Qinghai Province has undergone an overall increase. Investments from both the RGLR project and the REPSR scheme contributed positively to the vegetation restoration in project areas, suggesting the effectiveness of both livestock management and ecological engineering measures. Such findings are in line with previous studies (Chen et al., 2017; Liu et al., 2019a). Through the targeted implementation of grazing bans, rangeland-livestock balance, and ecological engineering measures, the two programs effectively promoted the restoration of rangeland vegetation. Meanwhile, local governments also provided subsidies for herders' resettlement, constructing sheds for confined feeding, and artificial forage planting, etc., which effectively helped herders to transform livestock farming from the open and nomadic grazing mode to a concentrated and sedentary captivity mode and reduced their dependence on natural rangelands without decreasing the herd size, therefore further strengthening the vegetation recovery and enhancing the resilience of the rangeland SESs (Du et al., 2016; Gao et al., 2016; Yu et al., 2023).

Discrepancy of investment efficiency existed between the two ERPs. The marginal return of the REPSR scheme was larger than that of the RGLR project by 37.86%, although the investment intensity of the former was nearly three times that of the latter (18,318 vs. 6421 Yuan/km^2, Table 2), implying that the investment efficiency of the REPSR was much higher than that of RGLR. The RGLR was mainly carried out at certain designated project areas by applying a combination of livestock management (e.g., grazing ban, grazing rest) and ecological engineering measures (e.g., supplementary grass seed sowing, planting artificial grass), whereas the REPSR predominantly applied livestock management measures (grazing ban and rangeland-livestock balance) and set grazing intensity requirements according to local natural conditions, so as to motivate herdsmen to utilize rangelands in a sustainable way. Three insights can be gained from such discrepancy. First, reducing grazing pressure by livestock management based on local natural conditions appears more efficient than ecological engineering measures. In other words, natural recovery relying on the natural growth of plants under zero to moderate grazing pressure is better than human engineering interventions. Second, moderate grazing intensity may be more beneficial for vegetation recovery than a strict no-grazing strategy. By continuous moderate-intensity disturbance, animal foraging helps to inhibit the growth of poisonous weeds, promote soil respiration, and improve rangeland species diversity and resilience, whereas grazing exclusion with fences unintentionally limits the animal mobility, artificially breaks the compensatory and mutual-beneficial link between plants and animals, and harms the integrity and long-term health of rangeland ecosystems, which is detrimental to the long-term resilience of the system (Li et al., 2015; Lu et al., 2017; Sun et al., 2020). Moderate grazing disturbance also contributes to ecosystem adaptability to weather extremes, such as drought (Li et al., 2021b). Third, in contrast with the one-size-fits-all approach adopted by the RGLR, the REPSR applies a more localized and targeted approach in terms of promoting the two livestock management measures. Under the REPSR, grazing ban largely aims at seriously degraded rangelands and ecological function areas such as water or biodiversity conservation areas, while rangeland-livestock balance is typically applied to areas with more benign soil and water conditions and without significant vegetation degradation. The flexible approach tailored to fit with local contexts fares better than the rigid one-size-fits-all approach.

4.2 Negative Spatial Spillover Effects of the RGLR Project and the REPSR Scheme

A negative spatial spillover effect was found in the investment of the REPSR scheme. However, the decomposition of total marginal effect via the partial differential method indicates that the indirect marginal effects of the two rangeland ERPs were not statistically significant. Nonetheless, both coefficients were negative and practically significant (−0.0845 and −0.1081, respectively, Table 6), larger than those of direct effects, indicating a possible existence of a spatial externality problem. The statistical insignificance of the two coefficients may be a result of a great discrepancy in the spatial spillover problem across different regions. That is to say, the negative spatial externality from the investment of the rangeland ERPs is severe and poorly managed in some regions, despite some others managing to properly control it. Spatial spillover of grazing pressure might occur in some regions but not in others.

The shift of grazing pressure from project areas to non-project areas, as well as from strict to slack rule-enforcement areas, has been frequently reported by earlier government documents and academic literature (Chen et al., 2017; Sun et al., 2020; Yin et al., 2019). First, given the vastness of the rangelands, detecting and enforcing rules on illegal grazing is difficult.

Furthermore, considering the significant economic benefit gained from animal husbandry, as well as its essential role in sustaining local herders' livelihoods, it is understandable that the local authority is reluctant to strictly regulate the illegal livestock grazing. The rangeland socio-ecological systems are complex and interconnected and are not constrained by administrative boundaries (Rathi, 2022). Therefore, management measures need to appropriately address cross-scale issues in order to enhance the resilience of rangeland SESs. As such, we suggest a collaborative institutional arrangement in which neighboring counties could join forces to monitor and enforce rules against illegal grazing activities in order to achieve the sustainability of the rangeland ecosystem as a whole (Qiu et al., 2020; Wang & Lo, 2022).

Second, the failure of monetary incentives to stop illegal grazing may also be rooted in the unreasonable compensation scheme that applies a uniform, area-based compensation standard to enrolled rangelands by households. Given the considerable variance in rangeland area owned by herders, natural conditions, family demography, and market environment across counties, significant differences in the benefits and costs incurred by local herders should be expected when they participate in the programs (Behrendt et al., 2022; Zhang et al., 2019). Therefore, such a relatively uniform compensation scheme causes a notable inequity problem in accrued compensation amounts between households and across regions (Aguilar-Gómez et al., 2020; Liu et al., 2021; Sheng & Wang, 2022). Herders with limited rangeland endowment end up receiving trivial compensation that can not cover the income loss from forgoing livestock production or cutting herd size, and therefore have no choice but to resort to illegal grazing. Last, the fundamental cause of the illegal grazing activities is the lack of alternative livelihood options and occupational skills. The primary goal of NBSs is to safeguard human well-being (Cohen-Shacham et al., 2019). Hence, greater emphasis needs to be placed on the cultivation and development of industries beyond animal husbandry, and on capacity building for the adaptation and transformation of the livelihoods of small-holder farmers in the context of multi-source risks, through, for example, occupational skill training, developing production cooperatives, as well as weaving safety nets.

5 Conclusions

Enormous investment has been made through the nature-based rangeland ERPs in China to restore rangeland vegetation during the last two decades. However, an important knowledge gap remains in understanding the comparative effectiveness of measures between grazing restriction and ecological engineering. Using a hybrid longitudinal dataset of climate, demography, economics, and two rangeland ERPs investments from 2000 to 2017, this chapter empirically examined the long-term investment effectiveness and efficiency of the two programs on vegetation restoration and the potential spatial spillover effects in the implementation process across counties in Qinghai. (1) Temporal changes in NDVI suggest that an overall vegetation greening was evident during the study period, although significant discrepancies across years and space were found. Three distinct phases were identified, and vegetation browning was largely found at the southeast of the province, which has relatively better soil, water, and vegetation conditions. (2) The modeling results of the SDM show that investment in the two rangeland ERPs yielded the desired return, although discrepancy of investment efficiency existed between them. The marginal return of the REPSR scheme was larger than that of the RGLR project, implying that localized and targeted management of grazing intensity might be more efficient than ecological engineering measures. (3) A cross-region heterogeneity of the spatial externality problem was speculated. That is, spatial spillover of grazing pressure might be severe and poorly managed in some regions but not at others. Above findings help narrow the knowledge gap in understanding the effectiveness and spatial externality of ERPs on rangeland vegetation rehabilitation as well as the varied performances that result from differential intervention measures and compensation designs.

To improve the investment effectiveness and efficiency of the rangeland ERPs and thus enhance rangeland ecosystem resilience, it is suggested that more locally tailored, spatially holistic intervention measures and a flexible compensation scheme should be developed. (1) A localized and targeted approach to grazing pressure management should be promoted. In particular, moderate-intensity grazing disturbance should be allowed and strict grazing exclusion should be limited in order to maintain the compensatory and mutually beneficial link and virtuous cycles between plants and animals, and to improve the resilience of the ecosystem. (2) To deal with the illegal livestock grazing problem, collaboration between neighboring counties on monitoring and rule enforcement should be strengthened. (3) To address the compensation inequity problem, a flexible, regressive compensation scheme that favors smallholder herders should be applied. (4) To enhance the adaptability and transformability of herders' livelihoods, more attention should be paid to providing alternative livelihood options and building individual and community livelihood capacity.

Research findings and policy suggestions gained by the current study provide important insights that may inform the refinement of rangeland ERPs being implemented in China and other developing countries. Future research should be extended to wider regions over a longer time span, so as to draw a more generalized and holistic understanding of the influences of the rangeland ERP investments. To this end, a finer and longer dataset across space and time should be collected, which should cover a wider geographical range, integrate climate and ecological indicators beyond vegetation coverage, accurately record demographic and economic evolutions, and combine observations from painstaking field experiments on grazing disturbance and rigorous household surveys.

References

Aguilar-Gómez, C. R., Arteaga-Reyes, T. T., Gómez-Demetrio, W., Ávila-Akerberg, V. D., & Pérez-Campuzano, E. (2020). Differentiated payments for environmental services: A review of the literature. *Ecosystem Services, 44*, 101131.

Behrendt, K., Brown, C., Qiao, G., & Zhang, B. (2022). Assessing the opportunity costs of Chinese herder compliance with a payment for environmental services scheme. *Ecological Economics, 193*, 107313.

Bianciardi, A., Becattini, N., & Cascini, G. (2023). How would nature design and implement nature-based solutions? *Nature-Based Solutions, 3*, 100047.

Bureau of Statistics, Qinghai Province. (2022). *Qinghai Statistical Yearbook 2021*. China Statistics Press (in Chinese).

Bustamante, M. M. C., Silva, J. S., Scariot, A., Sampaio, A. B., Mascia, D. L., Garcia, E., Sano, E., Fernandes, G. W., Durigan, G., Roitman, I., Figueiredo, I., Rodrigues, R. R., Pillar, V. D., de Oliveira, A. O., Malhado, A. C., Alencar, A., Vendramini, A., Padovezi, A., Carrascosa, H., Freitas, J., Siqueira, J. A., Shimbo, J., Generoso, L. G., Tabarelli, M., Biderman, R., de Paiva Salomo, R., Valle, R., Junior, B., & Nobre, C. (2019). Ecological restoration as a strategy for mitigating and adapting to climate change: Lessons and challenges from Brazil. *Mitigation and Adaptation Strategies for Global Change, 24*, 1249–1270.

Chen, C., Park, T., Wang, X., Piao, S., Xu, B., Chaturvedi, R. K., Fuchs, R., Brovkin, V., Ciais, P., Fensholt, R., Tommervik, H., Bala, G., Zhu, Z., Nemani, R. R., & Myneni, R. B. (2019). China and India lead in greening of the world through land-use management. *Nature Sustainability, 2*, 122–129.

Chen, H., Ding, R., & Shao, L. (2022a). Cognitive antecedents and formation pathways of confined feeding mode adoption by herders in China. *Journal of Rural Studies, 94*, 326–335.

Chen, H., Shao, L., Zhao, M., Zhang, X., & Zhang, D. (2017). Grassland conservation programs, vegetation rehabilitation and spatial dependency in Inner Mongolia, China. *Land Use Policy, 64*, 429–439.

Chen, X., Yu, L., Du, Z., Xu, Y., Zhao, J., Zhao, H., Zhang, G., Peng, D., & Gong, P. (2022b). Distribution of ecological restoration projects associated with land use and land cover change in China and their ecological impacts. *Science of the Total Environment, 825*, 153938.

Cohen-Shacham, E., Andrade, A., Dalton, J., Dudley, N., Jones, M., Kumar, C., Maginnis, S., Maynard, S., Nelson, C. R., Renaud, F. G., Welling, R., & Walters, G. (2019). Core principles for successfully implementing and upscaling Nature-based Solutions. *Environmental Science & Policy, 98*, 20–29.

Cumming, G. S., & Peterson, G. D. (2017). Unifying research on social–ecological resilience and collapse. *Trends in Ecology & Evolution, 32*, 695–713.

Dong, Q., Zhao, X., Wu, G., Shi, J., & Ren, G. (2013). A review of formation mechanism and restoration measures of "black-soiltype" degraded grassland in the Qinghai-Tibetan Plateau. *Environmental Earth Sciences, 70*, 2359–2370.

Du, B., Zhen, L., Yan, H., & de Groot, R. (2016). Effects of government grassland conservation policy on household livelihoods and dependence on local grasslands: Evidence from Inner Mongolia, China. *Sustainability, 8*, 1314.

Du, Y., Ke, X., Dai, L., Cao, G., Zhou, H., & Guo, X. (2020). Moderate grazing increased alpine meadow soils bacterial abundance and diversity index on the Tibetan Plateau. *Ecology and Evolution, 10*, 8681–8687.

Espoir, D. K., & Sunge, R. (2021). Co2 emissions and economic development in Africa: Evidence from a dynamic spatial panel model. *Journal of Environmental Management, 300*, 113617.

Fan, F., Liang, C., Tang, Y., Harker-Schuch, I., & Porter, J. R. (2019). Effects and relationships of grazing intensity on multiple ecosystem services in the Inner Mongolian steppe. *Science of the Total Environment, 675*, 642–650.

Folke, C., Carpenter, S., Walker, B., Scheffer, M., Chapin, T., & Rockström, J. (2010). Resilience thinking: Integrating resilience, adaptability and transformability. *Ecology and Society, 15*.

Fu, W. J., Jiang, P. K., Zhou, G. M., & Zhao, K. L. (2014). Using Moran's I and GIS to study the spatial pattern of forest litter carbon density in a subtropical region of southeastern China. *Biogeosciences, 11*, 2401–2409.

Gao, L., Kinnucan, H. W., Zhang, Y., & Qiao, G. (2016). The effects of a subsidy for grassland protection on livestock numbers, grazing intensity, and herders' income in inner Mongolia. *Land Use Policy, 54*, 302–312.

Goret, T., Janssens, X., & Godefroid, S. (2021). A decision-making tool for restoring lowland grasslands in Europe. *Journal for Nature Conservation, 63*.

Gu, Y., Pang, B., Qiao, X., Xu, D., Li, W., Yan, Y., Dou, H., Ao, W., Wang, W., Zou, C., Zhang, X., & Cao, B. (2022). Vegetation dynamics in response to climate change and human activities in the Hulun Lake basin from 1981 to 2019. *Ecological Indicators, 136*.

He, J., Shi, X., & Fu, Y. (2021). Identifying vegetation restoration effectiveness and driving factors on different micro-topographic types of hilly Loess Plateau: From the perspective of ecological resilience. *Journal of Environmental Management, 289*, 112562.

Hu, Y., Huang, J., & Hou, L. (2019). Impacts of the grassland ecological compensation policy on household livestock production in China: An empirical study in Inner Mongolia. *Ecological Economics, 161*, 248–256.

Jiang, H., Xu, X., Guan, M., Wang, L., Huang, Y., & Jiang, Y. (2020). Determining the contributions of climate change and human activities to vegetation dynamics in agro-pastural transitional zone of northern China from 2000 to 2015. *Science of the Total Environment, 718*, 134871.

Kooch, Y., Amani, M., & Abedi, M. (2022). Vegetation degradation threatens soil health in a mountainous semi-arid region. *Science of the Total Environment, 830*, 154827.

Li, D., Hou, L., & Zuo, A. (2021a). Informal institutions and grassland protection: Empirical evidence from pastoral regions in China. *Ecological Economics, 188*, 107110.

Li, P., & Bennett, J. (2019). Understanding herders' stocking rate decisions in response to policy initiatives. *Science of the Total Environment, 672*, 141–149.

Li, X., Lyu, X., Dou, H., Dang, D., Li, S., Li, X., Li, M., & Xuan, X. (2021b). Strengthening grazing pressure management to improve grassland ecosystem services. *Global Ecology and Conservation, 31*.

Li, Y., Fan, M., & Li, W. (2015). Application of payment for ecosystem services in China's rangeland conservation initiatives: A social-ecological system perspective. *The Rangeland Journal, 37*.

Liu, C., & Nie, G. (2022). Spatial effects and impact factors of food nitrogen footprint in China based on spatial durbin panel model. *Environmental Research, 204*, 112046.

Liu, J., Chen, H., Hou, X., Zhang, D., & Zhang, H. (2021). Time to adopt a context-specific and market-based compensation scheme for a new round of the Grain for Green Program. *Land Use Policy, 108*, 105675.

Liu, M., Dries, L., Huang, J., Min, S., & Tang, J. (2019a). The impacts of the eco-environmental policy on grassland degradation and livestock production in Inner Mongolia, China: An empirical analysis based on the simultaneous equation model. *Land Use Policy, 88*, 104167.

Liu, X., Ma, Z., Huang, X., & Li, L. (2020). How does grazing exclusion influence plant productivity and community structure in alpine grasslands of the Qinghai-Tibetan Plateau? *Global Ecology and Conservation, 23*.

Liu, Y., Zhang, Z., Tong, L., Khalifa, M., Wang, Q., Gang, C., Wang, Z., Li, J., & Sun, Z. (2019b). Assessing the effects of climate variation and human activities on grassland degradation and restoration across the globe. *Ecological Indicators, 106*.

Liu, Y., Zhao, L., Liu, Y., Huang, Z., Shi, J., Wang, Y., Ma, Y., Esteban Lucas-Borja, M., López-Vicente, M., & Wu, G.-L. (2022). Restoration of a hillslope grassland with an ecological grass species (Elymus tangutorum) favors rainfall interception and water infiltration and reduces soil loss on the Qinghai-Tibetan Plateau. *Catena, 219*.

Lu, C., Zhao, T., Shi, X., & Cao, S. (2018). Ecological restoration by afforestation may increase groundwater depth and create potentially large ecological and water opportunity costs in arid and semiarid China. *Journal of Cleaner Production, 176*, 1213–1222.

Lu, X., Kelsey, K. C., Yan, Y., Sun, J., Wang, X., Cheng, G., & Neff, J. C. (2017). Effects of grazing on ecosystem structure and function of alpine grasslands in Qinghai–Tibetan Plateau: A synthesis. *Ecosphere, 8*.

Luo, Y., Lu, Y., Fu, B., Zhang, Q., Li, T., Hu, W., & Comber, A. (2019). Half century change of interactions among ecosystem services driven by ecological restoration: Quantification and policy implications at a watershed scale in the Chinese Loess Plateau. *Science of the Total Environment, 651*, 2546–2557.

Maria Teixeira Lins, C., Rodrigues de Souza, E., Emanuelle Monteiro dos Santos Souza, T., Katharinne Silva Souza Paulino, M., Rodrigues Monteiro, D., Severino de Souza Júnior, V., Rugero Magalhães Dourado, P., Ernesto de Andrade Rego Junior, F., Jacques Agra da Silva, Y., & Schaffer, B. (2023). Influence of vegetation cover and rainfall intensity on soil attributes in an area undergoing desertification in Brazil. *Catena, 221*, 106751.

McNaughton, S. J. (1983). Compensatory plant growth as a response to herbivory. *Oikos, 40*, 329–336.

McNaughton, S. J. (1985). Ecology of a grazing ecosystem: The Serengeti. *Ecological Monographs, 55*, 259–294.

Miao, L., Sun, Z., Ren, Y., Schierhorn, F., & Müller, D. (2020). Grassland greening on the Mongolian Plateau despite higher grazing intensity. *Land Degradation & Development, 32*, 792–802.

Ning, X., Zhu, N., Liu, Y., & Wang, H. (2022). Quantifying impacts of climate and human activities on the grassland in the Three-River Headwater Region after two phases of Ecological Project. *Geography and Sustainability, 3*, 164–176.

Piao, S., Mohammat, A., Fang, J., Cai, Q., & Feng, J. (2006). NDVI-based increase in growth of temperate grasslands and its responses to climate changes in China. *Global Environmental Change, 16*, 340–348.

Piao, S., Nan, H., Huntingford, C., Ciais, P., Friedlingstein, P., Sitch, S., Peng, S., Ahlstrom, A., Canadell, J. G., Cong, N., Levis, S., Levy, P. E., Liu, L., Lomas, M. R., Mao, J., Myneni, R. B., Peylin, P., Poulter, B., Shi, X., Yin, G., Viovy, N., Wang, T., Wang, X., Zaehle, S., Zeng, N., Zeng, Z., & Chen, A. (2014). Evidence for a weakening relationship between interannual temperature variability and northern vegetation activity. *Nature Communications, 5*, 5018.

Qiu, H., Su, L., Feng, X., & Tang, J. (2020). Role of monitoring in environmental regulation: An empirical analysis of grazing restrictions in pastoral China. *Environmental Science & Policy, 114*, 295–304.

Rathi, A. (2022). Is Agrarian Resilience limited to Agriculture? Investigating the "farm" and "non-farm" processes of Agriculture Resilience in the rural. *Journal of Rural Studies, 93*, 155–164.

Sheng, J., & Wang, H. (2022). Participation, income growth and poverty alleviation in payments for ecosystem services: The case of China's Wolong Nature Reserve. *Ecological Economics, 196*, 107433.

Shi, P., Bai, L., Zhao, Z., Dong, J., Li, Z., Min, Z., Cui, L., & Li, P. (2023). Vegetation position impacts soil carbon losses on the slope of the Loess Plateau of China. *Catena, 222*, 106875.

Simonson, W. D., Miller, E., Jones, A., García-Range, S., Thornton, H., & McOwen, C. (2021). Enhancing climate change resilience of ecological restoration — A framework for action. *Perspectives in Ecology and Conservation, 19*, 300–310.

Sun, J., Liu, M., Fu, B., Kemp, D., Zhao, W., Liu, G., Han, G., Wilkes, A., Lu, X., Chen, Y., Cheng, G., Zhou, T., Hou, G., Zhan, T., Peng, F., Shang, H., Xu, M., Shi, P., He, Y., Li, M., Wang, J., Tsunekawa, A., Zhou, H., Liu, Y., Li, Y., & Liu, S. (2020). Reconsidering the efficiency of grazing exclusion using fences on the Tibetan Plateau. *Science Bulletin, 65*, 1405–1414.

Teng, Y., Zhan, J., Liu, S., Agyemanga, F. B., Li, Z., Wang, C., & Liu, W. (2022). Integrating ecological and social vulnerability assessment in Qinghai Province, China. *Physics and Chemistry of the Earth, Parts A/B/C*, 126.

Wang, L., Li, Z., Wang, D., Chen, J., Liu, Y., Nie, X., Zhang, Y., Ning, K., & Hu, X. (2021). Unbalanced social-ecological development within the Dongting Lake basin: Inspiration from evaluation of ecological restoration projects. *Journal of Cleaner Production, 315*.

Wang, X., & Lo, K. (2022). Pastoralism and conservation: The politics and notions of environmental justice under the grazing ban policy in Inner Mongolia, China. *Political Geography, 99*, 102779.

Yang, X., Sun, W., Li, P., Mu, X., Gao, P., & Zhao, G. (2018). Reduced sediment transport in the Chinese Loess Plateau due to climate change and human activities. *Science of the Total Environment, 642*, 591–600.

Yin, Y., Hou, Y., Langford, C., Bai, H., & Hou, X. (2019). Herder stocking rate and household income under the Grassland Ecological Protection Award Policy in northern China. *Land Use Policy, 82*, 120–129.

Yu, Y., Yan, J., & Wu, Y. (2023). Review on the socioecological performance of grassland ecological payment and award policy with the consideration of an alternate approach for nonequilibrium ecosystems. *Rangeland Ecology & Management, 87*, 105–121.

Zhai, D., Gao, X., Li, B., Yuan, Y., Li, Y., Liu, W., & Xu, J. (2022). Diverse chronic responses of vegetation aboveground net primary productivity to climatic changes on Three-River Headwaters region. *Ecological Indicators, 139*.

Zhang, J., Brown, C., Qiao, G., & Zhang, B. (2019). Effect of eco-compensation schemes on household income structures and herder satisfaction: Lessons from the grassland ecosystem subsidy and award scheme in Inner Mongolia. *Ecological Economics, 159*, 46–53.

Zhang, R., Tan, S., Hannaway, D., & Dai, W. (2020). Multi-household grassland management pattern promotes ecological efficiency of livestock production. *Ecological Economics, 171*, 106618.

Zhang, X., & Jin, X. (2021). Vegetation dynamics and responses to climate change and anthropogenic activities in the Three-River Headwaters Region, China. *Ecological Indicators, 131*.

Zuo, Y., Li, Y., He, K., & Wen, Y. (2022). Temporal and spatial variation characteristics of vegetation coverage and quantitative analysis of its potential driving forces in the Qilian Mountains, China, 2000–2020. *Ecological Indicators, 143*.

Smart Villages—An Instrument for Building Rural Resilience in the European Context: The Example of Poland

Monika Stanny, Łukasz Komorowski, and Anna Rosa

Abstract

Rural areas in Europe, while characterised by a diversity of landscapes, cultures, and traditions, face common challenges: depopulation, limited access to public services, and climate and economic challenges. These are global challenges. In response, the EU implements policies for sustainable rural development, including through the Common Agricultural Policy (CAP), which serves as the foundation of support for farmers and rural areas. A new element in supporting rural development in the EU is the concept of "smart villages," based on leveraging local resources, technological innovations, and community engagement. Smart villages align with key European initiatives such as the European Green Deal and the Long-Term Vision for Rural Areas by 2040. This article focuses on analyzing the concept of smart villages and its significance in enhancing the resilience of European rural areas, with a particular emphasis on Poland. It examines the evolution of the concept and its role in EU policies and presents specific examples from Poland that demonstrate how innovative approaches can contribute to the development of local economies. The initiatives analyzed provide evidence that integrating modern technologies with traditional agriculture and local resources can improve the quality of life in rural areas and increase their attractiveness to younger generations. The analysis findings indicate that the development of smart villages requires long-term financial support and coordinated actions across various administrative levels. The analysis is based on EU documents, Eurostat data, and case studies, providing a comprehensive perspective on the opportunities and challenges of implementing smart villages in rural development. Smart villages can become a catalyst for change, leading to more cohesive, sustainable, and resilient rural development in Europe and globally.

Keywords

Smart villages · Rural resilience · Common agricultural policy · Poland

M. Stanny (✉) · Ł. Komorowski · A. Rosa
Institute of Rural and Agricultural Development of the Polish Academy of Sciences (IRWiR PAN), Warsaw, Poland
e-mail: mstanny@irwirpan.waw.pl; lkomorowski@irwirpan.waw.pl; arosa@irwirpan.waw.pl

Y. Li (ed.), *Global Perspectives on Building Resilience for Sustainable Rural Development*, Sustainable Development Goals Series, https://doi.org/10.1007/978-981-96-6149-7_8

1 Introduction

Rural areas in Europe account for 83% of the continent's total land area and are home to about 30% of the European Union's (EU) population.[1] While these areas are characterised by a diversity of landscapes, cultures, and traditions, they also face a number of common challenges related to demographics, the economy, the environment, and the availability of public services. The EU is undertaking a variety of rural development measures aimed at reducing regional disparities and improving the quality of life for residents (European Commission, 2021).

Rural development serves as the "second pillar" of the Common Agricultural Policy (CAP), complementing the "first pillar" of income support and market measures by promoting the social, environmental, and economic sustainability of rural areas. EU countries have created various interventions for the CAP 2023–2027 programme period, including agriculture and non-area-based initiatives, such as investment, cooperation, and knowledge sharing, in their CAP strategic plans to meet their diverse needs. Instruments such as LEADER/CLLD (Community-Led Development), the European Agricultural Fund for Rural Development (EAFRD), and the Smart Rural Initiative support innovative projects that combine traditional approaches to agriculture with modern technologies and sustainability. According to Eurostat, under the 2023–2027 CAP, more than 36% of the total EU budget has been allocated to support rural development. The purpose of these funds is not only to improve agricultural competitiveness, but also to promote sustainable development, environmental protection, and social inclusion (European Commission, 2021). However, the effectiveness of these measures depends on cooperation at various administrative levels and the involvement of local communities in decision-making processes.

In this chapter, we undertake a discussion of the smart-village concept and its role in enhancing rural resilience in Europe, taking Poland as an example. Smart villages are initiatives that leverage local resources, digital innovations, and community engagement to foster sustainable rural development and address key challenges such as depopulation, limited access to public services, and climate change risks. The purpose of this article is to answer the question: how can the concept of smart villages contribute to the long-term resilience of rural areas in Europe, especially in the context of Poland?

The article aims to trace the evolution of the smart-village concept, its implementation in EU policies, and to present practical examples of its application in Poland. The discussion is based on official EU documents (such as European Commission reports on rural development policies, including the CAP and the European Green Deal), Eurostat data (on demography, economic trends, and access to public services in rural areas), and case studies from Poland exemplifying the practical application of the smart-village concept. Reference to specific examples illustrates both the opportunities and challenges of integrating the concept into rural development strategies. Academic literature on rural development, the issue of smart villages, and the resilience of local communities to development challenges was also used.

2 The Current State and Challenges of Europe's Rural Areas

2.1 Population Decline and Ageing

One of the most important challenges facing Europe's rural areas is population ageing and decline. According to Eurostat, the population of rural areas in many regions of Europe is shrinking, especially in the east and south of the continent (Eurostat, 2021a). This trend is the result of

[1] Calculated by JRC Eurostat demographic data by modelling at LAU (local administrative unit) level in 2018 considering all municipalities and cities in Europe with low population or population density; Atlas of Demography: https://migration-demography-tools.jrc.ec.europa.eu/atlas-demography (accessed: 26.08.2024).

both negative birth rates and the migration of young people to cities in search of better job prospects, education, and health services (European Commission, 2021). In 2021, for example, every rural region in Bulgaria, Estonia, Croatia, Latvia, Lithuania, Hungary, and Romania recorded a negative raw birth rate. Of these, Bulgaria and Romania recorded the highest depopulation rates, reaching over 20 per thousand in some places. In the case of the Czech Republic, Italy, Poland, Slovenia, and Slovakia, negative rates were recorded in most regions. The opposite is true for Ireland, where every region had a positive raw rate of natural population change in 2021. (Eurostat, 2023a). There are several reasons why Eastern European countries lead the list of countries with the fastest population decline. First, birth rates in these countries fell dramatically after the collapse of the USSR. While in 1988 the average woman had 2.1 children, by 1998 the number had already dropped to 1.2. Second, emigration played a significant role. Due to the expansion of EU borders between 2004 and 2007, Eastern European countries saw emigration equal to 6.3 million people by 2016 (DevelopmendAid, 2024).

Population ageing poses an additional challenge, as rural areas already have a higher percentage of older people than the EU average. In 2023, according to Eurostat, people aged 65 and older accounted for an average of 22% of the rural population, and in some regions, such as central Greece and western Spain, this percentage was even higher (Eurostat, 2021b). An ageing population brings with it the need to adapt infrastructure, health, and social services, which is often difficult due to limited financial resources (OECD, 2020).

2.2 Slow De-agrarianisation of Livelihoods

Rural areas in Europe often face economic challenges, such as limited employment opportunities and lower incomes compared to urban areas. Eurostat indicates that in 2023, the unemployment rate in rural areas averaged 7% (including, for example, more than 11% in Spain and less than 3% in Poland), while it was one percentage point lower in urban areas (Eurostat, 2023b). In addition, many people working in rural areas are employed in low-wage sectors such as agriculture, forestry, and fishing, which translates into lower average household income. In 2021, the median equivalised net income[2] of individuals residing in rural areas of the EU was 87.5% of the average income observed for those living in urban areas.

Agriculture remains a key sector of the economy in rural areas, but in many regions it faces challenges from globalisation, climate change, and increasing international competition. In 2023, the agriculture sector accounted for 1.7% of the EU's GDP, the same as in 2004, but in some countries, the share was much higher, such as Poland's 2.8%, Greece's 3.8% (2022), and Romania's 4.2% (2021).[3] These challenges are prompting farmers to seek more sustainable production methods and diversify their economic activities. In 2023, agricultural labour input in the EU corresponded to 7.6 million full-time workers, with the majority being non-salaried, equivalent to 5.3 million full-time workers. The EU's agricultural sector has seen a long-term decline in workforce numbers, with an average annual reduction of 2.6% between 2008 and 2023. This downward trend persisted in 2023, though at a slower estimated rate of −1.3% (Eurostat, 2024a).

2.3 Services Further Away

Another major problem in rural areas is limited access to public services and infrastructure. It is common and universal in global terms to find that rural residents often have difficult access to healthcare, education, public transport and high-speed Internet (Crescenzi & Rodríguez-Pose,

[2]Equivalised disposable income refers to a household's total income, after taxes and deductions, available for spending or saving. This amount is divided by the number of household members, adjusted into "equivalised adults" using the modified OECD equivalence scale, which assigns weights to members based on their age.

[3]Data from Statista.com (accessed: 03.09.2024).

2012; Cañal Fernández & Álvarez, 2021). Let us look at the data characterising access to Internet connections. In 2020, 8.5% of rural households were not yet covered by any fixed broadband network. In addition, less than 40% had no access to high-speed broadband (Perpiña Castillo et al., 2023). Admittedly, according to Eurostat (2024b), in 2023 as many as 90% of rural households had access to the Internet, but the quality of digital infrastructure often limits the development of new forms of economic activity, such as remote work or e-commerce (Philip & Williams, 2019; Wolski, 2019; Janc & Komorowski, 2024). A good-quality Internet connection (access to high-speed broadband) is therefore a key feature to take full advantage of the digital transformation.

Many rural residents also struggle with limited access to public transport, which hinders mobility and access to work, education and services. One hundred twenty-nine million EU citizens live in sparsely populated areas that face a lack of or low supply of public transport services. For residents without access or ability to drive, the lack of public transport options means limited economic opportunities and increased social isolation (European Parliament, 2021). This exacerbates inequality and accelerates rural depopulation (Interreg Europe, 2024). For example, in Poland, according to the report "Rural Development Monitoring. Stage IV" (MROW), almost 26% of villages (11,000) are not connected (without transfers) by public transport, rail or road, to the municipal seat. Many municipalities simply cannot afford to run such connections. In rural localities, the substitute for public transport is the car. Statistically, every person of working age already owns one, but most of them are more than 10 years old (Stanny et al., 2023).

2.4 Climate Change

Rural areas play a key role in environmental protection and sustainable development. They are home to many biodiverse ecosystems that are threatened by human activities, climate change, and agricultural intensification. According to Eurostat, EU rural areas occupy about 40% of the total area protected under the Natura 2000 network, which aims to protect natural habitats and endangered species (European Commission, 2018). At the same time, these areas are vulnerable to the effects of climate change, such as droughts, floods, and extreme weather events. In some regions, such as southern Europe, these impacts are expected to be more severe, which could negatively affect agriculture and local economies (Bindi & Olesen, 2011). In 2021, the OECD introduced the "Rural Agenda for Climate Action" (OECD 2021) to emphasise the critical role of rural areas in driving essential reforms for achieving net-zero emissions. The agenda calls for a stronger integration of rural policies into climate change mitigation strategies while fostering long-term collaboration. It aims to enhance the support provided by rural policies to subnational actors, private stakeholders, and civil society in managing the transition to net-zero economies. Additionally, the agenda advocates for incorporating rural development opportunities into national and regional climate strategies. Recognizing the importance of the green transition, governments at both national and international levels are allocating substantial resources to support rural areas. The Rural Agenda provides a framework to help OECD member states effectively leverage these funds and implement sustainable strategies.

Recent commitments aimed at accelerating the transition to net-zero emissions in rural areas include:

- European Union: Under the new Common Agricultural Policy (2023–2027), €387 billion has been allocated over 7 years to help farmers and rural communities address climate change, protect the environment, and transition toward sustainable and resilient food systems (European Commission, 2021).
- United States: The USDA earmarked over $914 million in new discretionary funding for fiscal year 2022 to support climate-smart agriculture and forestry initiatives. An additional $564 million was allocated for clean energy activities across USDA operations (United States Department of Agriculture, 2021).
- Australia: In 2020, the federal government announced a AU$1.9 billion investment in a

new energy technology package. This initiative includes the creation of Australia's first regional hydrogen export hub, a King Review Co-Investment Fund, a carbon capture and storage (CCS) Deployment Fund, and a Future Fuels Fund to promote innovative technologies (The Sydney Morning Herald, 2020).

These investments reflect the growing commitment of governments to prioritise rural areas in the global push for net-zero emissions.

3 From Diagnosis to Vision: A Vision for Europe's Rural Areas

"*After the experience of pandemic lockdowns and austerity, Russian aggression in Ukraine and war in Europe, there is no longer any doubt that the reality around us is different in many respects. Many previous 'certainties' have been shaken—at the macro level, such as the shape of markets, production chains, or, finally, the availability of products, marketability and, perhaps above all, the fundamental predictability of prices; and at the individual level, even that most basic issue of security from military aggression, violence, total lawlessness and barbarism*" (Buzek, 2023). The present, not only that presented above by the former Prime Minister of Poland (in 1997–2001) and the President of the European Parliament (in 2009–2012), but any present (cf. Fig. 1; Meijers & van der Wouw, 2019), especially when it is characterised by a great sense of uncertainty, the need to adapt to change, determines our idea of potential possibilities, and therefore our conception of the future. Being a conscious subject of development, every human being wants to influence their environment and shape their future, making their life better and their surroundings friendlier. It is an understandable aspiration for one to try to increase control over one's environment and to shape the future in such a way that, on the one hand, meets their needs and aspirations, and, on the other, will be resistant to crises.

To achieve this, one needs to have a clear vision of what one wants to accomplish—both in the short term and in the long term. Such a vision is based on our knowledge, experiences, aspirations, and hopes. While many of these desires and plans can be realised by an individual on their own, much of it requires cooperation and collective action, carried out at various levels—from the family, to small social groups and local communities, to regional, national, and even groups of affiliated states. The subjectivity of the individual must thus be in harmony with the subjectivity of groups and organisations, based on the principle of subsidiarity,[4] which promotes collective action where the individual or smaller groups cannot meet challenges alone (Wilkin, 2023). Let us emphasise here the social role of vision, the importance of which particularly increases in situations of uncertainty.

Constructing a Vision Is the Starting Point for the Following Sequence of Proceedings: Vision—Strategy—Plan—Action

Rural organisations working together in the European Rural Parliament[5] have long advo-

[4]The principle of subsidiarity indicates that the authority should be auxiliary, supportive of the actions of the individuals who established it. Where it is not necessary, the state should let civil society, families, the free market act. Instead, in a situation that requires intervention, it should solve the problem at the level as close to the citizens as possible, i.e. in the municipalities rather than in the regions, and in the regions rather than in the center. Subsidiarity is one of the basic constitutional principles of the European Union [Article 5(3) of the Treaty on the Functioning of the European Union. (2004)]. This means that only those actions should be taken at the Community level whose transfer to a higher level means greater efficiency and effectiveness than if the conduct of the relevant actions were left to the exclusive competence of the governments of individual member states.

[5]The European Rural Parliament is not a formal part of the European Union government, nor is it a legislative or decision-making body. It is an apolitical initiative that works "from the bottom up" focusing on achieving practical, agenda-driven policy outcomes that meet the challenges and opportunities of rural residents. The ERP campaign is based on the belief that the interests of rural communities are under-represented in national and European debates and in the formulation of policies and programmes; that rural residents themselves best understand their own strengths, opportunities and needs; and that rural communities should be encouraged to take action for their own well-being and be partners with governments at all levels [retrieved 01/08/2024: https://europeanruralparliament.com/].

Fig. 1 An artist's impression of the successive catastrophes threatening the modern world: from the COVID-19 pandemic, to recession, to climate change, and biodiversity loss—author Graeme MacKay. Source: https://mackaycartoons.net/tag/waves/ (accessed 6.06.2024)

cated the need for an active rural policy at the EU level that goes beyond CAP issues. In recent years, it can be said that the voice of rural organisations has been heard and taken seriously. In 2020, the European Commission invited individuals and organisations interested in rural issues to jointly build a long-term vision for European rural areas until 2040. This process included:

- public consultations, during which citizens, organisations, local governments and others could submit their ideas on the future of the countryside; the Commission's appeal received more than 2300 responses;
- discussions, workshops and other events to gather the voices of villagers in member states; more than 3000 people from 19 countries participated;
- a survey conducted by Eurobarometer on the role of rural areas and citizens' willingness to allocate public funds for investment in these areas (Budzich-Tabor, 2023).'

In 2021, the three EU Commissioners in charge of democracy and population issues (Dubravka Šuica), rural affairs and agriculture (Janusz Wojciechowski), and cohesion and reform (Elisa Ferreira) jointly presented the European Commission's Communication on a long-term vision for EU rural areas (European Commission, 2021). This communiqué outlines the vision pursued by and for these areas, according to which in 2040 European rural areas are to be stronger, connected, resilient and prosperous (Fig. 2). How these adjectives are understood is explained below.

Stronger rural areas: should foster resilient local communities where women and men alike have equal opportunities to engage actively in policymaking and decision-making. The aim is to enhance societal recognition of the countryside's importance and empower its residents through political processes that instill a sense of agency, ensuring rural areas remain appealing places to live and work. To achieve this, innovative approaches to service delivery should be adopted, leveraging digital tools and promoting robust social innovation.

Connected rural areas: public transport services (including connections to suburban and urban areas) should be maintained or improved, and investments should be made in digital infrastructure to better connect rural EU areas. It is important to strive to improve access to technology, such as expanding high-speed Internet coverage and developing digital competencies, making rural public transport more accessible, and introducing innovative solutions to facilitate mobility.

More resilient rural areas for prosperity: conservation of natural resources, landscape restoration, greening of agricultural practices, and the shortening of supply chains can enhance the resilience of rural areas to climate change, natural hazards, and economic crises. Resilience to shocks means that rural areas should better cope with the consequences of climate change and environmental degradation, and rural communities should be more inclusive, making use of people's diverse skills and creativity.

Prosperous rural areas: diversifying economic activities and adding value to agriculture and agri-food activities should result in sustainable local economic strategies that will have a positive impact on employment. Rural areas can become more prosperous through diversification of economic activities. Prosperous rural areas are an achievable goal when diverse sources of income are based on sustainable economies, tourism, and other sectors, increase their innovative potential, and keep young people in the countryside. It is the further development of entrepreneurship and the social economy that is expected to create competitive jobs in the countryside.

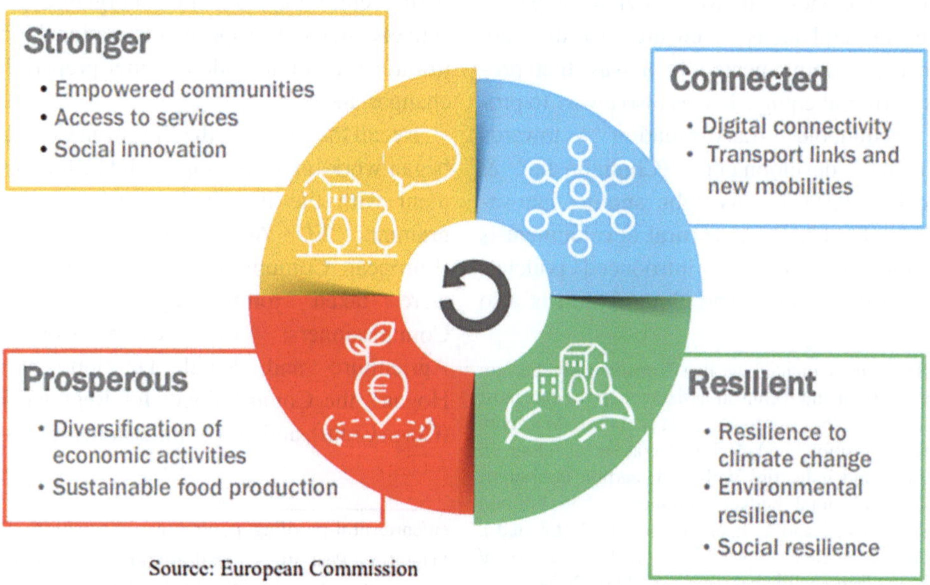

Fig. 2 The main objectives of the "EU Long-Term Vision for Rural Areas". Source: European Commission (2021, p. 10)

Although the vision, as intended, is general in nature and does not speak directly about how to achieve it, at the same time, at the EU level within the CAP, a number of programmes and activities are proposed to the member states coinciding with the four points quoted. One example is a classical instrument such as LEADER, which is evolving into a CLLD (Community-Led Development) instrument, in which the local population is the main decision-maker and disposer of funds for grassroots activities within the framework of European funds. The second example is already a new idea called *smart villages*, in which technological and social innovations, development of services in non-standard ways, and a broader quality of life in the countryside at the "micro" scale are supported with local resources (Komorowski, 2022). It is this scale (places) that is seen as one level of envisioning the future, and conversations between representatives of different stakeholder groups related to the future development of rural communities and development priorities are its key. An important element of this consensus is the activation of local communities and their real influence on decision-making processes. Widespread difficulties related to access to services, as well as infrastructure shortages, call for a review of public policies in rural areas to ensure that they are designed and implemented in a way that promotes territorial equity. This is also a way to promote "equitable transformation"[6] towards sustainable development (Chmieliński & Wieliczko, 2021). In this light, another innovative example of looking at rural development is predicting the effects of introduced policies, called *rural proofing*.[7] The Commission is also proposing a "Rural Pact,"[8] which can be joined by public authorities, organisations, companies, and other entities willing to cooperate for the benefit of the countryside within the scope of their competencies.

From Vision to Strategy: The Idea of a Smart Village in the European Union

Many factors, such as natural and climatic conditions, geographic features, historical and cultural development, demographic and social changes, national and regional peculiarities, and economic prosperity determine the diversity of rural areas. Each of these elements shapes the unique character of individual rural areas and requires that responses and solutions for the development of these areas be tailored to their specific needs and opportunities. This implies the need to move from vision to designing development strategies that take into account the individual characteristics of each area and its environment.

Hopefully, local communities will be able to take advantage of various forms of support through which they will become *smart*, building resilience and flexibility to face the coming threats. At the same time, through activities in agreements such as the European Rural Parliament, villagers and rural organisations will actively shape European and national policies towards the countryside to better prepare for the changes ahead.

Recall that in 2017, the European Commission began work on a new approach to rural development with the introduction of the pilot programme "EU Action for Smart Villages" (European Commission, 2017). The first steps were taken jointly by three European Commissioners: the Commissioner for Agriculture and Rural Development (Phil Hogan), the Commissioner for Regional Policy (Corin Crețu), and the Commissioner for Mobility

[6]The EU has implemented the Just Transition Mechanism to offer financial and technical assistance to regions most impacted by the shift to a low-carbon economy. At its core is the Just Transition Fund, designed to aid areas reliant on fossil fuels and industries with high carbon emissions. The initiative's total budget amounts to €55 billion.; European Council and Council of the EU Portal (n.d.), European Green Deal; https://www.consilium.europa.eu/pl/policies/green-deal/#what (accessed 20.01.2024).

[7]Rural proofing entails assessing public policies across various sectors, including education, health, and taxation, to determine their effects on rural areas. https://ec.europa.eu/enrd/rural-proofing_pl.html (accessed: 20.08.2024).

[8]The Rural Pact offers a platform for collaboration among public authorities, civil society, businesses, academia, and citizens at European, national, regional, and local levels; https://ruralpact.rural-vision.europa.eu/rural-pact_en (accessed: 20.08.2024).

and Transport (Violeta Bulc). In 2017, they provided a definition of the concept of smart villages and outlined the way forward to support this initiative. The document specified that the concept refers to rural areas and local communities, which should build on available resources while striving to expand them, especially in digital technologies, innovation, and knowledge. They were to aim not only to improve traditional and new networks and services, but also to raise rural living standards, including the quality of public services, make better use of resources, reduce negative environmental impacts, and create new opportunities for rural value chains. In short, an overall improvement in the quality of life of rural residents was sought.

In parallel, a development programme called the "Smart Villages Initiatives," developed by Terry van Gevelt and John Holmes (Van Gevelt & Holmes, 2015), was taking shape in Africa and Asia, outlining the features and principles of the smart-village vision. At the center of this vision was villagers' access to modern energy sources, which would act as a catalyst for development in other areas of socio-economic life and ultimately improve the residents' quality of life. And one of the main features of smart villages was considered to be their "connection" to cities through the digital infrastructure. A smart village understood in this way would become an alternative to city life.

Parallels can be drawn between the seemingly different visions of smart villages—the European concept and those in developing countries; both visions share a common goal: improving the quality of life in rural areas. In both cases, the key element is the territorial approach, which in the developing-country concept focuses on singling out remote, peripheral areas, while the European concept takes into account "territorial sensitivity"—that is, relying on the needs and potential of individual areas and strategies implemented by local communities, rather than universal, top-down solutions. In contrast, in both visions, the idea of smart villages proposes a wide range of activities that can be implemented to improve the quality of life of rural residents (Fig. 3).

However, there are important differences between the two approaches. The *Smart Villages* Initiative was designed to address the developmental (civilisational) backwardness of developing countries and therefore focused mainly on the problems of access to electricity, which were essentially solved in Europe in the late twentieth century. In addition, the concept assumes that smart villages will be the equivalent of smart cities, which the European approach rejected from the outset.

Taking into account the European context, when considering smart villages in the EU, we therefore focus in particular on the evolution of the smart-villages approach within the EU structures, especially its integration process with CAP for 2023–2027. "Path" developed in Asia or Africa, on the other hand, provides a background that reminds us that the idea is discussed globally.

The origins of the concept under discussion can be traced back to the 1990s. The first Cork Declaration (1996), entitled "A Vibrant Countryside," a result of the European Conference on Rural Development in Ireland, pointed out the key role of integrating advances in information technology in sustainable rural development. In addition to this, the importance of adequate infrastructure, training, and education was emphasised. Even then, attention was given to public participation and bottom-up activities as important factors in local and community-driven rural development.

Twenty years later, in the second Cork Declaration of 2016, entitled "Better Life in Rural Areas," which set new directions for rural and agricultural policies, one of the ten points was devoted to investing in the viability and vitality of rural areas (p. 5): "Investment in the development of economic activity, public and private services, basic infrastructure and capacity building should contribute to the achievement of common EU goals, particularly with regard to jobs and inclusive green growth." A key action in this regard was to overcome the digital divide between rural and urban areas, both in terms of access to a good-quality Internet network and the digital competence of rural residents. The declaration

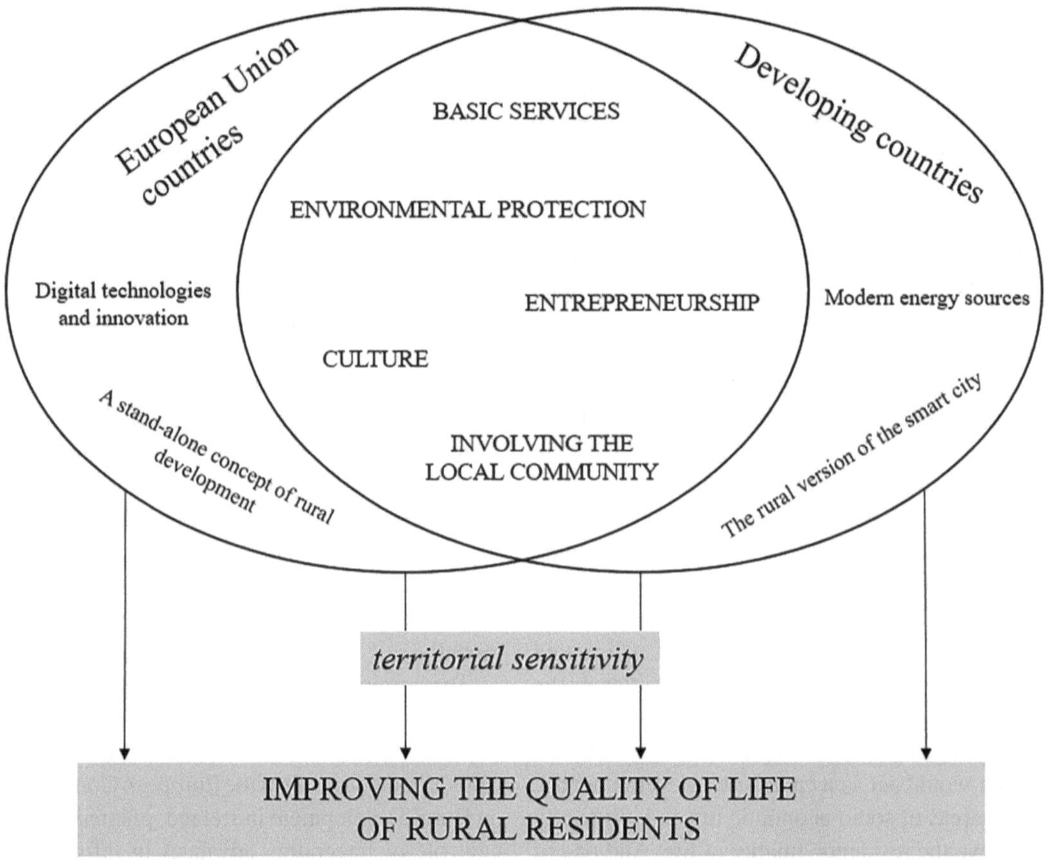

Fig. 3 Key assumptions of the *smart-village* concept in developing countries and the European Union. Source: Komorowski (2024, p. 510)

had a huge impact on the subsequent work on the concept of a smart village in the EU.

Important events for the development of the concept include the 2018 meeting on the future of smart villages in Bled, Slovenia. Its outcome was the Bled Declaration (2018) for a smart future for EU rural areas. The signatories of this declaration stated that innovative, integrated, and community-inclusive development of the digital economy in rural areas has the potential to improve the quality of life of their residents and increase economic and social cohesion between rural and urban areas. The body of the document also lists technological potentials that could be the flywheel of smart rural development, including precision agriculture, digital platforms, the sharing economy, the closed-loop economy, and rural tourism, among others. The authors of the declaration stressed that while digital technologies are an important tool for the development of smart villages, they are not the only one. Equally important are human resources, which require mobilisation and synergy of efforts. In order to achieve this synergy, coordinated investments in digital infrastructure and in the competencies of rural residents are necessary (as according to the signatories of the second Cork Declaration of 2016).

Another noteworthy "voice" in the evolution of the concept analysed is the European Rural Parliament. Following its deliberations in Candás, Spain, in 2019, the Manifesto of the European Rural Parliament (2019) was published. The document updates the content of previous documents developed during the parliament's deliberations in 2015 in Austria and 2017 in the Netherlands.

The update included, among other things, the addition of the concept of "smart villages" under the headings of policy framework, support for rural villages, and development of communications and digitisation. Then, for the first time, the concept was treated holistically, combining the leading technological sphere with the political (financial and systemic support) and social spheres (responses to the socio-economic challenges of villages). The need for multi-level work to concretise the concept's assumptions in the (then) upcoming programming period was emphasised, including through the operation of national working groups on the smart countryside. The task of these groups would be to raise awareness among rural residents of their needs and challenges and to develop ways to respond to them, taking into account the use of existing institutional structures, such as LEADER. Among these challenges, the following were specifically mentioned: population emigration, the disappearance of services, and the lag in the development of information and communication technologies. Significantly, it places the responsibility for addressing the last-mentioned challenge in the hands of entities with real power to do so, such as governments, international funds, and telecommunications providers.

The latest declaration of the European Rural Parliament was announced in 2022 after deliberations in the Polish city of Kielce (Declaration of the Villagers from Kielce, 2022), taking place during the recent agreements between EU member states and the European Commission on the shape of the CAP Strategic Plan. Each member state was developing its CAP Strategic Plan, taking account of the assumptions of the *European Green Deal*, a set of EU policy initiatives with the overarching goal of achieving climate neutrality in Europe by 2050.[9] It notes that smart village activities: "Because the role of rural communities is so important, they require integrated support conducted with special attention. The goals and vision for the development of these areas should be reflected in a goodwill and viable programme for rural areas at the national and European levels, with adequate funding for this and the integration of their development into a broader political framework. An integrated, strategic approach to this issue will require a new approach from the public sector. Cooperation is needed among all those with responsibility for rural areas at the European, national and regional levels. In order to respond to the real situation of rural areas, policy programmes at every level and in every field must be characterised by a holistic and integrated approach." It should be noted that the declaration was made in a new geopolitical reality—the accumulation of crises that began with the COVID-19 pandemic (Fig. 1). Adding to the effects of the pandemic were restrictions on the supply of energy resources from Russia in Europe, which meant problems with their availability and rising prices. Another crisis came with Russia's aggression against Ukraine, as it meant a full-scale war at the gates of the EU, and countries such as Poland became a frontline country. The facts cited therefore explain why there was an additional role assigned to smart villages—namely, strengthening the resilience of rural communities to crises.

The declarations and manifestos cited were mainly declaratory and guided the actions taken related to the smart-village concept. However, they were not the only important points in the work on the concept. Among the most important initiatives is the activity of the Thematic Group on Smart and Competitive Rural Areas at the European Network for Rural Development (ENRD),[10] which met from 2017 to 2020. This group was a "centre" for the exchange of views and experiences from various EU countries and one of the main advisory bodies for the design of solutions to support smart villages in the CAP. The results of the group's work can be attributed

[9]The principles of the European Green Deal are set out in the 2019 "Communication from the Commission to the European Parliament, the European Council, the Council, the Economic and Social Committee and the Committee of the Regions entitled the European Green Deal" European Commission. (2019)—https://eur-lex.europa.eu/legal-content/PL/TXT/?uri=CELEX%3A52019DC0640&qid=1653851156228

[10]https://ec.europa.eu/enrd/smart-and-competitive-rural-areas_en.html (accessed: 25.08.2024).

to five key spheres of smart villages (Komorowski et al., 2023): infrastructure, institutions, services, communities, and initiatives. This shows that practical activities with regard to the concept coincided with its theoretical assumptions and took into account current socio-economic processes in rural areas. Indeed, it has been pointed out that the factors driving the emergence of smart villages are (ENRD, 2018):

- Responding to demographic changes, particularly depopulation
- Seeking local solutions to ease the burden on the public budget
- Utilising and strengthening links between villages and cities
- Enhancing the contribution of rural areas to the transition towards a low-carbon, circular economy

Promoting and facilitating digital transformation. A key role in the discussion of creating support for smart villages was played by recommendations developed by representatives of various sectors—social, governmental, and scientific—at an international workshop entitled "Smart Villages as a Solution to Key Rural Challenges," which took place in 2019 in Warsaw, Poland. The recommendations focus mainly on the practical implementation of the idea of smart villages, rather than defining what exactly should be introduced. According to them, key activities should include:

- Using existing experience, such as that gained from the LEADER initiative or activities related to the Village Renewal programme. It is important to avoid creating new bureaucratic structures, as it is better to focus on developing existing solutions and involving local leaders in decision-making processes;
- Starting with small projects (e.g., in one locality) and then expanding to work with external partners to solve broader problems that may be difficult to overcome on a limited scale;
- Paying attention to the need for intensive digitisation of rural areas and the development of digital competence of residents, treating this as a priority and long-term task;
- Using the experience of local leaders and available human resources, and, in the absence of such resources, seeking to attract and engage them
- Promote innovative and unusual solutions to local problems that can serve as a model for other communities and reward active and committed groups

These recommendations are an important element in the process of designing support for smart villages, emphasising the need for a practical approach to implementing the smart-village concept, based on past experience, engaging local leaders and resources, and promoting innovative solutions to local problems.

4 Smart Villages—Examples from Poland

Thematic villages are becoming a popular and effective way to build the resilience of local rural communities (Sala, 2016). By basing their activities on local cultural, natural or historical resources, these villages not only strengthen the identity and attractiveness of the region but also create new economic development opportunities for the area.

An example of a thematic village in which the involvement of residents played a key role is Stara Białka, known as "Mleczny Zakątek." This small village in Lower Silesia province, Kamienna Góra county, has a population of less than 200. In 2021, a survey was conducted among the local community to determine the characteristic features of the village. When asked about the nouns that best described Stara Białka, residents most often pointed to words such as "cow," "milk," "peace," and "quiet." These responses reflected the village's strong ties to cow's milk production and the natural, peaceful nature of the place. The survey also asked about residents' willingness to engage in the process of change in the village. Although opinions were divided, almost half of the respondents expressed a willingness to participate in village activities. As a result of this interest, a meeting between residents and a national animator of thematic villages was organised to discuss the development of Stara Białka as a thematic village. During this meeting, the village community collectively recognised that the greatest asset of their village is cow's milk production. As reported, the village raises about 700 dairy cattle, which annually produce about 2 million litres of milk, making Stara Białka one of the largest milk producers in Kamienna Góra county. Surprisingly, the number of cows in the village is six times larger than the number of residents. As a result of joint activities, the residents unanimously decided that their village should become a thematic village called "Stara Białka—Mleczny Zakątek." The initiators of the project developed an action plan to develop the thematic village and create an attractive tourist offer.

Despite the short time it has been operating as a thematic village, since 2021, Stara Białka is showing dynamic development, with each successive activity involving more and more residents, including people who previously did not actively participate in the life of the local community. This involvement is turning into a grassroots initiative—residents themselves are coming up with new ideas and proposals for the development of this small village. Stara Białka—"Mleczny Zakątek" can be considered a village in line with the idea of smart villages for several key reasons:

1. Leveraging local resources and capital. The village is developing on the basis of its strengths, such as cow's milk production, which is central to the local economy. Dairy farming is the foundation of the business, which fits in with the smart villages idea of sustainable use of local resources.
2. Community involvement. The increase in the activity of local residents, who are increasingly willing to get involved in village life and bring their own initiatives, reflects the model of smart villages, in which the local community actively participates in the decision-making process and in creating the future of the village.
3. Diversification of income sources. Activities related to the establishment of the thematic village, such as the development of rural tourism and the promotion of local products (milk, dairy products), enable the diversification of residents' income, which is an important aspect of building economic resilience.
4. Preserving tradition and cultural heritage. Stara Białka integrates traditional forms of agriculture with modern approaches, preserving local heritage while introducing innovative solutions, which promotes sustainable development and cultural identity building.
5. Leveraging innovation and local knowledge. Strengthening the village's offerings through innovations in farm management, dairy processing and promotion of local products is an element in line with the idea of smart villages, where innovative approaches are key to effective village development.
6. Socio-economic sustainability. Mleczny Zakątek's activities are based on sustainable development, which combines economic, social and environmental aspects.

Stara Białka promotes the concept of smart villages by leveraging its strengths and local capital, seeking modern solutions, and strengthening traditions and networks among local stakeholders. These activities contribute to strengthening services through innovation and making better use of local knowledge, allowing the village to develop effectively in a sustainable way that responds to modern challenges. The involvement of local residents in the process of creating a new identity for the village was a key success factor, highlighting the importance of active community participation in building village resilience and creating innovative forms of development.

Another example of solutions that are part of the smart villages concept is the "Pszczelandia" educational farm, located in Barciany, in the Warmian-Masurian province, Kętrzyn county. It is an example of activities aimed at improving the quality of life, stopping the exodus of residents, implementing digitisation, and developing and strengthening social capital. The owners of the homestead have been running an apiary of 1000 bee stems since 1988, using natural and traditional beekeeping methods. Their decades of experience in beekeeping, honey processing, ecology, and apitherapy make them an example of a modern approach to traditional crafts in rural areas. The activities of "Pszczelandia" not only promote entrepreneurship in areas with difficulties related to the political transformation and create new jobs, but also show that even a small village can be innovative and open to tourists. The homestead plays a significant role in building sustainable village development, focusing on four key spheres:

1. Social—through educational activities, including training, workshops and study visits, "Pszczelandia" engages residents and tourists, increasing their knowledge and competence.
2. Cultural—the homestead nurtures local traditions and crafts, which helps build a strong identity for the rural community. Preserving cultural heritage strengthens residents' ties with the place where they live, which in turn increases their resilience to migration and demographic change.
3. Green—owners focus on protecting the environment, raising environmental awareness and using renewable energy sources. A sustainable approach to natural resources promotes the long-term stability of rural areas, as well as their resilience to threats of environmental degradation or climate change.
4. Economic—offering a rich programme of workshops and other activities, "Pszczelandia" increases farm income and creates stable jobs for the local community.

The initiative contributes to the resilience of the local community for several reasons. First of all, "Pszczelandia" engages the local community which fosters social ties and a community based on cooperation and mutual support. In the context of smart villages, social capital plays a key role, as communities with strong ties are more resilient to crises. The activity of the educational homestead promotes the creation of new jobs in rural areas (diversification of production). Increased employment and the development of local entrepreneurship reduce the risk of depopulation, while at the same time raising residents' standard of living, which is key to building long-term rural resilience. In addition, "Pszczelandia" offers a variety of services, such as education, workshops and rural tourism, which allows the local community to diversify its sources of income. This makes the village less dependent on single economic sectors, such as agriculture, which increases its economic resilience to changing market conditions or crises in specific industries. In conclusion, "Pszczelandia" not only promotes the development of the local economy and environmental protection, but also builds a cooperative community. This is crucial for rural resilience, as it enables rural communities to respond more effectively to the challenges of the modern world, both economically and environmentally.[11]

5 Summary

Intensifying climate and environmental change, the energy crisis, the economic and social impact of the COVID-19 pandemic and related constraints, rising waves of migration and armed conflicts, including the war in Ukraine—all of this is causing people in Europe today to think about the future of the countryside not so much in terms of taking advantage of opportunities, but in terms of facing challenges—although some of these challenges bring new opportunities that rural areas will be able to take advantage of. Going forward, rural communities must not only respond to current threats, but also actively shape their future.

The conclusions of the study of the smart-village concept indicate that it responds to the pressing challenges facing rural areas in Europe, as well as around the world. One of the most important conclusions of the study is that the development of smart villages requires close cross-sector cooperation. Involving local com-

[11] More examples of smart solutions in Poland are presented in Kalinowski et al. (2022).

munities, government, science and the private sector is essential to creating effective, sustainable solutions. Community empowerment "bring a new vision for a hopeful future as they shape new sustainable living values, infrastructure, and the interaction between these living values and infrastructure incubators which challenge the current structures" (ERCA, 2024, p. 7).

The smart-village concept offers a sustainable development model that combines technological innovation, local resources and community engagement to create stronger and more resilient local structures. One of the key elements of smart villages' success is their bottom-up nature. Local communities that engage in decision-making processes and jointly elaborate development strategies show greater flexibility and efficiency in adapting to changing conditions. These activities lead not only to an improvement in residents' quality of life, but also to the strengthening of local identity, which promotes better preservation of cultural heritage and traditions.

Analysing specific cases from Poland, such as "Mleczny Zakątek" in Stara Białka or the "Pszczelandia" educational farm in Barciany, we can see that the concept in question can significantly affect the creation of new jobs and diversification of residents' sources of income, and thus the development of local economies even in regions facing difficult challenges, such as depopulation or lack of infrastructure. Grassroots initiatives, in which both local natural resources and innovative management approaches play a key role, show that rural communities can become more independent and resilient to global crises (see Li, 2023). The development of thematic villages or the promotion of organic forms of agricultural production are good examples of how tradition and modernity can be combined. With the use of modern tools, such as digitisation or precision agriculture, they can not only revitalise the local economy, but also contribute to building more integrated, self-sufficient and resilient communities.

The smart-village concept also shows that rural development does not have to be based solely on traditional agriculture. Diversification of economic activities, including the development of rural tourism, renewable energy or digital services, opens up new opportunities for these areas. In the long term, this can contribute to making the countryside more attractive as a place to live and work, especially for the younger generation, who often choose to emigrate to cities.

It is also worth noting that the concept is part of broader European policy goals, such as the European Green Deal, the Common Agricultural Policy 2023–2027 and the long-term vision for rural areas until 2040. Smart villages are therefore part of a larger transformation that aims not only to improve the quality of life in rural areas, but also to narrow the rural-urban gap and adapt to the changes brought by development challenges including countering crisis situations such as the effects of climate change (see Czarnecki et al., 2023).

In the European context, smart villages are at the heart of a long-term rural development strategy. The document, which was written in 2021, identifies four key areas of action to achieve the goals of building stronger, connected, resilient and pro-poor rural areas by 2040. As part of this vision, smart villages are seen as the foundation of community-based local development. Their success therefore depends not only on financial support, but above all on the involvement of the communities themselves and coordinated action at various administrative levels. It is important that rural development policies are integrated, and that both national governments and international organisations and local initiatives are involved in their implementation.

Ultimately, smart villages can become a catalyst for sustainable change in rural areas, provided that public policies are properly tailored to their specific needs and challenges, and are therefore territorially sensitive. Investing in digitisation, education and local economic initiatives, combined with long-term financial support, creates real opportunities for rural areas in Europe and beyond to become more cohesive, sustainable, and resilient.

References

Bindi, M., & Olesen, J. E. (2011). The responses of agriculture in Europe to climate change. *Regional Environmental Change, 11*(Suppl 1), 151–158. https://doi.org/10.1007/s10113-010-0173-x

Budzich-Tabor, U. (2023). Polish countryside: Resilient and flexible. In M. Halamska, M. Kłodziński, & M. Stanny (Eds.), *Polish Countryside 2044. Vision of Development* (pp. 15–24). IRWiR PAN.

Buzek, J. (2023). Polish Countryside 2044: Towards energy self-sufficiency. In M. Halamska, M. Kłodziński, & M. Stanny (Eds.), *Polish Countryside 2044. Vision of Development* (pp. 25–28). IRWiR PAN.

Cañal Fernández, V., & Álvarez Pinilla, A. (2021). Are infrastructures important to stop rural depopulation? In *XIII Congreso de Economía Agroalimentaria* (pp. 239–242). Universidad Politécnica de Cartagena.

Chmieliński, P., & Wieliczko, B. (2021). Long-term vision for rural development and critical enablers for achieving it: Case of Mazowieckie Region (Poland). In J. Černiauskienė (Ed.), *Proceedings of the 10th International Scientific Conference "Rural Development 2021"* (pp. 345–349). https://doi.org/10.15544/RD.2021.060

Crescenzi, R., & Rodríguez-Pose, A. (2012). Infrastructure and regional growth in the European Union. *Papers in Regional Science, 91*(3), 487–513.

Czarnecki, A., Dacko, A., & Dacko, M. (2023). Extreme weather event risk awareness among second home owners and their economic and non-economic response strategies. In B. A. Adie & C. M. Hal (Eds.), *Second homes and climate change*. Routledge. https://doi.org/10.4324/9781003091295

Declaration from Bled. (2018). *For a smarter future of the rural areas in EU*. Meeting at Bled, Slovenia, April 13. https://pametne-vasi.info/wp-content/uploads/2018/04/Bled-declaration-for-a-Smarter-Future-of-the-Rural-Areas-in-EU.pdf

Declaration from Cork. (1996). *A living countryside*. The European Conference on Rural Development, November 7–9. http://www.andreamarescotti.it/Materiali/5-Politiche/1996_UE_Cork_declaration.htm

Declaration from Cork. (2016). *Better life in rural areas*. European Conference on Rural Development Cork 2.0, September 5–6. Publications Office of the European Union. https://ec.europa.eu/enrd/sites/default/files/cork-declaration_pl.pdf

Declaration of the Villagers from Kielce. (2022). *12-15.09.2022*. https://ksow.pl/aktualnosc/europejski-parlament-wiejski-2022-deklaracja-z-kielc

DevelopmendAid. (2024). *Countries with the highest population decline*. https://www.developmentaid.org/news-stream/post/152772/top-20-countries-with-the-highest-rate-of-population-decline

ENRD. (2018). Smart villages. Revitalization of services in rural areas. *EU Rural Review*, No. 26. https://ec.europa.eu/enrd/sites/default/files/enrd_publications/publi-enrd-rr-26-2018-pl.pdf

ERCA. (2024). *Compass for Local Empowerment in rural areas*. Co-production of the European Partnership for Citizen-empowerment (SBE) and the European Rural Community Alliance (ERCA). https://cdn.ruralcommunities.eu/wp-content/uploads/2024/07/ERCA-SBE-Compass-Local-empowerment.pdf

European Commission. (2017). *EU action for smart villages*. https://ec.europa.eu/enrd/news-events/news/eu-action-smart-villages_en.html

European Commission. (2018). *Farming for Natura 2000*. https://doi.org/10.2779/85823. https://www.miteco.gob.es/content/dam/miteco/es/biodiversidad/temas/ecosistemas-y-conectividad/biodiv-medios-agrarios/farming-for-natura-2000.pdf

European Commission. (2019). *Green deal*. https://eur-lex.europa.eu/legal-content/PL/TXT/?uri=CELEX%3A52019DC0640&qid=1653851156228

European Commission. (2021). *European strategies for rural development*. https://agriculture.ec.europa.eu/common-agricultural-policy/rural-development_en

European Council and Council of the EU Portal. (n.d.). *European Green Deal*. Retrieved January 20, 2024, from https://www.consilium.europa.eu/pl/policies/green-deal/#what

European Parliament. (2021). *Transport infrastructure in low-density and depopulating areas*. Policy Department for Structural and Cohesion Policies Directorate-General for Internal Policies (PE 652.227). https://www.europarl.europa.eu/RegData/etudes/STUD/2021/652227/IPOL_STU(2021)652227_EN.pdf

European Rural Parliament Manifesto. (2019). Candás, Spain, November 6–9. https://lgd-srws.pl/images/biblioteka/zalaczniki/20191120925206848.pdf

Eurostat. (2021a). *Population change in rural areas: Dynamics and trends*. https://ec.europa.eu/eurostat/statistics-explained/index.php?title=Population_and_population_change_statistics

Eurostat. (2021b). *Population structure and ageing*. https://ec.europa.eu/eurostat/statistics-explained/index.php?title=Population_structure_and_ageing#The_share_of_elderly_people_continues_to_increase

Eurostat. (2023a). *Natural population decrease in most EU regions in 2021*. https://ec.europa.eu/eurostat/web/products-eurostat-news/w/ddn-20231121-4

Eurostat. (2023b). *Unemployment rates by sex, age and other typologies*. https://ec.europa.eu/eurostat/databrowser/view/urt_lfu3rt__custom_12726571/default/table?lang=en

Eurostat. (2024a). *Performance of the agricultural sector.* https://ec.europa.eu/eurostat/statistics-explained/index.php?title=Performance_of_the_agricultural_sector#Value_of_agricultural_output

Eurostat. (2024b). *Digital economy and society statistics - households and individuals.* https://ec.europa.eu/eurostat/statistics-explained/index.php?title=Digital_economy_and_society_statistics_-_households_and_individuals#Internet_access_of_households.2C_2010_and_2023

Interreg Europe. (2024). *Rural mobility. Improving rural mobility across Europe through novel transport solutions and innovative policy approaches.* Project summary; Retrieved September 01, 2024, from https://www.interregeurope.eu/rural-mobility

Janc, K., & Komorowski, Ł. (2024). From analogue to digital – Polish countryside 2004-2024. In P. Chmieliński & G. Gorzelak (Eds.), *Polish countryside and Polish agriculture. 20 years in the European Union* (pp. 287–300). Institute of Rural and Agricultural Development, Polish Academy of Sciences. https://doi.org/10.53098/978-83-89900-78-4

Kalinowski, S., Komorowski, Ł., & Rosa, A. (2022). *The smart village concept. Examples from Poland.* IRWiR PAN, CDR w Brwinowie, o. w Warszawie, Warszawa.

Komorowski, Ł. (2022). Smart initiatives in a suburban community: An example from the holy cross mountains in Poland. *Mountain Research and Development, 42*(1), D1–D9. https://doi.org/10.1659/MRD-JOURNAL-D-21-00037.1

Komorowski, Ł. (2024). *Smart villages as a concept for rural development.* Ph.D. thesis in social sciences, economics, and finance. Institute of Rural and Agricultural Development, Polish Academy of Sciences, Warsaw.

Komorowski, Ł., Rosa, A., & Kalinowski, S. (2023). The concept of smart villages – is it the next stage of rural renewal? *Village and Agriculture, 3*(200), 187–207. https://doi.org/10.53098/wir032023/09

Li, Y. (2023). A systematic review of rural resilience. *China Agricultural Economic Review, 15*(1), 66–77. https://doi.org/10.1108/CAER-03-2022-0048

Meijers, E., & van der Wouw, D. (2019). Struggles and strategies of rural regions in the age of the 'urban triumph'. *Journal of Rural Studies, 66*, 21–29. https://doi.org/10.1016/j.jrurstud.2019.01.027

OECD. (2020). *Regions and cities facing ageing.* In *OECD Regions and Cities at a Glance 2020.* OECD Publishing. https://doi.org/10.1787/9a6f8396-en

OECD. (2021). *Rural agenda for climate action.* https://www.oecd.org/content/dam/oecd/en/topics/policy-sub-issues/environmental-transition-in-regions/oecd-rural-agenda-for-climate-action.pdf

Perpiña Castillo, C., Jacobs-Crisioni, C., Barranco, R., Curtale, R., Kompil, M., Vallecillo, S., Auteri, D. and Dijkstra, L. (2023). Opportunities and challenges for remote rural areas in the European Union, European Commission, Ispra. JRC135398.

Philip, L., & Williams, F. (2019). Remote rural home-based businesses and digital inequalities: Understanding needs and expectations in a digitally underserved community. *Journal of Rural Studies, 68*, 306–318. https://doi.org/10.1016/j.jrurstud.2018.09.011

Sala, K. (2016). Wioski tematyczne jako przykład innowacyjności w turystyce wiejskiej. *Zeszyty Naukowe Małopolskiej Wyższej Szkoły Ekonomicznej w Tarnowie, 30*(2), s. 117-126.

Stanny, M., Rosner, A., & Komorowski, Ł. (2023). *Monitoring rural development: Stage IV. Decade of socio-economic changes. Foundation European Fund for Rural Development, Institute of Rural and Agricultural Development.* Polish Academy of Sciences.

The Sydney Morning Herald. (2020). *'Unlocking new tech': $1.9 billion for low-emission energy projects.* Retrieved August 25, 2024, from https://www.smh.com.au/politics/federal/unlocking-new-tech-1-9-billion-for-lowemission-energy-projects-20200916-p55was.html

Treaty on the Functioning of the European Union. (2004). Dz.U.2004.90.864/2. Retrieved August 20, 2024, from https://lexlege.pl/traktat-o-funkcjonowaniu-unii-europejskiej/art-5/

United States Department of Agriculture. (2021). *FY Budget Summary 2022.* Retrieved August 29, 2024, from https://www.usda.gov/sites/default/files/documents/2022-budget-summary.pdf

Van Gevelt, T., & Holmes, J. (2015). *A vision for smart villages* (Briefing No. 5). CMEDT – Smart Villages Initiative.

Wilkin, J. (2023). On the need and principles for creating a vision for the development of the Polish countryside. In M. Halamska, M. Kłodziński, & M. Stanny (Eds.), *Polish Countryside 2044: Vision of development.* Institute of Rural and Agricultural Development, Polish Academy of Sciences.

Wolski, O. (2019). Digitalisation of rural areas and agriculture in the EU debate: How far from what research says? *Village and Agriculture, 2*(183), 7–28. https://doi.org/10.7366/wir022019/01

Rural Development Challenges and Resilience Building in Northern and Central Europe: A Comparison of Sweden and Slovakia

Hans Westlund and Kamila Borsekova

Abstract

In line with the emergence of the post-urban world, a general development trend is that rural areas are being divided into two opposite parts: the city-close countryside, which is being integrated into the growing city-regions, and a peripheral countryside outside the growth influences of the city-regions. (Between these areas, there might be intermediate zones interacting with the city-regions, but not enough to transition fully into integration). This means that "rural areas" cannot be treated as one unit, but as two units with very different development tendencies. This chapter discusses the situation and tendencies for these two types of rural areas in Sweden and Slovakia, respectively. Both are members of the European Union, and their rural policies are determined by the Union's Common Agricultural Policy. However, the countries' histories and institutional legacies are very different. Postwar Sweden has been a welfare market economy, while Slovakia, after World War II, became a part of the socialist Czechoslovakia and did not become an independent state until 1993. Against this background, this chapter presents and analyzes current challenges for rural development in the two countries, presents and discusses two successful cases of development in the rural peripheries, and, based on these, draws some conclusions on building rural resilience.

Keywords

Rural development · Challenges · Resilience building · Sweden · Slovakia

1 Introduction

As late as the year 1800, about 97% of the world's population was living in rural areas. Today, just over 200 years later, the share is under 50% and is expected to be down to 40% in the 2030s. In a process that started with the agricultural and industrial revolutions in England and that slowly spread across the world, the population of rural areas has steady shrunk at the expense of cities, towns, and smaller central places. In most of the developed world, the process of classic urbanization—migration from rural to urban areas—reached an end about 1970. However, with the start of the ICT revolution and the emerging knowledge economy in the 1980s, it became

H. Westlund (✉)
KTH Royal Institute of Technology, Stockholm, Sweden
e-mail: hanswes@kth.se

K. Borsekova
Matej Bel University, Banska Bystrica, Slovakia
e-mail: kamila.borsekova@umb.sk

Y. Li (ed.), *Global Perspectives on Building Resilience for Sustainable Rural Development*, Sustainable Development Goals Series, https://doi.org/10.1007/978-981-96-6149-7_9

clear that the process of urbanization had reached a completely new stage. The traditional urban-rural dichotomy had been dissolved and replaced with cohesive, multi-nuclei city-regions that comprised dense centers but also sparse areas and a mosaic of land use. Densification of suburbs, as well as regional enlargement across municipal boundaries through better transportation infrastructure and public transit, were other characteristics of the functional city-regions along with a re-urbanization from places at lower levels in the urban hierarchies to national and regional centers. We have, in earlier research, denominated these changes as a transformation to a "post-urban world" (Westlund & Borsekova, 2023; Westlund, 2014, 2018).

For rural areas, this transformation had drastic consequences. On the one hand, rural areas within the emerging city-regions were increasingly interwoven into "the urban fabric" (Lefebvre, 2003, p. 3) and became an integrated part of the city-region. On the other hand, the vast rural areas outside the positive economic influence of the city-regions gradually lost importance as the knowledge economy's growth changed the city-regions' exchange with the outside world. The raw materials and natural resources of the outer hinterlands declined in relative importance, while the exchange of knowledge-intensive services and products increased in relative importance. In addition, ever-increasing capital intensity in the exploitation of natural resources meant a steadily decreasing number of jobs in the peripheral areas.

This means that the area that traditionally has been called "the countryside" is being divided into two parts with completely different development paths. One part is a city-close countryside that becomes a part of expanding city regions. The other part is a peripheral countryside that slowly fades away—unless it is able to supply something that there is sufficient urban demand for, which makes it possible for these areas to survive. Studies that ignore this circumstance will produce misleading results—results that cannot serve as foundations for rural development policies.

The two very different types of rural areas that characterize the post-urban world are general attributes that must be taken into consideration when forming rural development policies. Another issue is that the prerequisites for rural development differ across the world. This chapter deals with European conditions, but as shown in other chapters of this book, rural areas in other parts of the world face partly other problems. However, also in Europe, the situations might vary due to differences in natural resources, climate, human and social capital, culture, and historical heritage. For almost half the twentieth century, the "iron curtain" divided Europe into democratic market economies and communist-governed state economies. Borsekova et al. (2021) have shown that the transition of post-socialist regions is not over. Capitalist regions in Europe perform better than post-socialist regions in eight of nine pillars of regional competitiveness. This is also a circumstance that must not be neglected when forming regional and rural development policies for Europe.

A relatively new feature in local and regional development policies worldwide is the focus on resilience. One of the most cited works on resilience and regional development is Martin and Sunley (2015). They show that the concept of resilience in the research literature has been given three different interpretations. The first is that resilience is defined as the ability of a system to withstand a shock and return to its original state without any changes. The second interpretation can be described as a system's ability to receive a shock by reorganizing and changing but still essentially retaining the same functions, structure, identity, and feedbacks. This interpretation implies that "bouncing back" to a previous state of equilibrium cannot in many cases be possible, but that the system can respond to the shock by changing but still maintaining its original functions and structures.

The third interpretation, called "adaptive resilience," is the most far-reaching, according to which resilience is the ability of a system to undergo structural and qualitative changes in the event of a shock but still maintain its core functions. This can be described in terms of "bouncing forward" instead of "bouncing back." This interpretation is very close to the concept of robustness. Martin and Sunley (2015) point out that, in recent years, robustness has come to be used for systems' ability to withstand shocks by, if necessary, changing their structures and components in order to maintain or restore core activities or functions. When we use the term "resilience" in this chapter, we do so in line with this third interpretation, i.e., as a synonym for robustness.

This chapter has three interconnected aims: (a) to map the two types of countryside in two European countries, Sweden and Slovakia; (b) to discuss the challenges for rural development in the two countries' peripheral parts; and (c) to give examples of how peripheral places in the two countries have formed strategies for building local rural resilience. Section 2 gives short overviews of the two types of rural regions in Sweden and Slovakia and their challenges. Section 3 presents two case studies of rural development in peripheral areas in each country. Section 4 discusses the conclusions and whether any general lessons for rural development can be learned from the case studies.

2 Two Types of Rural Regions in Sweden and Slovakia and Their Challenges

We analyze rural regions of two European countries with significant differences in location, size, population density, and history. Sweden is located on the Scandinavian Peninsula in the northernmost part of Western Europe, whereas Slovakia is located in Central Europe. The population of Sweden was 10.55 million in 2022, while Slovakia's population was 5.46 million. In size, Sweden, with its 528,447 km^2, is one of the largest countries in Europe, while Slovakia is 49,035 km^2. Sweden's population density is 26 persons per km^2, but the population is very unevenly distributed, with a relatively dense population in the south and very sparse in the northwestern parts of the country. Slovakia's population density is almost six times higher, with 121 persons per km^2.

The histories of the two countries are very different as well. Sweden has been an independent country since 1523, when it left the union with Denmark and Norway, even though the areas that have been part of Sweden have varied. From the 1930s to the 1970s, Sweden was governed by Social Democrats, and a welfare society was built up. Sweden joined the European Union 1995. After having had a long period of shrinking social gaps, they started to increase after 1980 (Henning et al., 2022). Sweden is currently the country with the fastest growing social gaps in Europe.

Slovakia, for most of its history until the last 30 years, predominantly existed not as a separate and independent state but as part of larger political entities. Nestled in Central Europe, the region's early history was shaped by the settlement of Celtic, Germanic, and Slavic tribes. In the eleventh century, Slovakia became part of the Kingdom of Hungary, a status that persisted for nearly a millennium. The sixteenth century witnessed the region falling under the influence of the Habsburg Monarchy and later the Austro-Hungarian Empire, where it remained until the conclusion of World War I. The post-war period heralded a significant geopolitical shift, with Slovakia merging with the Czech lands to form Czechoslovakia in 1918. However, this union was disrupted during World War II, when Slovakia briefly became a separate state allied with Nazi Germany. Following the war, Slovakia rejoined Czechoslovakia, which transitioned into a communist state under Soviet influence in 1948. The peaceful Velvet Revolution in 1989 ended communist rule, leading to the 'Velvet Divorce' in 1993, after which Slovakia emerged as an independent nation. Since then, Slovakia has integrated into the European community, joining NATO and the European Union in 2004, marking a significant chapter in its history as a sovereign state. These historical events and developments, influenced by a mix of multicultural and multinational elements, have shaped the settlement structure of Slovakia, characterized by a large number of smaller cities, except for the capital, surrounded by rural areas.

2.1 Sweden

Figure 1 shows a division of Sweden into four municipality types.[1] The black (metropolitan) and the dark grey (regional centers) are cities with a positive population development. The white (sparse rural) areas all belong to the peripheral group we have discussed above. Most municipalities (164 of 290) belong to the light grey (rural) group, and it is within this group that we find the most divergent population trends.

Summaries of central variables for the four municipality types are found in Table 1. The metropolitan areas have had the strongest population growth, followed by the regional centers. The rural group had on average a certain growth, while the sparse rural group lost population. The other variables express the same pattern. The metropolitan group has the lowest average age, the highest population share with higher education and the highest accessibility,[2] while the sparse rural group has the most negative values.

Figure 2 shows population development in Swedish municipalities from 2000 to 2019. A comparison with Fig. 1 confirms that all metropolitan municipalities and the regional centers (with the exception of four suburb municipalities) increased their population during the period. Among the rural municipalities, 47% increased their population by more than 2.5%. Twenty-five percent of the rural municipalities were stagnating (±2.5% population change), and 28% lost more than 2.5% of their population. Most of the stagnating and declining municipalities have peripheral locations, in Sweden as a whole or in their home region. Only one of the sparse rural municipalities increased its population, the ski tourism center of Åre.[3]

2.2 Slovakia

Since the beginning of the twenty-first century, a number of transformative changes have been ongoing in Slovakia's residential and population development, the initiation phase of which can be traced back to after the collapse of the previous political regime. Slovakia covers an area of 49,034 square kilometers and has a population of 5,431,026 (as of 31.12.2022). As illustrated in Fig. 3, its residential structure is considerably fragmented. The country comprises 141 cities and 2890 municipalities. Slovakia is relatively evenly populated, with higher population density in cities and their surrounding areas. Naturally, mountainous regions, rural areas, and the

[1] The municipalities are classified into four different municipal types (MT). (MT 1) metropolitan areas (N=46), (MT 2) urban areas/regional centers (N=47), (MT 3) rural areas/countryside (N=164), and (MT 4) sparsely populated rural areas (N=33). The four types of areas are defined as follows: Metropolitan areas (MT 1): Includes municipalities where 100% of the population lives within cities or within a 30 km distance from the cities. Using this definition, there are three metropolitan areas in Sweden: the Stockholm, Gothenburg and Malmö regions. Urban areas/regional centers (MT 2): Municipalities with a population of at least 30,000 inhabitants and where the largest city has a population of 25,000 people or more. Smaller municipalities adjacent to these urban municipalities are included in a local urban area if more than 50% of the labor force in the smaller municipality commutes to a neighbor municipality. In this way, a functional-region perspective is adopted. In practice, this group contains regional centers outside the metropolitan areas and their "suburb municipalities". Rural areas/countryside (MT 3): Municipalities that are not included in the metropolitan areas and urban areas/regional centers are classified as rural areas/countryside, given they have a population density of at least five people per square kilometer. Sparse populated rural areas (MT 4): Municipalities that are not included in the three categories above and have less than five people per square kilometer.

[2] The accessibility measure is based on Harris' (1954) classic measure of market potential. For a detailed explanation, see Johansson et al. (2003).

[3] An additional sparse rural municipality, bordering to Åre shows growth as well. However, in this case the growth took place right outside the border of the regional center, i.e., it is an expression of the regional center's growth.

Fig. 1 Sweden divided in four municipality types

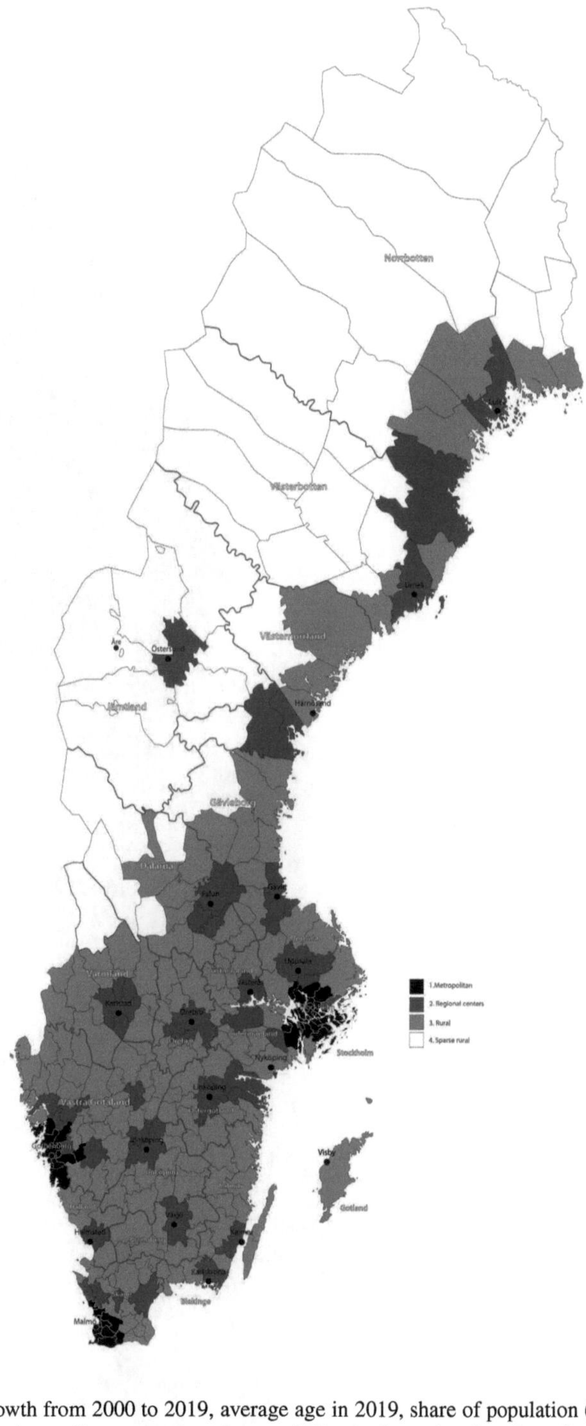

Table 1 Average population growth from 2000 to 2019, average age in 2019, share of population (16–64 years) with higher education (3 years' university studies or more), and accessibility in the national road net for the four groups of Swedish municipalities

Municipality type	% Population growth	Avg. age	% Higher education	Avg. accessibility
Metropolitan	28.6	40.1 years	33.2	195,904
Regional centers	18.0	41.4 years	26.2	56,364
Rural	5.5	44.4 years	18.3	27,103
Sparse rural	−10.5	46.2 years	16.7	6354

Source: Statistics Sweden

Fig. 2 Population development in Sweden's municipalities from 2000 to 2019. Green: more than +2.5%; Yellow: +2.5% to −2.5%; Red: more than −2.5%. Source: Statistics Sweden

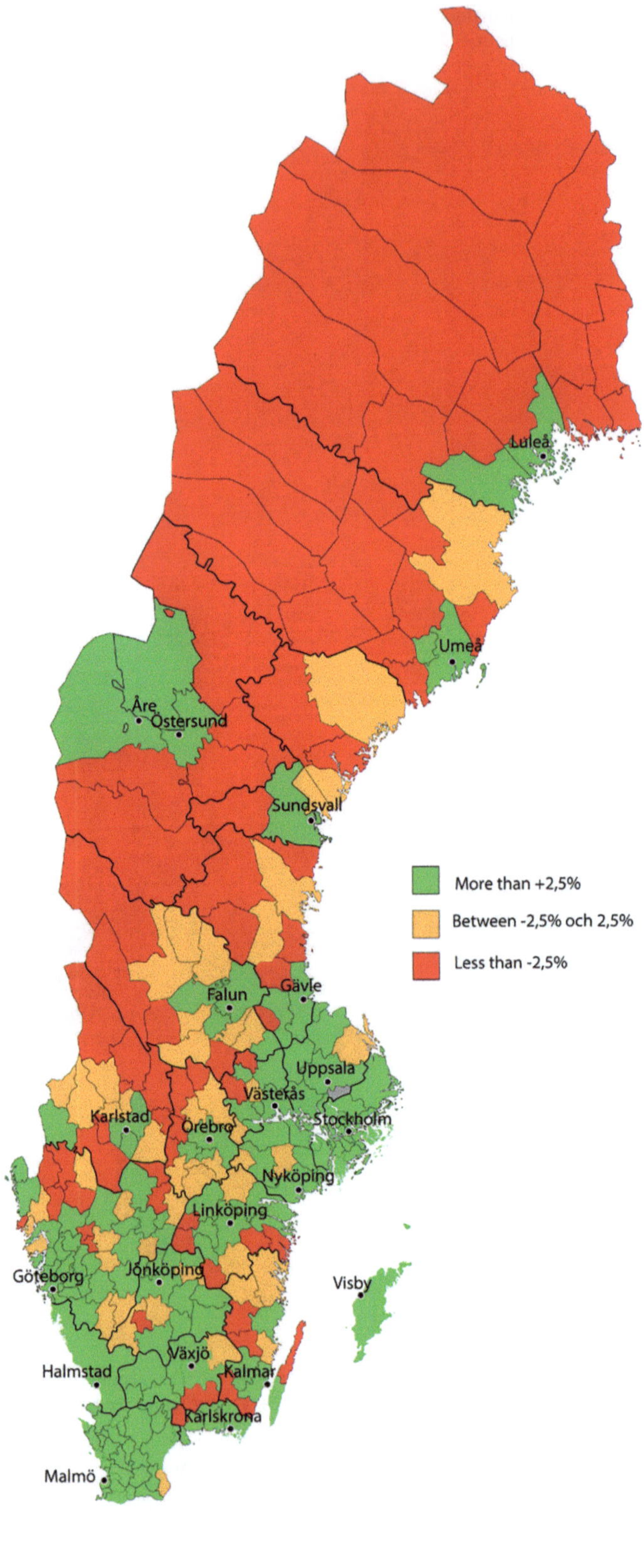

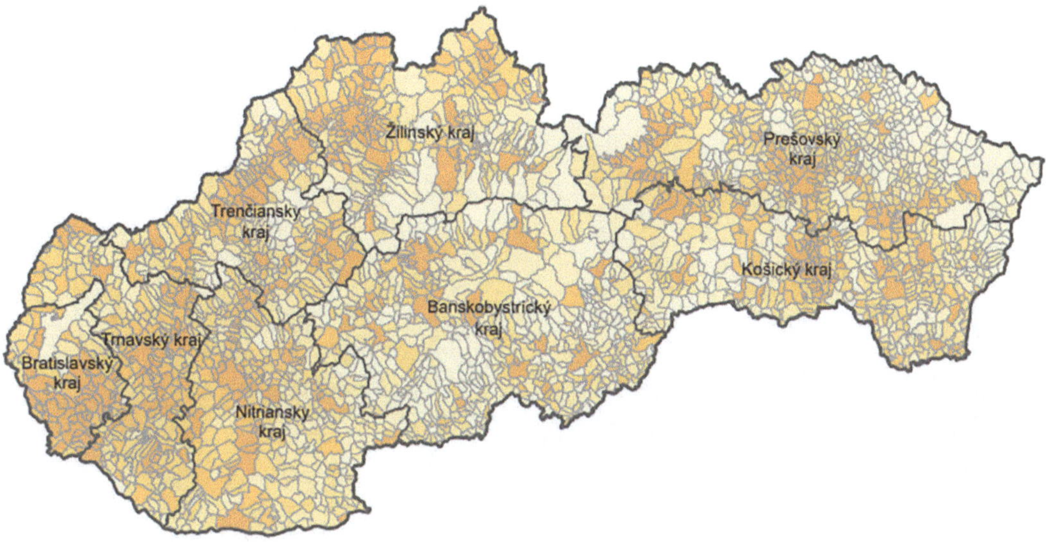

Fig. 3 Number and population density (per sq km) in Slovakia in 2021. The lightest color: 0–20, the darkest color: 155–5000. Source: Statistical Office of the Slovak Republic

Table 2 Average population growth from 2000 to 2019, average age in 2021, share of population (16–64 years) with higher education (3 years' university studies or more), and accessibility in the national road net

	% Population growth	Average age (2021)	% Higher education (2021)
Metropolitan areas	8.82	41.55%	27.89
Big and mid-size cities without suburbs	0.85	42.06%	27.67
Big and mid-size cities incl. suburbs	6.65	41.65%	26.84
Smaller cities and rural areas	−1.20	41.42%	14.50
Slovakia	0.96	41.39%	18.52

Source: Own calculations based on data from Statistical Office of the Slovak Republic

peripheral zones in the south and east of the country are less populated.

In this chapter, we will focus in more detail on urban and rural classification of Slovakia. Table 2 provides deeper insights into the demographic dynamics across Slovakia.

The table presents a comprehensive overview of Slovakia's demographic and educational landscape in 2021, categorized by different urban and rural areas. The Metropolitan Areas, encompassing Bratislava and Košice along with their suburbs, show significant population growth of 8.82%, an average age of 41.55 years, and 27.89% of the population with higher education. In contrast, Big and Mid-Size Cities Without Suburbs, which include the main districts of Slovakia's eight largest cities, exhibit a modest population growth of 0.85%, an average age slightly higher at 42.06 years, and 27.67% of the population holding higher education degrees. When suburbs are included in this category, population growth increases to 6.65%, the average age slightly decreases to 41.65 years, and the percentage with higher education drops marginally to 26.84%. Smaller Cities and Rural Areas, however, face a population decline of −1.20%, have

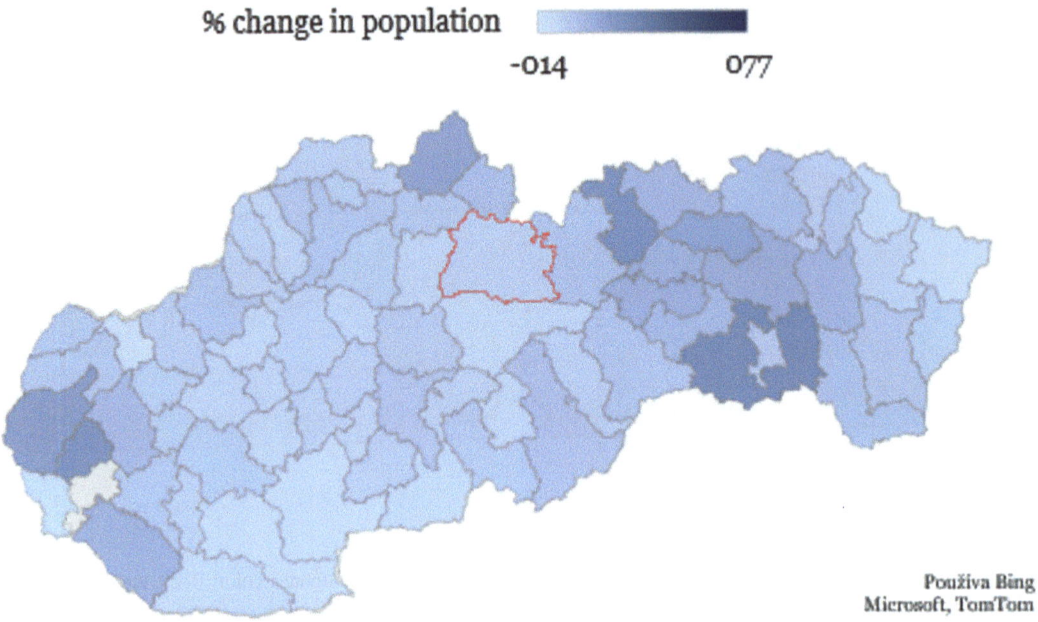

Fig. 4 Percentual change in population in Slovak counties between 2000 and 2019. Source: Own processing based on data from the Statistical Office of the Slovak Republic

an average age of 41.42 years, and only 14.50% of their population has higher education. Overall, for Slovakia as a whole, population growth stands at 0.96%, the average age is 41.39 years, and 18.52% of the population possesses higher education qualifications. This disparity highlights the concentration of educational opportunities and attainment in urban settings. Significantly, and to Slovakia's disadvantage, the level of higher education attainment is much lower than that of Sweden, both in urban and rural areas.

Figure 4 shows the percentual change in population in Slovak counties between 2000 and 2019.

The map presents the percentage change in population for various counties in Slovakia between 2000 and 2019. The county marked in red is the one where Jasná is located (see next chapter). The data reveals a mix of population increases and decreases across the country. Notably, Senec County experienced the most significant increase, with a remarkable 76.62% rise in population, highlighting it as a peak in population growth. This county is attractive for living not only because of its proximity to the capital city, Bratislava, but also due to its high-quality environment and the recreational area of the Sunny Lakes. On the other hand, Bratislava V County saw a substantial decrease, with a 13.67% reduction in its population, marking it as one of the most significant declines. Other notable increases include Pezinok County (approximately 30 km from the capital city) and Košice—surroundings (located as a suburban area of the second-largest city in Slovakia), with 20.22% and 23.09% growth, respectively. In conclusion, the demographic and educational landscape of Slovakia in 2021, as well as the population changes between 2000 and 2019, reveal significant regional disparities. Metropolitan areas, particularly Bratislava and Košice, are experiencing robust population growth and higher levels of educational attainment, in contrast to smaller cities and rural areas, which are facing population decline and lower educational levels. This trend underscores the urban-rural divide in Slovakia, with urban areas attracting more residents and offering greater educational opportunities. These trends reflect the broader socio-economic shifts within Slovakia, emphasizing the concentration of growth and development in and around major urban centers.

3 Case Studies of Two Successful Peripheral Places

3.1 Åre

Here we focus on the peripheral municipality that had the strongest population growth: Åre, best known as Sweden's leading ski resort. Åre is located close to the geographical center of Sweden, about 600 km from the closest metropolitan region, Stockholm. In 1882, the railway between Stockholm and Trondheim (in Norway) reached Åre, and a station was built. With the railway came tourists, first for walking in the mountains and to guest the health resorts that were established in Åre and its vicinities. In the early twentieth century, ski tourism in the winter took over as tourism activity. The excellent accessibility—tourists could take the night train from Stockholm and literally strap on the skis when they stepped off the train and take the ski lift up the mountain—made Åre the leading ski resort in Sweden (even though alpine skiing remained a limited sport until the end of the 1960s). Åre hosted the World Championship in Alpine Skiing in 1954, and although this event made Åre well known in Sweden, its impact on the local development was negligible. Most of the local population earned their living in forestry and agriculture, and with shrinking employment in these sectors, the area that nowadays forms Åre municipality experienced population losses from 1950 to 1975. Thus, Åre's development seemed to follow the same negative path as the rest of the peripheral, rural Sweden, but in contrast to other peripheral areas, the 1970s became a turning point for Åre municipality and in particular for the village. What was the reason?

During the post-war period, the Swedish economy flourished. The living standard rose, and the number of paid vacation days increased. More people could afford to take their families skiing in the mountains during the schools' winter holidays—but a rising share of these trips headed abroad, to the European Alps. The Swedish government reacted to this outflow of money and decided to support the development of modern Swedish ski resorts by including tourism among the industries that were subject to regional policy. Åre was one of the main recipients of this support. Among other investments, a cable car to the top of the Åre Mountain was built 1975–1976 with the help of government support corresponding to ca 90 million Euro 2024. New hotels were built in the same period.

The probably most important factor behind the boom in alpine skiing in Sweden (and the boom in Åre) during the 1970s and 1980s was, however, not the state-supported investments but the legendary Ingemar Stenmark, double Olympic gold medalist, five World Championship golds, and 86 World Cup victories in the disciplines of slalom and giant slalom. Interestingly, it was the Jasná ski resort in Slovakia that paved the way to the world of elite skiing for this athlete, as his first major success was achieved in Jasná (see next Sect. 3.2) in 1974 when he became the junior European champion in the giant slalom. During a number of years around 1980, Sweden was in a Stenmark-mania, and "everybody" should go alpine skiing. Åre experienced a strong growth of small tourism businesses, and the population in the village increased with ca 20% during the period 1975–2000.

The economic crisis of the early 1990s meant a severe blow to Åre, with bankruptcies and falling real estate prices. The crisis also meant the entry of new, international, professional owners and managers into Åre. Initially, these actors focused solely on their own business, but over time came an insight that local collaboration was necessary for strengthening the brand and attractiveness of Åre (Nordin & Westlund, 2009). The municipality had a leading role in this collaboration that set a "Vision 2011" and thereafter a "Vision 2020" (Fig. 5).

Expressed in terms of social capital, i.e., social networks and trust and other values supported by these networks, before the boom, the local ski club and local businesses dominated the village. When new major actors, foreign and domestic, emerged, partly new values of professional commercialization came to dominate. The new actors overtook the leading business positions in Åre, but the experiences of the "old" actors and existing values and practices could not be neglected.

Fig. 5 Example of marketing pictures on Åre. Source: Google.com

Table 3 Population of Sweden, Åre Municipality, and Åre village in the years 2000 and 2020 and change in percent

	2000	2010	2020	Change in %
Sweden	8,872,000	9,378,000	10,350,000	17
Åre Municipality	9745	10,274	12,049	24
Åre village[a]	1574	2334	4311	174

Source: Statistics Sweden

[a]The last row in the table shows the merged population for Åre, Duved, and Björnänge villages. From 2015, Duved was officially merged with Åre in the statistics, as the two villages had expanded and become a joint unit. Björnänge is located just south of Åre and is an integrated part of the bigger village

The municipality took the initiative to gather old and new actors in an informal public-private collaboration that, over time, set up a vision for the future development of the village. Hence, Åre reproduced its social capital in its transformation from a small mountain village to an international ski resort (Nordin & Westlund, 2009). Two of the most salient results were that Åre hosted the World Championship in Alpine Skiing in 2007 and 2019. Both events resulted in considerable investments in new ski lifts, hotels, adventure pools, housing, second homes, shops, and in the entertainment industry—and new actors and their social networks.

Table 3 shows that Åre's population growth indeed is remarkable and that the growth after 2000 was much higher than in the period 1975–2000. The municipality as a whole had a higher percentage population growth than Sweden as s whole from 2000 to 2020, which in itself is unique for a peripheral inland municipality. However, the real boom took place in Åre village and its two close neighboring villages, which have grown with 174% so that they have formed one integrated village.

There is very little research on the transformation of Åre village, from a railway station with some surrounding houses to a dense, international ski resort with pubs, bars, dance venues, and restaurants in all price ranges. However, there are many testimonies that current Åre is something very different from the village that existed during Ingemar Stenmark's heydays. It is sometimes said that Åre is the most remote suburb of Stockholm. Åre has simply become a modern, urban spot in the wilderness that makes it possible to enjoy both urban and rural amenities simultaneously. It is of course not the 4311 inhabitants in Åre village that make this possible, but the tourists that come there to consume these amenities and attractions. The number of commercial guest beds in Åre was 42,000 in 2023, i.e., almost ten times the number of inhabitants (https://aresweden.com). This is an increase of 15,000 beds (55.5%) since 2011.

Even if it still is merely a hypothesis, there are strong reasons to believe that this rather unique combination of amenities is a main explanation behind the exceptional growth of both tourists and inhabitants in the Åre village, in particular after 2010. Regarding the inhabitants, the group attracted by these amenities is a rather heterogeneous one and stretches, among others, from seasonal workers in the tourism industry that ski and party in their leisure, to lifestyle entrepreneurs that run their own businesses, often connected to the tourism industry and combine it with outdoor life, and individuals and families simply attracted by the sporty lifestyle in Åre.

3.2 Jasná

According to the Slovak Tourism Development Strategy 2020, mountain destinations in Slovakia are major tourism attractions, appealing not only to domestic but also to international tourists. In total, there are 107 mountain destinations, which vary in scale from small local resorts to large destinations of international significance (Gajdošík et al., 2017). Jasná, recognized as Slovakia's largest and best ski resort, is situated within the geomorphological confines of the Low Tatras, specifically in the NAPANT protected area at the terminus of the Demänovská Valley. This is confirmed by Bučeková et al. (2019), who conducted a classification of ski resorts in Slovakia. Their study aimed to establish a comprehensive classification system for these resorts, taking into account various factors. These included the physical geographic characteristics of the ski resorts, the level of ski infrastructure, and the availability of other services. According to this classification, the highest quality is represented by Category I ski resorts. Notably, Jasná Low Tatras is the sole resort in this top category, distinguishing itself significantly from other resorts. It boasts exceptional natural conditions, the most extensive range of slopes, the highest transport capacity, and the broadest and most diverse array of additional amenities and services.

Jasná is located in the county of Liptovský Mikuláš, within the Demänovská Dolina Municipality. This resort, strategically positioned on the northern and southern aspects of Chopok mountain, presents a unique geographical and recreational setting, with altitudes ranging from 900 to 2024 m above sea level.

The winter season at Jasná is distinguished by its comprehensive skiing and snowboarding facilities. The resort features over 45 km of slopes, designed to accommodate a wide spectrum of skill levels, from beginners to experts. A notable aspect of Jasná is its sophisticated artificial snow system, which covers over 26 km, ensuring reliable snow conditions for at least 5 months each year. The resort operates from early December to late April and includes provisions for night skiing on selected slopes.

Jasná's infrastructure is comprised of 29 lifts and cable cars, with a combined transport capacity of 30,744 individuals per hour, making it the largest and most efficient ski resort in Slovakia. Beyond traditional skiing and snowboarding, the resort is a venue for various seasonal events and offers activities like ski alpinism, freeriding (part of the World Freeride Tour), cross-country skiing, and other winter sports.

The resort employs a skipass system that utilizes contactless technology, facilitating ease of use and operational efficiency. These skipasses, valid for multiple days, are also accepted at other resorts in Liptov and High Tatras and at Aquapark Tatralandia, thereby extending the recreational options available to visitors. This interconnectivity between the Low and High Tatras contributes to the development of a significant international tourist destination. Jasná includes Slovakia's largest skiing and snowboarding school, along with rental services and a selection of branded sports shops. For children, there are available amenities such as a winter playground, animation programs, and professional childcare services (Fig. 6).

The culinary offerings at Jasná are characterized by a variety of dining establishments, ranging from self-service facilities to à la carte restaurants, strategically located along the ski slopes. The region is also notable for its après-ski culture, wellness and spa centers within the resort, and the nearby Aquapark Tatralandia, which complement the offer and create a complex tourist destination. In the warmer months, the area around Chopok mountain within the National Park Low Tatras undergoes a transformation into a green environment. It becomes a

Fig. 6 The landscape of Low Tatras in winter, dominated with Chopok mountain. Source: Authors

Table 4 Population of Slovakia, county Liptovský Mikuláš and Demänovská Dolina (both home to the Jasná resort) in the years 2003 and 2022, and change in percent

	2003	2012	2022	Change in %
Slovakia	53,78,821	54,05,338	54,31,026	1
Liptovský Mikuláš	73,739	72,537	71,599	−3
Demänovská dolina	195	268	356	83

Source: Statistical Office of the Slovak Republic

hub for outdoor activities, including a network of hiking trails, cycling paths, and scenic cable car rides. Educational trails on both sides of Chopok offer insights into the local flora and fauna, enhancing the visitor experience. Additionally, the resort area provides opportunities for various outdoor activities, such as paragliding and the Rope Park.

In the context of the Slovak Tourism Development Strategy 2020, mountain destinations within Slovakia have been identified as pivotal tourism attractions, drawing interest from both domestic and international tourists (Gajdošík et al., 2017). Among these, Jasná stands out due to its dynamic evolution in network structures over time. Initially, the network within Jasná was primarily localized, catering to a smaller, more domestic audience. However, as the destination gained prominence, there was a marked shift in its network structure.

This evolution can be characterized by a transition from a simple, locally focused network to a more complex, internationally oriented one. The expansion of Jasná's network is evident in several aspects. Firstly, there has been a significant increase in the connectivity between Jasná and major urban centers, both within Slovakia and in neighboring countries. This is facilitated by improved transportation links and marketing strategies targeting international tourists.

Furthermore, the internal network of Jasná has also undergone substantial changes. This is underlined by population changes as depicted in Table 4.

In examining the population trends from 2003 to 2022, a notable increase is observed in Demänovská Dolina, the home municipality of Jasná. Over this period, the population of Demänovská Dolina has risen significantly by 83%. This growth starkly contrasts with the overall population trends in Slovakia, as the country experienced a modest increase of 1% in its population. The substantial population growth in Demänovská Dolina, set against the backdrop of these broader demographic trends, highlights its unique development, influenced by factors such as tourism growth due to the popularity of Jasná.

The methodological framework applied in the network analysis of Jasná's tourism structure (for more information, see Gajdošík et al., 2017; Gajdošíková et al., 2019), utilizing retrospective data from two distinct periods (1995 and 2015), reveals a significant transformation in the destination's network structure. This transformation is contextualized within the broader historical and legislative landscape of Slovakia's tourism industry.

In 1995, Jasná's network structure was predominantly community-based, centralized around local tourism associations (public-private stakeholders) and municipalities (public stakeholders). This period was marked by major structural changes in the resort's offerings, stimulated by new legislative and institutional frameworks that supported entrepreneurial activities within the business environment.

By 2015, a notable shift towards a corporate-based model was observed. This shift was influenced by the Tourism Support Act no. 91/2010 Coll., which initiated the top-down development of destination management organizations in Slovakia's micro-regions. The network analysis, employing metrics such as degree centrality, closeness centrality, and betweenness centrality, revealed a transition from public leadership to a more pronounced engagement of private stakeholders in destination leadership. Specifically, the company TMR emerged as a significant private stakeholder, gaining a powerful position within the network. Degree centrality indicated that stakeholders with more ties (connections) had greater opportunities and autonomy, making them less dependent on any specific stakeholder in the destination and hence more powerful. Over time, a shift from public leaders to the engagement of private stakeholders, particularly TMR, in destination leadership became evident.

Closeness centrality further underscored the power dynamics within the network. Stakeholders closer to a larger number of other stakeholders could exert influence through direct bargaining and exchange. TMR's rise was facilitated by its structural advantage in the network, characterized by shorter path lengths between cooperating stakeholders and the interdependence of financial capital, activities, and innovations (e.g., new mountain transport facilities, product packages, and loyalty programs).

Betweenness centrality revealed the strategic positioning of stakeholders within the network. While local tourism associations initially had the highest betweenness centrality, TMR's position as a key intermediary stakeholder grew significantly. This centrality measure indicated an increasing dependency of other stakeholders on TMR, particularly in the context of ski tourism infrastructure.

The analysis of network centralization indices and expert opinions further confirmed the shift from a community-based to a corporate-based structure. TMR's control over key resources, including all cable cars and ski lifts, ski rental facilities, ski schools, summer equipment rental, ski-in/ski-out hotels, and après-ski bars, allowed it to create a comprehensive and competitive tourism product. This structural ownership underlines the corporate model of the destination, where TMR determines the strategic orientation and entrepreneurial objectives, relegating the role of Destination Management Organizations (DMOs) to a more limited scope (Gajdošík et al., 2017; Gajdošíková et al., 2019).

In conclusion, Jasná's significance as a ski destination is augmented by its status as the home resort of Olympic champion Petra Vlhová and as a venue for the Alpine Skiing World Cup. Boasting over 50 km of expertly groomed pistes and a comprehensive educational complex for winter sports, Jasná is a pivotal element in Slovakia's skiing landscape and a notable destination in the European context. Its diverse après-ski entertainment, led by top-tier DJs and artists, cements its status as a premier destination for outdoor enthusiasts and researchers in recreational geography and sports tourism.

4 Discussion: Unique Examples or Possible Lessons for Other European Rural Areas?

In a Swedish perspective, the population growth in the village of Åre is unique. There is no other village in Sweden's peripheral countryside that can show such growth. There are a number of other successful ski resorts in Sweden, but it is only in Åre that tourism has led to such growth of the permanent, local community. The main domestic competitor, the Sälen area in the southernmost part of the Swedish mountains, with ca 55,000 commercial guest beds, is in fact larger than Åre. In Sälen village, the number of inhabitants grew from 489 to 842 (72%) between 2000 and 2020, while the population of the whole Malung/Sälen municipality decreased by 6%. The growth of the Sälen village is an impressive result but still very far from Åre's corresponding figure: 174%.

A major reason for Åre's unparalleled growth is probably its denseness—its concentration of amenities in one spot. In Åre village, "everything" is within walking distance. In Sälen, the ski slopes are located along a ca 20 km long mountainside, and the village is located in a valley ca 10 km from the closest slopes. In Åre, the natural conditions made it possible to concentrate investments in ski tourism *and* to create a dense urban village with a multitude of actors in "the middle of wilderness," where residents and visitors can simultaneously enjoy both urban and rural amenities. This denseness does not exist in Sälen, and the multitude of activities and actors does not exist in the smaller ski resorts. For a visitor that stays in these resorts a week for a week of skiing, the lack of urban density and diversity does not matter very much, but for luring residents, density and diversity seem to be more important. We suggest that these tentative conclusions should serve as hypotheses for future research.

From a Slovak perspective, the population growth in the municipality of Jasná is unique, akin to trends observed in Sweden. Its growth in population is comparable only to that of attractive municipalities near the capital city.

A major factor contributing to Jasná's growth and its status as a leader among Slovak ski destinations is its strategic location in the picturesque and culturally rich Liptov region.

Liptov, situated in the northern part of Slovakia and extending to the Slovak-Polish border, is recognized for its natural landscapes and its cultural and historical heritage. The region encompasses several natural reserves, including the Tatra National Park, the Low Tatras National Park, and the Veľká Fatra National Park. These protected areas highlight Liptov's commitment to environmental preservation and maintaining its ecological integrity. The region's diverse terrain and natural features make it an attractive destination for outdoor activities. In the summer, it offers opportunities for hiking and cycling, while in winter, it becomes a hub for downhill and cross-country skiing. Additional natural attractions include numerous caves, as well as healing mineral and thermal springs, and the Liptovská Mara water reservoir, highlighting the area's ecological diversity (Fig. 7).

Fig. 7 The landscape of Liptov region with Liptovská Mara water reservoir. Source: Authors

The cultural and historical heritage of Liptov includes castles, manor houses, and the ruins of medieval castles. The villages in Liptov exhibit examples of folk architecture, with Vlkolínec being a UNESCO World Heritage Site. The region is also distinguished by its traditional shepherd culture.

Jasná's location in the attractive tourism region of Liptov, which employs targeted place marketing and its tools (for more information see Vaňová et al., 2017a, 2017b), has significantly contributed to boosting Jasná's attractiveness as part of a comprehensive natural, historical, and cultural destination. This strategic positioning has been instrumental in enhancing Jasná's prominence and appeal as a top ski destination.

Drawing parallels to Sweden's notable achievements in hosting major skiing events, Jasná has demonstrated a similar capacity. It has hosted the Alpine Ski World Cup six times, marking its presence on the international skiing stage. These events were held in 1979, 1982, and 1984, a period when Jasná was part of Czechoslovakia, and more recently in 2016, 2021, and 2024, reflecting its enduring appeal and relevance in the world of alpine skiing. However, it's worth noting that Jasná is not the only ski resort in Slovakia with a rich history in hosting significant skiing competitions. Another famous resort, located in the High Tatras, hosted the Alpine Ski World Cup in 1935 and 1970. Despite its historical significance, this resort in the High Tatras currently finds itself slightly overshadowed by Jasná, especially in the realm of marketing and global recognition. This contrast highlights Jasná's successful blend of heritage and modern marketing strategies, which have effectively bolstered its position as a leading ski destination, not just in Slovakia, but on the international stage as well.

A tentative conclusion is that the success of Åre will be very hard to replicate. The particular combination of historic and new buildings and establishments, and, over time, recurrent waves of investments in infrastructure and facilities cannot be copied to another place. Moreover, a substantial part of the investments that took place before the two World Championships in 2007 and 2019 would probably not have happened without these unique events, and Åre is the only Swedish place scheduled to organize future World Cups and Championships.

Even if Åre, ever since the World Championships in 1954, has been defined as the No. 1 Swedish ski resort—and thereby by definition as a unique place—this does not mean that other places have nothing to learn from Åre.

In principle, every tourist destination is an example of the application of Alonso's concept "borrow size" (Alonso, 1973, p. 200), as the tourists represent an external demand that makes it possible to increase the local supply of products and amenities to a level it would not have had without the additional demand. This additional supply attracts visitors, but it gives the destination a competitive advantage in competing for residents as well. In addition, it should be stressed that "borrow size" is not only applicable to the local supply and demand for products and amenities. It can also be applied to issues like access to knowledge, competence, and financial capital. Successful tourism destinations might start with local capital solely, but sooner or later, as in the examples of Åre and Jasná, it is external capital that makes it possible to take another step to a higher level. This holds for knowledge and competence as well.

Another general principle that in the case of tourism destinations also is connected to the "borrow size" concept, is spatial agglomeration economies. Åre's and Jasná's high urban densities are excellent examples of agglomeration economies that thanks to the visitors, makes it possible to offer a diversity of amenities and entertainment as well as specialized services, which normally only exists in much bigger places.

The "creative destruction" that hit Åre after the financial crisis of the early 1990s ended up in a fairly informal public-private collaboration between the municipality and leading private actors. The collaboration aimed at setting up long-term visions and goals on the way to fulfilling the visions. This local public-private collaboration, based on a joint interest in strengthening the brand of Åre and increasing the place's attractiveness for tourism and for permanent residence, became a method to form social capital among the leading actors. New actors that originally only focused on the interests of their own business gradually became involved in this informal group and in the local, social environment. There is, of course, no official information on the role of this informal group in attracting external investors, competence, and management, but it should not be underestimated.

The evolution of Jasná's network structure from a community-based model in 1995 to a corporate-based model by 2015, as detailed by Gajdošík et al. (2017) and Gajdošíková et al. (2019), underscores a significant transformation in the resort's management and strategic direction. This shift, influenced by legislative changes and the emergence of private stakeholders like TMR, has led to a more dynamic and competitive tourism product. However, it is imperative that this development respects the protection of natural areas to ensure the sustainability of Jasná's growth. The resort's future success hinges not only on maintaining its appeal as a ski destination but also on its integration into the broader tourism landscape of the Liptov region. Strengthening collaboration with local stakeholders and incorporating Jasná into the overall tourism strategy of Liptov will enhance its attractiveness and contribute to a more balanced and sustainable tourism ecosystem. This approach will not only preserve the natural beauty and cultural heritage of the region but also support the long-term viability and prosperity of Jasná as a leading destination in Slovakia.

5 Routes to Build Rural Resilience

In this chapter, we have shown how the post-urban development has resulted in two types of rural areas in Sweden and Slovakia with very different development trends: growth in the city-close rural areas that are being integrated into growing city regions, and decline in most of the rural areas that are located outside the city regions. Moreover, we have shown that the disadvantages of the peripheral rural municipalities not only consist of that that they are depopulating but that they have a problematic age structure, lower education levels, and lower accessibility. However, we have also shown that certain rural areas have been able to "swim against the tide," deal with the challenges, and in fact have been able to grow when they have appeared to supply urban customers with a unique combination of amenities that have appealed to their tourism preferences. These rural areas have shown examples of "adaptive resilience" or "robustness" in that they have developed new employment in new industries when the traditional employment in the primary sector decreased.

It should be noted that the adapted resilience in Åre has been a protracted process. Even if the population started to grow in 1975, after 25 years of downturn, the tourism boom meant only a 20% population increase during the following 25 years. It was not until after the millennium shift that the remarkable increase took place. It is highly probable that the large investments for the World Championships in 2007 and 2019, and the urban transformation they meant for the village, played a significant role in increasing Åre's attractiveness as a place of residence and not only for tourism. If these experiences can be generalized, they indicate that there is no quick fix for peripheral areas in the post-urban world, but also that if investments are long-term and followed up gradually, they might in the long run give a good yield.

Drawing on the resilience journey of Åre, Jasná's path to resilience can be similarly envisioned as a gradual and strategic process. The development of Jasná, much like Åre, is not a quick fix but a long-term endeavor. The transformation of Jasná from a predominantly community-based network structure in 1995 to a more corporate-based model marks the beginning of this journey. This shift, while enhancing the resort's competitiveness, also underscores the need for sustainable development that respects the natural environment and integrates with the broader Liptov region's tourism strategy.

The resilience of Jasná, similar to Åre, will likely hinge on substantial and sustained investments, not only in its ski infrastructure but also in its overall urban transformation. These investments should aim to enhance Jasná's appeal both as a tourist destination and as a place of residence. The key lies in balancing the development of tourism facilities with the preservation of natural and cultural heritage, ensuring that growth is both environmentally sustainable and beneficial to the local community.

Furthermore, the experiences from Åre suggest that long-term investments, if strategically planned and gradually implemented, can yield significant benefits. For Jasná, this could mean not only enhancing its status as a top ski destination but also contributing to the socio-economic development of the Liptov region. In the long run, such an approach could transform Jasná into a resilient, attractive, and sustainable destination, thriving in the post-urban world and setting a benchmark for other peripheral areas.

Seen from a more general perspective, the development of the countryside in Sweden and Slovakia is an excellent illustration of the post-urban thesis of the division of rural areas into a growing group, close to the metropolitan regions or regional centers, and a peripheral, declining group outside the positive influence of the growing cities. We could also observe a stagnating, intermediate group of municipalities. The experiences from Åre and Jasna show that peripheral location does not necessarily mean decline, but the experiences from all other peripheral places indicate that it most often does. This brings us to the question of what policy can do to strengthen rural resilience. In our opinion, rural resilience cannot be built in local isolation. Resilience is a regional process and, as we have shown, it is the regional centers that are the engines of regions' growth. A national policy for local resilience should include the regional centers and their role in strengthening the region.

To enhance the resilience of rural areas in the face of post-urban development, as exemplified by Sweden and Slovakia, a comprehensive, multifaceted strategy is essential. This strategy should focus on leveraging unique local amenities to attract tourism, balancing development with the preservation of natural and cultural resources, and making long-term investments in infrastructure and services. Diversifying the local economy beyond traditional sectors and integrating rural areas into broader regional development plans are also crucial. This approach necessitates active community involvement and empowerment in decision-making processes, ensuring that development aligns with local needs and values. Additionally, national and regional policies should support rural resilience through financial incentives, grants, and sustainable practices. Embracing technology and innovation can help overcome geographical limitations, while education and training programs can equip residents with skills for emerging sectors. Effective marketing and branding strategies are also vital in rebranding rural areas as attractive destinations. By combining sustainable development, community engagement, regional collaboration, and strategic investment, at least some rural areas can transform into vibrant, sustainable communities, contributing significantly to broader regional and national growth.

Acknowledgments The first author gratefully acknowledges financial support from the Swedish Research Council Formas, Grant No. FR 2021/0004. The second author gratefully acknowledges financial support from the BRRIDGE project, funded by the European Union's Horizon Europe research and innovation program under grant agreement No. 101079219 and the project SECreTour, funded by the European Union's Horizon Europe research and innovation program under grant agreement No. 101132584. Views and opinions expressed are those of the author(s) only and do not necessarily reflect those of the European Union.

References

Alonso, W. (1973). Urban zero population growth. *Daedalus, 102*(4), 191–206. https://www.jstor.org/stable/20024174

Borsekova, K., Korony, S., & Nijkamp, P. (2021). Traces of the iron curtain: A multivariate analysis of regional cohesion in Europe. *Socio-Economic Planning Sciences, 78*, 101040. https://doi.org/10.1016/j.seps.2021.101040

Bučeková, I., Erbert, L., & Klobučník, M. (2019). Klasifikácia lyžiarskych stredísk na Slovensku.

Geografický časopis, 71, 363–382. https://doi.org/10.31577/geogrcas.2019.71.4.19

Gajdošík, T., Gajdošíková, Z., Maráková, V., & Flagestad, A. (2017). Destination structure revisited in view of the community and corporate model. *Tourism Management Perspectives, 24*, 54–63. https://doi.org/10.1016/j.tmp.2017.07.009

Gajdošíková, Z., Gajdošík, T., & Kučerová, J. (2019). Slovak winter tourism festinations: Future playground for tourists in the Carpathians. *Tourism and hospitality management, 25*(1), 161–178. https://doi.org/10.20867/thm.25.1.7

Harris, C. D. (1954). The market as a factor in the localization of industry in the U.S. *Annals of the Association of American Geographers, 44*, 315–348.

Henning, M., Westlund, H., & Enflo, K. (2022). Urban–rural population changes and spatial inequalities in Sweden. *Regional Science Policy & Practice*, 1–15. https://doi.org/10.1111/rsp3.12602

Johansson, B., Klaesson, J., & Olsson, M. (2003). Commuters' non-linear response to time distances. *Journal of Geographical Systems, 5*(3), 315–329.

Lefebvre, H. (2003). *The urban revolution*. University of Minnesota Press. (French original first published 1970).

Martin, R., & Sunley, P. (2015). On the notion of regional economic resilience: conceptualization and explanation. *Journal of Economic Geography, 15*(1), 1–42.

Nordin, S., Westlund, H. (2009). Social Capital and the Life Cycle Model: The Transformation of the Destination of Åre. *Tourism, 57*(3), 259–284.

Vaňová, A., Vitálišová, K., & Borseková, K. (2017a). *Place marketing*. Belianum.

Vaňová, A., Vitálišová, K., & Borseková, K. (2017b). *Place marketing, public and nonprofit marketing case studies*. Belianum.

Westlund. (2014). Urban futures in planning, Policy and Regional Science: Are We Entering a Post-Urban World? *Built Environment, 40*(4), 447–457.

Westlund, H. (2018). Urban-rural relations in the post-urban world. In T. Haas & H. Westlund (Eds.), *In the post-urban world: Innovative transformations in Global City regions* (Vol. 2018, pp. 70–81). Routledge.

Westlund, H., & Borsekova, K. (2023). Rural problems, policies and possibilities in a post-urban world. *Regional Science Policy & Practice, 15*(4), 717–728. https://doi.org/10.1111/rsp3.12664

Index

Y. Li (ed.), *Global Perspectives on Building Resilience for Sustainable Rural Development*, Sustainable Development Goals Series, https://doi.org/10.1007/978-981-96-6149-7

MIX
Papier aus verantwortungsvollen Quellen
Paper from responsible sources
FSC® C105338

If you have any concerns about our products,
you can contact us on
ProductSafety@springernature.com

In case Publisher is established outside the EU,
the EU authorized representative is:
Springer Nature Customer Service Center GmbH
Europaplatz 3, 69115 Heidelberg, Germany

Printed by Libri Plureos GmbH
in Hamburg, Germany